ENTROPY OPTIMIZATION AND MATHEMATICAL PROGRAMMING

INTERNATIONAL SERIES IN OPERATIONS RESEARCH & MANAGEMENT SCIENCE

Frederick S. Hillier, Series Editor
Department of Operations Research
Stanford University
Stanford, California

Saigal, Romesh
The University of Michigan
LINEAR PROGRAMMING: A Modern Integrated Analysis

Nagurney, Anna/ Zhang, Ding
University of Massachusetts @ Amherst
PROJECTED DYNAMICAL SYSTEMS AND VARIATIONAL INEQUALITIES WITH APPLICATIONS

Padberg, Manfred/ Rijal, Minendra P.
New York University
LOCATION, SCHEDULING, DESIGN AND INTEGER PROGRAMMING

Vanderbei, Robert J.
Princeton University
LINEAR PROGRAMMING: Foundations and Extensions

Jaiswal, N.K.
Ministry of Defense, INDIA
MILITARY OPERATIONS RESEARCH: Quantitative Decision Making

Gal, Tomas / Greenberg, Harvey J.
FernUniversität Hagen/ University of Colorado @ Denver
ADVANCES IN SENSITIVITY ANALYSIS AND PARAMETRIC PROGRAMMING

Prabhu, N.U.
Cornell University
FOUNDATIONS OF QUEUEING THEORY

ENTROPY OPTIMIZATION AND MATHEMATICAL PROGRAMMING

S.-C. Fang
North Carolina State University
Raleigh, North Carolina, USA

J.R. Rajasekera
International University of Japan
Urasa, Niigata, Japan

H.-S. J. Tsao
University of California
Berkeley, California, USA

KLUWER ACADEMIC PUBLISHERS
Boston/London/Dordrecht

Distributors for North America:
Kluwer Academic Publishers
101 Philip Drive
Assinippi Park
Norwell, Massachusetts 02061 USA

Distributors for all other countries:
Kluwer Academic Publishers Group
Distribution Centre
Post Office Box 322
3300 AH Dordrecht, THE NETHERLANDS

Library of Congress Cataloging-in-Publication Data

Fang, Shu-Cherng.
Entropy optimization and mathematical programming / S.-C. Fang, J. R. Rajasekera, H.-S. J. Tsao.
p. cm. -- (International series in operations research & management science ; 8)
Includes bibliographical references and index.
ISBN 0-7923-9939-0
1. Operations research. 2. Entropy (Information theory) 3. Mathematical optimization. I. Rajasekera, J. R. II. Tsao, H.-S. J. III. Title. IV. Series.
T57.6.F35 1997
519.7--dc21 97-17742
CIP

Printed on acid-free paper.

Printed in the United States of America

Dedicated to those we love, in particular,

Chi-Hsin,

Lilly,

Shu-Wen, Hsiao-Tzu,
Hueylian, Allison and Jason.

CONTENTS

PREFACE

The term *entropy* was coined by Rudolf Clausius around 1865 as a measure of the amount of energy in a thermodynamic system. Claude Shannon gave a new meaning to this term in 1948 as a measure of uncertainty in the context of communication theory. With E. T. Jaynes' *maximum entropy principle* and its variations, the concept of using Shannon's entropy to select a probability distribution that complies with known observations about an unknown distribution and yet remains most noncommittal has penetrated a wide range of disciplines. Various *entropy optimization models* have proven their usefulness with successful applications in image reconstruction, pattern recognition, statistical inference, queueing theory, spectral analysis, statistical mechanics, transportation planning, urban and regional planning, input-output analysis, portfolio investment, information analysis, linear and nonlinear programming, etc.

While the entropy optimization models have been repeatedly used in different fields, corresponding solution methods have been invented and reinvented in similar but different forms for use in these fields. A good number of applicable solution methods have been loosely constructed without rigorous mathematical treatment. Some of them even lack convergence proofs. A systematic presentation with proper mathematical treatment of this material is needed by the practitioners and researchers alike in all application areas. The purpose of this book is to fill this gap. It is our intention to offer perspectives that meet the needs of diverse *user communities* in ways that the users would be comfortable with. With this consideration, we focus on the entropy optimization problems in finite dimensional Euclidean space such that only some basic familiarity with optimization is required of a reader.

This book is divided into seven chapters. Chapter 1 introduces the concept of entropy and entropy optimization principles. Chapter 2 reports some important practical applications of entropy optimization in a small number of selected fields. This hopefully helps the reader relate the concept of entropy optimization to his/her own application areas. Chapter 3 deals with methods for solving entropy optimization problems subject to linear constraints, while Chapter 4 handles problems with quadratic, entropic, and general convex constraints. In addition to reviewing other major existing methods, a recently developed *unconstrained dual-perturbation approach* is introduced and discussed in detail in both chapters. Capitalizing on results obtained in Chapter 3, Chapter 5 introduces the entropic perturbation approach to solving several important classes of mathematical programming problems, including linear and convex quadratic programs. Chapter 6 reexamines the entropic perturbation approach from a different but more general viewpoint. A relation between entropy optimization and Bayesian statistics is explained in Chapter 7. Extensions of results obtained earlier in the book as well as some other related results are also included in this last chapter.

The writing of this book would never have been completed without encouragement, information, and comments from those surrounding us. Part of this book has been derived from joint research performed with some of them. For example, Section 1.1 is based on a working paper by R. Kanoub. We gratefully acknowledge the contributions of E. L. Peterson, R. Kanoub, C.-F. Hu, H. L. W. Nuttle, S. D. Roberts, and J. R. Wilson of the North Carolina State University; T. Otsuki and U. Sumita of the International University of Japan; P. Varaiya and A. Kanafani of the University of California at Berkeley; C.-J. Lin and R. Saigal of the University of Michigan at Ann Arbor; S.-Y. Wu and R.-L. Sheu of the National Cheng-Kung University; M.-S. Chen, H.-F. Wang, and E. K. Yang of the National Tsinghua University; R. K.-C. So and C. Scott of the University of California at Irvine; R. W. Hall of the University of Southern California; D.-N. Lee of AT&T Laboratories; X. Li of the Dalian University of Technology; Y. Censor of the University of Haifa; and M. Yamada of Tokai International Securities Co. Ltd. We are also grateful to Professor Frederick S. Hillier of Stanford University, the Series Editor, and Gary Folven, the Managing Editor, of Kluwer Academic Publishers, for their continued encouragement and kind guidance. Some of the research effort leading to the results presented in this book was financially supported by the Cray Research Institute, National Textile Center, North Carolina Supercomputing Center, National Science Council of the Republic of China, Walter Clark Professorship Endowment, and the Foundation of International Education at Tokyo, Japan. We thank these institutions for their kind support.

Shu-Cherng Fang, Raleigh, North Carolina

J. R. Rajasekera, Niigata, Japan

H.-S. Jacob Tsao, Berkeley, California

1

INTRODUCTION TO ENTROPY AND ENTROPY OPTIMIZATION PRINCIPLES

This chapter provides a historical perspective of the concept of *entropy*, Shannon's reasoning, and the axioms that justify the principles of entropy optimization, namely, the maximum entropy and minimum cross-entropy principles. The mathematical forms of various entropy optimization problems are also discussed along with references to the existing literature. The chapter consists of two sections. Section 1.1 introduces the concept of entropy, and Section 1.2 classifies different entropy optimization problems to be studied in this book.

1.1 INTRODUCTION TO FINITE-DIMENSIONAL ENTROPY

The word *entropy* originated in the literature on thermodynamics around 1865 A.D. in Germany and was coined by Rudolf Clausius [6] to represent a measure of the amount of energy in a thermodynamic system as a function of the temperature of the system and the heat that enters the system. Clausius wanted a word similar to the German word *energie* (i.e., energy) and found it in the Greek word $\eta\ \tau\rho o\pi\eta$, which means transformation [1]. The word entropy had belonged to the domain of physics until 1948 when Claude Shannon, while developing his theory of communication at Bell Laboratories, used the term to represent a measure of information after a suggestion by John Von Neumann, who was then with Princeton University. Shannon wanted a word to describe his newly found *measure of uncertainty* and sought Von Neumann's help. Von Neumann's reasoning to Shannon [26] was that:

No one really understands entropy [in the sense of Physics]. Therefore, if you know what you mean by it and you use it when you are in an argument, you will win every time.

Whatever the reason for the name is, the concept of *Shannon's entropy* has penetrated a wide range of disciplines. They include statistical mechanics [16], thermodynamics [16], statistical inference [25], business and finance [7, 31], nonlinear spectral analysis [24], pattern recognition [28], transportation [12], urban and regional planning [22, 19], queueing theory [14], information theory [23, 13], parameter estimation, and linear and nonlinear programming [11, 21]. It is worth noting that, at the time when Shannon introduced his concept of entropy, no relationship, except for the similar mathematical expressions, was known to exist between *Shannon's entropy* and *thermodynamics entropy.* The relationship was only established later ([18], Chapter 3).

The concept of entropy is closely tied to the concept of *uncertainty embedded in a probability distribution.* In fact, entropy can be defined as a measure of probabilistic uncertainty. For example, suppose the probability distribution for the outcome of a coin-toss experiment is $(0.0001, 0.9999)$, with 0.0001 being the probability of having a tail. One is likely to notice that there is much more "certainty" than "uncertainty" about the outcome of this experiment and hence about the probability distribution. In fact, one is almost certain that the outcome will be a head. If, on the other hand, the probability distribution governing that same experiment were $(0.5, 0.5)$, one would realize that there is much less "certainty" and much more "uncertainty," when compared to the previous distribution. Generalizing this observation to the case of n possible outcomes, we conclude that the uniform distribution has the highest uncertainty out of all possible probability distributions. This implies that, if one had to choose a probability distribution for a chance experiment without any prior knowledge about that distribution, it would seem reasonable to pick the uniform distribution. This is because one would have no reason to choose any other and because that distribution maximizes the "uncertainty" of the outcome. This is called Laplace's Principle of Insufficient Reasoning [18]. Note that we are able to justify this principle without resorting to a rigorous definition of "uncertainty." However, this principle is inadequate when one has some prior knowledge about the distribution. Suppose, for example, that one knows some particular moments of the distribution, e.g., the expected value. In this case, a mathematical definition of "uncertainty" is crucial. This is the case where Shannon's measure of uncertainty, or Shannon's entropy, plays an indispensable role [23].

To define *entropy,* Shannon proposed some axioms that he thought any measure of uncertainty should satisfy and deduced a unique function, up to a multiplicative constant, that satisfies them. It turned out that this function actually possesses many more desirable properties. In later years, many researchers modified and replaced some of his axioms in an effort to simplify the reasoning. However, they all deduced that same function.

In this book, we focus on finite-dimensional entropy, i.e., Shannon's entropy defined on discrete probability distributions that have a finite number of outcomes (or states). Let $\mathbf{p} \equiv (p_1, p_2, \ldots, p_n)^T$ be a probability distribution associated with n possible outcomes, denoted by $\mathbf{x} \equiv (x_1, x_2, \ldots, x_n)^T$, of an experiment. Denote its entropy by $S_n(\mathbf{p})$. Among those defining axioms, Kapur and Kesavan stated the following [18]:

1. $S_n(\mathbf{p})$ should depend on all the p_j's, $j = 1, 2, \ldots, n$.

2. $S_n(\mathbf{p})$ should be a continuous function of p_j, $j = 1, 2, \ldots, n$.

3. $S_n(\mathbf{p})$ should be permutationally symmetric. In other words, if the p_j's are merely permuted, then $S_n(\mathbf{p})$ should remain the same.

4. $S_n(\frac{1}{n}, \frac{1}{n}, \ldots, \frac{1}{n})$ should be a monotonically increasing function of n.

5. $S_n(p_1, p_2, \ldots, p_n) = S_{n-1}(p_1 + p_2, p_3, \ldots, p_n) + (p_1 + p_2) S_2(\frac{p_1}{p_1+p_2}, \frac{p_2}{p_1+p_2})$.

Properties 1, 2, and 3 are obvious. Property 4 states that the maximum uncertainty of a probability distribution should increase as the number of possible outcomes increases. Property 5 is the least obvious but states that the uncertainty of a probability distribution is the sum of the uncertainty of the probability distribution that combines two of the outcomes and the uncertainty of the probability distribution consisting of only those two outcomes adjusted by the combined probabilities of the two outcomes.

It turns out that the unique family of functions that satisfy the defining axioms has the form $S_n(\mathbf{p}) = -k \sum_{j=1}^{n} p_j \ln p_j$, where k is a positive constant, ln represents the natural logarithmic function, and $0 \ln 0 = 0$ [18]. Shannon chose $-\sum_{j=1}^{n} p_j \ln p_j$ to represent his concept of entropy [23]. Among its many other desirable properties, we state the following:

(i) Shannon's measure is nonnegative and concave in $p_1, \ldots, p_n$.

(ii) The measure does not change with the inclusion of a zero-probability outcome.

(iii) The entropy of a probability distribution representing a completely certain outcome is 0, and the entropy of any probability distribution representing uncertain outcomes is positive.

(iv) Given any fixed number of outcomes, the maximum possible entropy is that of the uniform distribution.

(v) The entropy of the joint distribution of two independent distributions is the sum of the individual entropies.

(vi) The entropy of the joint distribution of two dependent distributions is no greater than the sum of the two individual entropies.

Property (i) is desirable because it is much easier to maximize a concave function than a non-concave one. Properties (ii) and (iii) are appealing because a zero-probability outcome contributes nothing to uncertainty, and neither does a completely certain outcome. Property (iv) was discussed earlier. Properties (v) and (vi) state that joining two distributions does not affect the entropy if they are independent and may actually reduce it if they are dependent.

We now revisit the problem of choosing a probability distribution for a chance experiment. We stated earlier that, in the absence of any information about the probability distribution, the best course of action is to choose the distribution that maximizes the uncertainty. In the absence of any information, this problem is relatively easy because the uniform distribution is clearly the one with maximum uncertainty. However, with the information about some of the moments of the distribution, it is not clear what distribution maximizes uncertainty while satisfying the moment constraints. Now that we can mathematically describe uncertainty in terms of entropy, we can generalize this reasoning and choose the distribution that maximizes uncertainty subject to the given moment constraints. In this way, we make full use of all the information given to us but avoid making any assumption about any information that is not available. Such reasoning leads to the following principle:

> *Out of all possible distributions that are consistent with the moment constraints, choose the one that has the maximum entropy.*

This principle was proposed by Jaynes ([18], Chapter 2) and has been known as the Principle of Maximum Entropy or Jaynes' Maximum Entropy Principle. It has often been abbreviated simply as **MaxEnt**. We now formulate the associated mathematical program.

Let X denote a random variable with n possible outcomes $\{x_1, x_2,..., x_n\}$. Let $\mathbf{p} = (p_1, p_2, \ldots, p_n)^T$ denote their probabilities, respectively. Let $g_1(X), g_2(X), \ldots, g_m(X)$ be m functions of X with known expected values $E[g_1(X)] = a_1$, $E[g_2(X)] = a_2$, $\ldots$, $E[g_m(X)] = a_m$, respectively. The distribution determination problem according to the Principle of Maximum Entropy can be formulated as the following mathematical problem:

$$\max \quad H_1(\mathbf{p}) = -\sum_{j=1}^{n} p_j \ln p_j \tag{1.1}$$

$$\text{s.t.} \quad \sum_{j=1}^{n} p_j g_i(x_j) = a_i, \quad i = 1, \ldots, m, \tag{1.2}$$

$$\sum_{j=1}^{n} p_j = 1, \tag{1.3}$$

$$p_j \geq 0, \quad j = 1, \ldots, n. \tag{1.4}$$

This is a concave programming problem with linear constraints. As will be shown in Chapter 3, the nonnegativity constraints are not binding for the optimal solution $\mathbf{p}^*$ because each p_j^* can be expressed as an exponential function in terms of the Lagrange multipliers associated with the equality constraints. Note that, in the absence of the moment constraints, the solution to the problem is the uniform probability distribution, whose entropy is $\ln n$. As such, the Maximum Entropy Principle can be viewed as an extension of Laplace's Principle of Insufficient Reasoning. The distribution selected under the Maximum Entropy Principle has also been interpreted as one that is the "most probable" in a certain sense. See Section 2.2 for details.

Recall that the whole discussion of the Maximum Entropy Principle was originally motivated by the task of choosing a probability distribution with known moment constraints. Suppose now that, in addition to those constraints, we have an *a priori* probability distribution $\mathbf{p}^0$ that we think our probability distribution $\mathbf{p}$ should be close to. In fact, in the absence of the moment constraints, we might choose $\mathbf{p}^0$ for $\mathbf{p}$. However, with the presence of the moment constraints, we would choose the probability distribution that is the "closest" to the *a priori* distribution among those that satisfy the moment constraints. But, to be able to do so, we need a precise definition of "closeness" or "deviation." In other words, we need to define some sort of deviation or, more precisely, "directed divergence" [18] on the space of discrete probability distributions that we are selecting the distribution from. Note that we are deliberately avoiding calling this measure a "distance." This is because a distance measure should be symmetric but such symmetry is not crucial in this case. Also, a distance measure should satisfy the triangular inequality, but this is not important in this context either. We can be content with a "one-way (asymmetric) deviation measure," $D(\mathbf{p}, \mathbf{p}^0)$, from $\mathbf{p}$ to $\mathbf{p}^0$. In fact, if a "one-way deviation measure" from $\mathbf{p}$ to $\mathbf{p}^0$ is not satisfactory, one can consider using a symmetric measure defined as the sum of $D(\mathbf{p}, \mathbf{p}^0)$ and $D(\mathbf{p}^0, \mathbf{p})$. What is important for this "directed divergence" measure is the following properties:

1. $D(\mathbf{p}, \mathbf{p}^0)$ should be nonnegative for all $\mathbf{p}$ and $\mathbf{p}^0$.
2. $D(\mathbf{p}, \mathbf{p}^0) = 0$ if and only if $\mathbf{p} = \mathbf{p}^0$.
3. $D(\mathbf{p}, \mathbf{p}^0)$ should be a convex function of $p_1, p_2, \ldots, p_n$.
4. When $D(\mathbf{p}, \mathbf{p}^0)$ is minimized subject to moment constraints but without the explicit presence of the nonnegativity constraints, the resulting p_j's should be nonnegative.

Property 1 is desirable for any such measure of deviation. If property 2 were not satisfied, then it would be possible to choose a $\mathbf{p}$ that has a zero directed divergence from $\mathbf{p}^0$, i.e., one that is as "close" to $\mathbf{p}^0$ as $\mathbf{p}^0$ itself, but differs from $\mathbf{p}^0$. Property 3 makes minimizing the measure much simpler, and property 4 spares us from explicitly considering n nonnegativity constraints. Fortunately, there are many measures that satisfy these properties. We may even be able to find one that satisfies the triangle inequality. But, simplicity of the measure is also desirable. The simplest and most important of those measures is the Kullback-Liebler measure ([18], Chapter 4), also known as the measure of cross-entropy, defined as $D(\mathbf{p}, \mathbf{p}^0) = \sum_{j=1}^{n} p_j \ln \frac{p_j}{p_j^0}$ with the convention that, whenever p_j^0 is 0, p_j is set to 0 and $0 \ln \frac{0}{0}$ is defined to be 0. In addition to the four properties discussed before, we state the following properties:

(i) $D(\mathbf{p}, \mathbf{p}^0)$ is a continuous function of $\mathbf{p}$ and $\mathbf{p}^0$.

(ii) $D(\mathbf{p}, \mathbf{p}^0)$ is permutationally symmetric, i.e., the measure does not change if the pairs of (p_j, p_j^0) are permuted among themselves.

(iii) $D(\mathbf{p}, \mathbf{p}^0)$ is convex in both $\mathbf{p}$ and $\mathbf{p}^0$.

(vi) $D(\mathbf{p}, \mathbf{p}^0)$ is not symmetric.

(v) If $\mathbf{p}$ and $\mathbf{q}$ are independent and $\mathbf{r}$ and $\mathbf{s}$ are also independent, then

$$D(\mathbf{p} * \mathbf{q}, \mathbf{r} * \mathbf{s}) = D(\mathbf{p}, \mathbf{r}) + D(\mathbf{q}, \mathbf{s}),$$

where $*$ denotes the convolution operation between two independent distributions.

(vi) In general, the triangular inequality does not hold. But, if distribution $\mathbf{p}$ minimizes $D(\mathbf{p}, \mathbf{p}^0)$ subject to some moment constraints and $\mathbf{q}$ is any other distribution that satisfies those same constraints, then

$$D(\mathbf{q}, \mathbf{p}^0) = D(\mathbf{q}, \mathbf{p}) + D(\mathbf{p}, \mathbf{p}^0).$$

Thus, in this special case, the triangular inequality holds but as an equality.

With cross-entropy interpreted as a measure of "deviation," the Kullback-Leibler's Minimum Cross Entropy Principle, or **MinxEnt**, can be stated as follows [18]:

> *Out of all possible distributions that are consistent with the moment constraints, choose the one that minimizes the cross-entropy with respect to the given a priori distribution.*

In mathematical terms, this principle is simply:

$$\min \qquad H_2(\mathbf{p}) = \sum_{j=1}^{n} p_j \ln \frac{p_j}{p_j^0} \tag{1.5}$$

$$\text{s.t.} \qquad \sum_{j=1}^{n} p_j g_i(x_j) = a_i, \quad i = 1, \ldots, m, \tag{1.6}$$

$$\sum_{j=1}^{n} p_j = 1, \tag{1.7}$$

$$p_j \geq 0, \quad j = 1, \ldots, n. \tag{1.8}$$

Note that the nonnegativity constraints are not binding as in the MaxEnt problem, for the same reason.

Now, if no *a priori* distribution is given, then one may use a maximum-entropy distribution in its place. This leads to the use of the uniform distribution, denoted by $\mathbf{u}$, as the *a priori* distribution. In this case,

$$D(\mathbf{p}, \mathbf{p}^0) = D(\mathbf{p}, \mathbf{u}) = \sum_{j=1}^{n} p_j \ln \frac{p_j}{1/n} = \ln n + \sum_{j=1}^{n} p_j \ln p_j.$$

Since minimizing $\sum_{j=1}^{n} p_j \ln p_j$ is equivalent to maximizing $-\sum_{j=1}^{n} p_j \ln p_j$, minimizing the cross-entropy with respect to the uniform distribution is equivalent to maximizing entropy and, therefore, MaxEnt is a special case of MinxEnt. These two principles can now be combined into a general principle:

> *Out of all probability distributions satisfying the given moment constraints, choose the distribution that minimizes the cross-entropy with respect to the given a priori distribution and, in the absence of it,*

choose the distribution that minimizes the cross-entropy with respect to the uniform distribution.

It is worth noting that the MaxEnt and MinxEnt principles could be generalized to the case of discrete probability distributions defined over a countably infinite number of possible outcomes or states. This will be discussed in Section 7.1. These two principles can also be generalized to the case of continuous probability distributions. However, the corresponding mathematical optimization problems are formulated with functionals as objectives and constraints. Section 7.2 contains a discussion of this generalization. Also note that many special cases of those problems, continuous or discrete, lead to well-known and widely used distributions that have been discovered independently of MaxEnt or MinxEnt. Again, the focus of this book will be on discrete probability distributions defined over a finite number of possible outcomes and their finite-dimensional variations.

In the next section, we classifies several types of entropy maximization and cross-entropy minimization problems.

1.2 ENTROPY OPTIMIZATION PROBLEMS

In addition to the MaxEnt and MinxEnt models introduced in the previous section, many more general entropy optimization models have proven useful in science and engineering. This section defines some of the better known models, which will be treated in detail in Chapters 2, 3 and 4.

Four classes of cross-entropy optimization models are defined in this section. They differ in the type of constraints. The four types of constraints are linear, quadratic, cross-entropic and general convex constraints. The corresponding cross-entropy optimization problems are discussed in the following four subsections. Applications of these models will be studied in Chapter 2. Theory and algorithms will be the subject of Chapters 3 and 4.

1.2.1 Cross-Entropy Minimization with Linear Constraints

Motivated by the applications of entropy in the areas of information theory [13], transportation planning [29], urban and regional planning [30], and queueing theory [14], Censor et al. [4] considered the following problem:

$$\min \quad \sum_{j=1}^{n} p_j \ln\left(\frac{p_j}{p_j^0}\right) \tag{1.9}$$

$$\text{s.t.} \quad \mathbf{p} \in Q, \tag{1.10}$$

$$\sum_{j=1}^{n} p_j = 1, \tag{1.11}$$

$$p_j \geq 0, \quad j = 1, 2, ..., n, \tag{1.12}$$

where $\mathbf{p}^0$ is the given *a priori* distribution and $Q \subset R^n$ is a constraint set of the following possible types:

$$Q_1 = \{\mathbf{p} \in R^n : \mathbf{Ap} = \mathbf{b}\},$$
$$Q_2 = \{\mathbf{p} \in R^n : \mathbf{Ap} \leq \mathbf{b}\},$$
$$Q_3 = \{\mathbf{p} \in R^n : \mathbf{c} \leq \mathbf{Ap} \leq \mathbf{b}\},$$

where $\mathbf{A}$ is an $m \times n$ matrix and $\mathbf{b}$ and $\mathbf{c}$ are m-dimensional column vectors.

Theory and algorithms for solving such a problem [3, 4, 5, 15, 20, 27] will be discussed in detail in Chapter 3.

1.2.2 Cross-Entropy Minimization with Quadratic Constraints

Motivated by studying the "graduation" problem in actuarial science and applications in many other diverse disciplines [17], the problem defined in the previous section was extended by Zhang and Brockett [32] to include quadratic constraints. The problem, as they defined, finds the probability distribution $\mathbf{p}$ that is the "closest" to a given *a priori* distribution $\mathbf{p}^0$ among those satisfying a given set of quadratic inequality constraints. It is defined as follows:

$$\min \quad \sum_{j=1}^{n} p_j \ln\left(\frac{p_j}{p_j^0}\right) \tag{1.13}$$

$$\text{s.t.} \quad g_i(\mathbf{p}) = \frac{1}{2}\mathbf{p}^T\mathbf{H}_i\mathbf{p} + \mathbf{b}_i^T\mathbf{p} + c_i \leq 0, \quad i = 1, 2, .., m, \tag{1.14}$$

$$\sum_{j=1}^{n} p_j = 1, \tag{1.15}$$

$$p_j \geq 0, \quad j = 1, 2, ..., n, \tag{1.16}$$

where $\mathbf{H}_i$ is an $n \times n$ symmetric positive semi-definite matrix, $\mathbf{b}_i$ is an n-dimensional column vector, and c_i is a constant, for $i = 1, 2, \ldots, m$.

Many applications can be found in [17] and [32]. One recent application of this model is in the area of portfolio "rebalancing" [31]. Theory and algorithms for solving such a problem [32, 9] will be discussed in detail in Sections 4.1 and 4.2.

1.2.3 Cross-Entropy Minimization with Cross-Entropic Constraints

This problem is to find a probability distribution $\mathbf{p}$ that is the "closest" to a given *a priori* distribution $\mathbf{p}^0$ among those that are within a "deviation measure" of e_i from a given set of distributions $\{\mathbf{p}^i | i = 1, 2, ..., m\}$. In mathematical terms, the problem is of the form:

$$\min \quad \sum_{j=1}^{n} p_j \ln\left(\frac{p_j}{p_j^0}\right) \tag{1.17}$$

$$\text{s.t.} \quad g_i(\mathbf{p}) = \sum_{j=1}^{n} p_j \ln\left(\frac{p_j}{p_j^i}\right) \leq e_i, \quad i = 1, 2, \ldots, m, \tag{1.18}$$

$$\sum_{j=1}^{n} p_j = 1, \tag{1.19}$$

$$p_j \geq 0, \quad j = 1, 2, \ldots, n. \tag{1.20}$$

Motivation to study this problem comes from information theory, where it has significant applications [2]. Theory and algorithms for solving such a problem [8] will be discussed in detail in Section 4.3.

1.2.4 Cross-Entropy Minimization with Convex Constraints

This problem generalizes all the problems defined earlier in this section. It has the following form:

$$\text{Min} \quad \sum_{j=1}^{n} p_j \ln\left(\frac{p_j}{p_j^0}\right) \tag{1.21}$$

$$\text{s.t.} \quad g_i(\mathbf{p}) = h_i\left(\mathbf{A}_i\mathbf{p}\right) + \mathbf{b}_i^T\mathbf{p} + c_i \leq 0, \quad i = 1, 2, .., m, \tag{1.22}$$

$$\sum_{j=1}^{n} p_j = 1, \tag{1.23}$$

$$p_j \geq 0, \quad j = 1, 2, ..., n, \tag{1.24}$$

where $\mathbf{A}_i$ is an $m_i \times n$ matrix with full row-rank, $\mathbf{b}_i$ is an n-dimensional column vector, c_i is a constant, and h_i is a convex function on R^{m_i}, for $i = 1, 2, \ldots, m$.

Note that when $h_i\left(\mathbf{A}_i\mathbf{p}\right) = 0$, the problem reduces to a linearly constrained cross-entropy problem; when $h_i\left(\mathbf{A}_i\mathbf{p}\right) = \frac{1}{2}\mathbf{p}^T\mathbf{A}_i^T\mathbf{A}_i\mathbf{p}$, it becomes quadratically constrained; when $\mathbf{A}_i = \mathbf{I}$, $h_i\left(\mathbf{p}\right) = \sum_{j=1}^{n} p_j \ln\left(\frac{p_j}{p_j^i}\right)$, and $\mathbf{b}_i = \mathbf{0}$, it involves cross-entropic constraints.

For many applications in engineering and social sciences, the embedded constraints are neither linear nor quadratic [17]. In this case, general convex constraints often need to be added to describe the underlying system. This motivates the study of the problem defined above. Theory and algorithms for solving such a problem [8, 10] will be discussed in detail in Section 4.4.

1.2.5 Concluding Remarks

The four types of problems identified in this section appear in many applications, some of which will be discussed in Chapter 2. The theory and algorithms for solving such problems are the subjects of Chapters 3 and 4. In addition to the entropy optimization problems defined in this section, entropy has played a major role in mathematical programming as a barrier function to many optimization problems. How to exploit the characteristics of entropy optimization in mathematical programming is the subject to be studied in Chapters 5, 6, and 7.

REFERENCES

[1] Baierlein, R., "How Entropy Got Its Name," *American Journal of Physics*, Vol. 60, 1992, p. 1151.

[2] Ben-Tal, A., Teboulle, M., and Charnes, A., "The Role of Duality in Optimization Problems involving Entropy Functionals with Applications to Information Theory," *Journal of Optimization Theory and Applications*, Vol. 58, 1988, pp. 209–223.

[3] Bregman, L.M., "The Relaxation Method of Finding the Common Point of Convex Sets and its Application to the Solution of Problems in Convex Programming," *U.S.S.R. Computational Mathematics and Mathematical Physics*, Vol. 7, 1967, pp. 200–217.

[4] Censor, Y., Elfving, T., and Herman, G.T., "Special Purpose Algorithms for Linearly Constrained Optimization," *Maximum-Entropy and Bayesian Spectral Analysis and Estimation Problems*, edited by C.R. Smith and G.J. Erickson, D. Reidel Publishing Company, 1987, pp. 241–254.

[5] Censor, Y., De Pierro, A.R., Elfving, T., Herman, G.T., and Iusem, A.N., "On Maximization of Entropies and Generalization of Bregman's Method of Convex Programming," Technical Report MIPG 113, Medical Image Processing Group, Department of Radiology, Hospital of the University of Pennsylvania, Philadelphia, PA., 1986.

[6] Clausius, R., "Ueber Verschiedene fur die Anwendung Bequeme Formen der Hauptgleichungen der Mechanischen Warmetheorie," *Annalen der Physik und Chemie*, Vol. 125, 1865, pp. 353–400.

[7] Cozzolino, J.M., and Zahner, M.J., "The Maximum Entropy Distribution of the Future Market Price of a Stock," *Operations Research*, Vol. 21, 1973, pp. 1200–1211.

[8] Fang, S.-C., Peterson, E.L., and Rajasekera, J.R., "Minimum Cross-Entropy Analysis with Entropy-Type Constraints," *Journal of Computational and Applied Mathematics*, Vol. 39, 1992, pp. 165–178.

[9] Fang, S.-C., and Rajasekera, J.R., "Quadratically Constrained Minimum Cross-Entropy Analysis," *Mathematical Programming*, Vol. 44, 1989, pp. 85–96.

[10] Fang, S.-C., and Rajasekera, J.R., "Minimum Cross-Entropy Analysis with Convex Constraints," *Information and Computation*, Vol. 116, 1995, pp. 304–311.

[11] Fang, S.-C., and Tsao, H.-S.J., "An Unconstrained Convex Programming Approach to Solving Convex Quadratic Programming Problems," *Optimization*, Vol. 27, 1993, pp. 235–243.

[12] Fang, S.-C., and Tsao, H.-S.J., "Linear-Constrained Entropy Maximization Problem with Quadratic Cost and its Applications to Transportation Planning Problems," *Transportation Science*, Vol. 29, 1995, pp. 353–365.

[13] Guiasu, S., *Information Theory with Applications*, McGraw-Hill, New York, 1977.

[14] Guiasu, S., "Maximum Entropy Condition in Queueing Theory," *Journal of Operational Research Society*, Vol. 37, 1986, pp. 293–301.

[15] Gordon, R., Bender, R., and Herman, G. T., "Algebraic Reconstruction Techniques (ART) for Three Dimensional Electron Microscopy and X-ray Photography," *Journal of Theoretical Biology*, Vol. 29, 1970, pp. 471–481.

[16] Jaynes, E.T., "Information Theory and Statistical Mechanics II," *Physics Review*, Vol. 108, 1957, pp. 171–190.

[17] Johnson, R.W., and Shore, J.E., "Minimum Cross-entropy Spectral Analysis of Multiple Signals," *IEEE Transactions on Acoustics, Speech, and Signal Processing*, Vol. ASSP-31, 1983, pp. 574–582.

[18] Kapur, J.N., and Kesavan, H.K., *Entropy Optimization Principles with Applications*, Academic Press, Boston, 1992.

[19] Kumar, V., Hoshino, K., and Kumar, U., "An Application of the Entropy Maximization Approach in Shopping Area Planning," *International Journal of General Systems*, Vol. 16, 1989, pp. 25–42.

[20] Lent, A., "A Convergent Algorithm for Maximum Entropy Image Restoration with a Medical X-Ray Application," in *Image Analysis and Evaluation*, edited by Shaw, R., Society of Photographic Scientists and Engineers (SPSE), Washington, D.C., 1977, pp. 249–257.

[21] Rajasekera, J.R., and Fang, S.-C., "Deriving an Unconstrained Convex Program for Linear Programming," *Journal of Optimization Theory and Applications*, Vol. 75, 1992, pp. 603–612.

[22] Scott, C.H., and Jefferson T.R., "Entropy Maximizing Models of Residential Location via Geometric Programming," *Geographical Analysis*, Vol. 9, 1977, pp. 181–187.

[23] Shannon, C.E., and Weaver, W., *The Mathematical Theory of Communication*, University of Illinois Press, Urbana, Illinois, 1962.

[24] Shore, J.E., "Minimum Cross-entropy Spectral Analysis," *IEEE Transactions on Acoustics, Speech, and Signal Processing*, Vol. ASSP-29, 1981, pp. 230–237.

[25] Tribus, M., *Rational Descriptions, Decisions, and Designs*, Pergamon Press, New York, 1969.

[26] Tribus, M., "An Engineer Looks at Bayes," in *Maximum-Entropy and Bayesian Methods in Science and Engineering, Vol. 1: Foundations,* edited by Gary J. Erickson and C. Ray Smith, Kluwer Academic Publishers, Dordrecht, The Netherlands, 1988, pp. 31–52.

[27] Tsao, H.-S.J., Fang, S.-C., and Lee, D.N., "On the Optimal Entropy Analysis," *European Journal of Operational Research*, Vol. 59, 1992, pp. 324–329.

[28] Wang, Y., and Lu, W., "Multicriterion Maximum Entropy Image Reconstruction from Projections," *IEEE Transactions on Medical Imaging*, Vol. 11, 1992, pp. 70–75.

[29] Wilson, A.G., "A Statistical Theory of Spatial Distribution Models," *Transportation Research*, Vol. 1, 1967, pp. 253–269.

[30] Wilson, A.G., *Entropy in Urban and Regional Planning*, Pion, London, 1970.

[31] Yamada, M., and Rajasekera, J.R., "Portfolio Re-balancing with the Entropy Criteria," Report No. 310, QUICK Research Institute Corp., Tokyo, 1993.

[32] Zhang, J., and Brockett, P.L., "Quadratically Constrained Information Theoretic Analysis," *SIAM Journal on Applied Mathematics*, Vol. 47, 1987, pp. 871–885.

2

ENTROPY OPTIMIZATION MODELS

Entropy optimization models have been successfully applied to practical problems in many scientific and engineering disciplines. As noted in Chapter 1, those disciplines include statistical mechanics, thermodynamics, statistical parameter estimation and inference, economics, business and finance, nonlinear spectral analysis, pattern recognition, transportation, urban and regional planning, queueing theory, and linear and nonlinear programming. Included in this book are example applications in the areas of (1) queueing theory, (2) transportation planning, (3) input-output analysis, (4) regional planning, (5) investment portfolio optimization, and (6) image reconstruction. They are discussed in six sections, from Sections 2.1 through 2.6.

2.1 QUEUEING THEORY

Queueing theory, whose origin goes back to the early 1900s [13], has many applications in such fields as telecommunications, manufacturing and computer systems. Within the context of queueing theory, a system consisting of servers, customers and queues is usually expressed in terms of arrival processes, service processes, queueing disciplines and system configurations. The resulting formulations are used to compute the probability distributions of various system attributes, e.g., queue length and waiting time. In many cases, finding an exact probability distribution could be difficult [42]. Since the early 1970's, an approximation scheme has been applied successfully to solving queueing problems [16, 40]. In that scheme, some attributes of the exact probability distribution are first obtained, e.g., the expected value, standard deviation, and quantiles, and then used as the constraints of an entropy optimization problem.

Consider, for example, a simple queueing system known as M/M/1 (exponential interarrival time, exponential service time, and single server model [18], page 9) in a steady state. Define:

$N \equiv$ random number of customers in the system (including service and queue),

$n \equiv$ maximum number of customers allowed in the system,

$p_j \equiv$ probability of having j customers in the system,

$\bar{N} \equiv$ expected number of customers in the system, i.e.,

$$\bar{N} = E(N) = \sum_{j=0}^{n} jp_j. \tag{2.1}$$

Now, let us assume that the expected number of customers in the system, i.e., $\bar{N}$, is known. To solve for $\{p_j\}$, the maximum entropy principle calls for the solution of the following problem [19]:

Model 2.1:

$$\max \quad -\sum_{j=0}^{n} p_j \ln p_j \tag{2.2}$$

$$\text{s.t.} \quad \sum_{j=0}^{n} jp_j = \bar{N}, \tag{2.3}$$

$$\sum_{j=0}^{n} p_j = 1, \tag{2.4}$$

$$p_j \geq 0, \quad j = 0, 1, 2, ..., n. \tag{2.5}$$

After converting Model 2.1 to a minimization form, we define the following Lagrangian function

$$L(\mathbf{p}, \alpha, \beta) = \sum_{j=0}^{n} p_j \ln p_j + \alpha \left(\sum_{j=0}^{n} p_j - 1 \right) + \beta \left(\sum_{j=0}^{n} jp_j - \bar{N} \right),$$

where $\alpha \in R$ and $\beta \in R$ are Lagrange multipliers associated with Constraints (2.3) and (2.4).

Taking and solving the first-order optimality conditions gives

$$p_j = \exp(-\alpha - \beta j - 1), \qquad j = 0, 1, ..., n. \tag{2.6}$$

From Equation (2.3) of Model 2.1, we have:

$$\exp(\alpha + 1) = \sum_{j=0}^{n} \exp(-\beta j) = \frac{1 - \delta^{n+1}}{1 - \delta}, \tag{2.7}$$

where

$$\delta = \exp(-\beta). \tag{2.8}$$

Then, Equations (2.6), (2.7) and (2.8) imply that, for $j = 0, 1, ..., n$,

$$p_j = \frac{\delta^j (1 - \delta)}{(1 - \delta^{n+1})}. \tag{2.9}$$

Note that δ is related to $\bar{N}$ in the following manner:

$$\bar{N} = \sum_{j=0}^{n} \frac{j\delta^j (1 - \delta)}{(1 - \delta^{n+1})} = \frac{\delta \left[1 - (n+1)\delta^n + n\delta^{n+1}\right]}{(1 - \delta^{n+1})(1 - \delta)}. \tag{2.10}$$

This is identical to the result derived in the classical queueing theory based on the theory of "birth and death process" [18].

The corresponding queueing problem with no capacity constraints is equivalent to Model 2.1 with $n = \infty$. It can be shown [19] that, when $n \to \infty$,

$$p_j = \frac{\bar{N}^j}{(1 + \bar{N})^{j+1}}, \tag{2.11}$$

where

$$\bar{N} = \frac{\delta}{1 - \delta}. \tag{2.12}$$

The well-known Pollaczek–Khinkchine formula ([18], page 254) derived by simply using a mean-value argument for a steady state M/M/1 queueing system states that

$$\bar{N} = \frac{\lambda}{\mu - \lambda}, \tag{2.13}$$

where λ is the arrival rate and $\mu > \lambda$ is the service rate. Substituting Equation (2.13) in Equation (2.11) gives

$$p_j = \left(1 - \frac{\lambda}{\mu}\right) \left(\frac{\lambda}{\mu}\right)^j, \qquad j = 0, 1, ..., \infty. \tag{2.14}$$

This result is also identical to the analytical result derived in the classical queueing theory based solely on the theory of "birth and death process" ([18], page 70).

Many other variations of Model 2.1 have been explored. For M/G/1 queueing systems without capacity constraints, it was shown in [8] that the corresponding entropy optimization problem became:

Model 2.2:

$$\max \quad -\sum_{j=0}^{\infty} p_j \ln(p_j/p_j^0) \tag{2.15}$$

$$\text{s.t.} \quad \sum_{j=0}^{\infty} jp_j = \rho(1+\rho y_s)/(1-\rho), \tag{2.16}$$

$$\sum_{j=0}^{\infty} p_j = 1, \tag{2.17}$$

$$p_j \geq 0, \quad j = 0, 1, 2, ..., \infty, \tag{2.18}$$

where $\rho = \lambda/\mu$, $y_s = (c_s^2 - 1)/2$ with c_s^2 being the squared coefficient of variation for the service time distribution, $p_0^0 = (1-\rho)$ and p_j^0 is a given constant, for $j = 1, 2, ..., \infty$.

For non-capacitated G/M/1 queueing systems, Constraint (2.16) is replaced by the following form ([8], page 351):

$$\sum_{j=0}^{\infty} jp_j = \rho/\{r_a(1-\rho)\} \tag{2.19}$$

where $r_a = 1/(1+y_a)$ and $y_a = (c_a^2 - 1)/2$ with c_a^2 being the squared coefficient of variation for the interarrival time distribution.

For GI/G/c queueing systems with no capacity constraint, an entropy optimization model has been formulated as follows ([50], page 817).

Model 2.3 :

$$\max \quad -\sum_{j=0}^{\infty} p_j \ln p_j \tag{2.20}$$

$$\text{s.t.} \quad \sum_{j=0}^{\infty} j p_j = \bar{Q}, \tag{2.21}$$

$$\sum_{j=0}^{\infty} p_j = 1, \tag{2.22}$$

$$p_j = a_j, \qquad j = 0, 1, ..., c-1, \tag{2.23}$$

$$p_j \geq 0, \qquad j = c, c+1, ..., \infty, \tag{2.24}$$

where $\bar{Q}$, the mean queue length, is estimated and given through analytical modeling, simulation, or empirical observation, and $a_j \geq 0$, $j = 0, 1, ..., c-1$, is also given ([50], page 817).

The application of entropy optimization to numerous queueing problems is a vibrant research area. Recent applications include analysis of open queueing networks with group arrival patterns [51] and queues with generalized blocking [33].

2.2 TRANSPORTATION PLANNING

A traditional transportation planning process consists of four major components: trip generation, trip distribution, modal split, and trip assignment. Many transportation planning problems can be formulated as a linearly constrained convex programming problem whose objective function consists of entropy functions and other cost-related terms [4, 10, 11, 25, 26, 34, 43, 44]. This section illustrates the usefulness of entropy optimization models in transportation planning with several well-known trip distribution models. These models will also be justified.

In a transportation planning process, a region is usually partitioned into a set of disjoint zones. An important task is to estimate the number of trips between all pairs of zones. We consider the problem of estimating these trip numbers within a 24-hour period. Usually, the total number of trips originated from any particular zone (also known as trip production for that zone) and the total number of trips destined for any particular zone (also known as trip attraction) can be estimated with sufficient accuracy either through surveys or other statistical means. However, accurate statistical estimation of the number of trips between a particular pair of origin and destination zones has been considered a difficult task. Therefore, different methods have been proposed to estimate the origin-destination trip numbers, based on the trip production, trip attraction, and other related information. One such piece of information is the average trip time, distance, or cost for the region under study. A widely used model is the following maximum entropy model:

Model 2.4:

$$\max \quad - \sum_{i=1}^{K}\sum_{j=1}^{L} x_{ij} \ln x_{ij} \tag{2.25}$$

$$\text{s.t.} \quad \sum_{j=1}^{L} x_{ij} = O_i, \quad i = 1, 2, \ldots, K, \tag{2.26}$$

$$\sum_{i=1}^{K} x_{ij} = D_j, \quad j = 1, 2, \ldots, L, \tag{2.27}$$

$$\sum_{i=1}^{K}\sum_{j=1}^{L} c_{ij} x_{ij} = C, \tag{2.28}$$

$$x_{ij} \geq 0, \quad i = 1, 2, \ldots, K, \text{ and } j = 1, 2, \ldots, L, \tag{2.29}$$

where x_{ij} denotes the trip number between zones i and j, O_i denotes the total number of trips originated from zone i, D_j denotes the total number of trips destined for zone j, K is the total number of origin zones, L is the total number of destination zones, c_{ij} is the travel time, distance, or cost between zones i and j, and C is the total cost incurred by all trips across the region. Note that K could be equal to L. The following three different reasons explain why the maximum entropy model above is especially appropriate for the trip distribution problem.

Given the total number T of trips from all origins to all destinations and treating each of the T trips as a distinctive entity, consider the problem of assigning these T trips to the $K \times L$ origin-destination pairs, or equivalently a $K \times L$ trip matrix. Each assignment leads to a trip matrix. Conceptually, one can exhaust all possible assignments and tally the number of times any particular trip matrix occurs. It seems natural to treat each assignment as equally likely and use the trip matrix that occurs the most number of times, i.e., the most probable trip matrix, as the estimated trip numbers. We now show that this process essentially leads to the maximization of the entropy function $-\sum_{i=1}^{K}\sum_{j=1}^{L} x_{ij}\ln x_{ij}$.

Let $\mathbf{M} = \{T_{ij}\}$ denote the $K \times L$ trip matrix and $\gamma(\mathbf{M})$ the number of assignments leading to the trip matrix $\mathbf{M}$. With some combinatorial analysis, it can be seen that

$$\gamma(\mathbf{M}) = \frac{T!}{\prod_{ij} T_{ij}!}. \tag{2.30}$$

Rather than maximizing $\gamma(\mathbf{M})$ directly, we solve an equivalent problem of maximizing $\ln\gamma(\mathbf{M})$. By Stirling's formula, we have

$$\ln T_{ij}! \approx T_{ij}\ln T_{ij} - T_{ij} + \frac{1}{2}(\ln 2\pi + \ln T_{ij}) \approx T_{ij}\ln T_{ij} - T_{ij}, \tag{2.31}$$

and

$$\ln\gamma(\mathbf{M}) \approx T\ln T - T - \sum_{ij}(T_{ij}\ln T_{ij} - T_{ij}), \tag{2.32}$$

$$= -\sum_{ij} T_{ij}\ln T_{ij} + T\ln T. \tag{2.33}$$

Since T is given, maximizing $\ln\gamma(\mathbf{M})$ is equivalent to maximizing entropy, as defined in (2.25). This is the first reason why the entropy optimization model is particularly suitable for the trip distribution problem.

The optimization problem with objective (2.25) and Constraints (2.26) - (2.29) produces an optimal solution in the following form:

$$x_{ij} = cr_i e^{-\alpha c_{ij}} s_j, \tag{2.34}$$

where $r_i = e^{\lambda_i}, s_j = e^{\mu_j}, c = e^{-1}$, and λ_i, μ_j and α are the Lagrange multipliers associated with the constraints. The exponential term is a decreasing function of c_{ij}, which may represent distance, cost, or some notion of generalized cost. This particular solution form has been referred to as the "gravity model" in the literature and the exponential term as the "exponential deterrence." Note that this deterrence is different from the "quadratic deterrence" in the Newtonian gravity law. This gravity model had been proposed and used for empirical estimates well before the advent of the entropy optimization formulation. The entropy optimization model provides a theoretical justification for a technique that was originally conceived based on empirical observation and validation. This is the second reason.

The third reason has to do with the interpretation of entropy as the negation of information, as proposed by Jaynes [23]. In this interpretation, least information implies maximum entropy. He proposed that, when faced with the problem of selecting a probability distribution to represent one's uncertainty about a random variable, one should select the one that satisfies all known properties of the distribution but contains the least information, or, equivalently, maximum entropy. This has been referred to as the maximum entropy principle, which calls for entropy maximization subject to the constraints representing the properties.

Normalizing the trip numbers by dividing them by the total number of trips T gives a bivariate probability distribution:

$$p_{ij} = \frac{T_{ij}}{T}. \tag{2.35}$$

It is easy to see that Constraints (2.26) – (2.29) can be converted to an equivalent set of probability constraints. The first two constraints specify the two marginal distributions while the third specifies the average cost. Finally, since

$$\sum_{ij} p_{ij} \ln p_{ij} = -\ln T + \frac{1}{T} \sum_{ij} T_{ij} \ln T_{ij}, \tag{2.36}$$

the trip distribution model defined by (2.25) – (2.29) is further justified by the maximum entropy principle. For more detail on the application of entropy optimization models or gravity models to transportation planning, the reader is referred to [48, 37] .

We now consider an equivalent entropy-optimization formulation and generalize this basic trip distribution model. It is well known that the gravity model defined by Equation (2.34) can be derived via several different approaches. We first define a general entropy optimization problem with quadratic cost terms, a special case of which is equivalent to the gravity model:

Model 2.5:

$$\max \quad -\sum_{i=1}^{K}\sum_{j=1}^{L} x_{ij}\ln x_{ij} - \alpha\sum_{i=1}^{K}\sum_{j=1}^{L} c_{ij}x_{ij} - \frac{\beta}{2}\sum_{i=1}^{K}\sum_{j=1}^{L} d_{ij}x_{ij}^2 \tag{2.37}$$

$$\text{s.t.} \quad \sum_{j=1}^{L} x_{ij} = O_i, \quad i = 1, 2, \ldots, K, \tag{2.38}$$

$$\sum_{i=1}^{K} x_{ij} = D_j, \quad j = 1, 2, \ldots, L, \tag{2.39}$$

$$x_{ij} \geq 0, \quad i = 1, 2, \ldots, K, \text{ and } j = 1, 2, \ldots, L. \tag{2.40}$$

It can be shown that, without the quadratic cost terms in the objective function, Model 2.5 is equivalent to the well-known gravity model. In other words, the optimal solution of Model 2.5, in the absence of the quadratic terms in its objective function, also takes the form of Equation (2.34).

Tomlin [43] pointed out that, in addition to the entropic terms, the inclusion of a linear cost (e.g., travel time) per unit flow from an origin to a destination alone in a trip distribution problem is an oversimplification. Assuming a convex quadratic cost that reflects the cost increase due to congestion in Model 2.5 should be a better approximation of the actual cost. The coefficients d_{ij}'s in the quadratic term of the objective function accommodate the possibility of difference in quadratic approximations for the $K \times L$ origin/destination pairs.

With the quadratic terms, the model is no longer a gravity model. Instead, its optimal solution satisfies the following equation:

$$x_{ij} = cr_i e^{-\alpha c_{ij} - \beta d_{ij} x_{ij}} s_j. \tag{2.41}$$

This differs from Equation (2.34) of the gravity model in that x_{ij} appears in the exponent in the right-hand-side expression and the optimal solution is no longer explicitly solvable as a function of the program parameters. When all βd_{ij}'s are very small, Model 2.5 produces approximately the gravity model. However, the quadratic terms have the following "self-deterrence" or "self-regulation" effect on the optimal solution. By increasing the value of one particular βd_{ij} but keeping all other parameters the same, the corresponding optimal trip number x_{ij} decreases. Moreover, for any particular positive value of βd_{ij}, the larger the solution x_{ij} of the gravity model (2.34) is, the larger the decrement from this solution to the optimal x_{ij} of the Model 2.5 is. This characteristic can be used to model deterrence due to congestion in the stage of trip distribution. Moreover, solving the combined trip distribution and assignment problem often involves a problem decomposition where the trip distribution is a component that is required to be solved repeatedly. This self-regulation feature of Model 2.5 could benefit the convergence rate due to its ability to reflect, at the stage of distribution, the congestion cost obtained in the assignment stage. For ease of discussion, we will refer to Model 2.5 as the "self-deterrent gravity model" in the rest of this book. We keep the notion of gravity in the name because the optimal solution x_{ij} is still a decreasing function of c_{ij}.

We now offer an alternative motivation for studying Model 2.5. Similar to the fact that the gravity model can be derived from the maximum entropy problem defined by (2.25) – (2.29), Equation (2.41) can be derived as the optimal solution of the following mathematical program:

Model 2.6:

$$\max \quad -\sum_{i=1}^{K}\sum_{j=1}^{L} x_{ij} \ln x_{ij} \tag{2.42}$$

$$\text{s.t.} \quad \sum_{j=1}^{L} x_{ij} = O_i, \quad i = 1, 2, \ldots, K, \tag{2.43}$$

$$\sum_{i=1}^{K} x_{ij} = D_j, \quad j = 1, 2, \ldots, L, \tag{2.44}$$

$$\sum_{i=1}^{K}\sum_{j=1}^{L} c_{ij} x_{ij} = C, \tag{2.45}$$

$$\frac{1}{2}\sum_{i=1}^{K}\sum_{j=1}^{L} d_{ij} x_{ij}^2 = D, \tag{2.46}$$

$$x_{ij} \geq 0, \quad i = 1, 2, \ldots, K, \text{ and } j = 1, 2, \ldots, L. \tag{2.47}$$

It is easy to verify that when the quadratic constraint in Model 2.6 is replaced by a slightly generalized constraint

$$\frac{1}{2}\sum_{i=1}^{K}\sum_{j=1}^{L} d_{ij}x_{ij}^2 + \sum_{i=1}^{K}\sum_{j=1}^{L} f_{ij}x_{ij} = D, \tag{2.48}$$

the self-deterrent gravity model has the following form:

$$x_{ij} = cr_i e^{-(\alpha c_{ij}+\beta f_{ij})-\beta d_{ij}x_{ij}} s_j. \tag{2.49}$$

Note that when the parameter D is not known with confidence or the equality constraint is too strong, we may consider using the corresponding inequality constraints. Also note that, as pointed out by Wilson [47] about parameter α, the parameter β of the self-deterrent gravity model does not have to be known exactly at the beginning. It could be found in the calibration process.

A particular application of Model 2.6 is as follows. In formulating and solving trip distribution problems, the planner often has information about the past trip distributions. This information often leads to certain "target values" for the current planning cycle. There are different ways to model such target values. One way is to use the target values as the "prior" and formulate a cross-entropy minimization problem, i.e., minimization of Kullback's discrimination information measure. Another way is to impose a constraint on a weighted sum of the squared deviation of the distribution x_{ij} from the target values p_{ij}, i.e.,

$$\sum_{i=1}^{K}\sum_{j=1}^{L} w_{ij}(x_{ij} - p_{ij})^2 = T, \tag{2.50}$$

or, equivalently,

$$\frac{1}{2}\sum_{i=1}^{K}\sum_{j=1}^{L} d_{ij}x_{ij}^2 + \sum_{i=1}^{K}\sum_{j=1}^{L} f_{ij}x_{ij} = D, \tag{2.51}$$

where $d_{ij} = 2w_{ij}$, $f_{ij} = -2w_{ij}p_{ij}$, and $D = T - \sum_{i=1}^{K}\sum_{j=1}^{L} w_{ij}p_{ij}^2$. Moreover, the quantity $\sum_{i=1}^{K}\sum_{j=1}^{L} w_{ij}(x_{ij} - p_{ij})^2$ has been used as a measure for the "deviation" between x_{ij} and p_{ij} (e.g., [7]).

Gravity models with additional constraints other than the production, attraction, and cost constraints have been proposed and studied (e.g., [12]). All those additional constraints are linear except Constraints (2.46) and (2.48), which were proposed in [15]. The quadratic cost term in the objective function of Model 2.5, i.e., Equation (2.37), or the quadratic constraint (2.46) of Model 2.6 provides an additional dimension of possible models for transportation planning. The additional parameter β of the self-deterrent gravity model provides one more degree of freedom in model identification and calibration. In addition, as will be shown in Section 5.3, Model 2.5 can be solved efficiently, in both theory and practice [14] . For a detailed discussion of the self-deterrent gravity model, the reader is referred to [15].

2.3 INPUT-OUTPUT ANALYSIS

The input-output analysis was first introduced by W. Leontief [28, 29] in the late 1930's to analyze the interdependence of industries in an economy. Assume that there are n industry sectors in an economy such as a country, state, or province. Let z_{ij} represent the sales *input* from sector i in the economy to sector j of the economy and let X_i represent the total *output* of sector i. Let Y_i be the sector i's "final demand," which represents the demand from those sources exogenous to the industry sectors under consideration, such as the consumer, the government and the overseas markets. Therefore, we have

$$X_i = \sum_{j=1}^{n} z_{ij} + Y_i, \quad i = 1, \ldots, n. \tag{2.52}$$

In the input-output analysis literature, the ratios

$$a_{ij} = z_{ij}/X_j, \qquad i, j = 1, ..., n, \tag{2.53}$$

are known as the *technical coefficients* ([32], page 11). The variable a_{ij} is interpreted as the amount of input from sector i to sector j per unit of the output of sector j. In the heart of input-output analysis is the square matrix $\mathbf{A} \equiv (a_{ij})$, $i, j = 1, ..., n$, which is known as the *technical coefficient matrix* or the *structural matrix* ([29], page 23).

The values z_{ij} are usually provided by data collected from various sources such as independent agencies, states, provinces, or countries [29]. However, collecting data can be time consuming and expensive [45, 32]. Hence a common practice in input-output analysis is to estimate their current values using previous estimates of these values and some aggregate information about their current values.

By defining U_i to be the total inter-industry sales by sector i, for $i = 1, ..., n$, and V_j to be the total inter-industry purchases by sector j, for $j = 1, ..., n$, we have

$$U_i = \sum_{j=1}^{n} z_{ij}, \qquad i = 1, ..., n, \tag{2.54}$$

and

$$V_j = \sum_{i=1}^{n} z_{ij}, \qquad j = 1, ..., n. \tag{2.55}$$

If an estimate $\mathbf{A}^0 \equiv (a_{ij}^0)$, $i, j = 1, ..., n$, of the technical coefficient matrix based on past data is available, it needs to be updated to accommodate the current information.

Assume that the current information available for the update includes U_i, V_j, and X_j, $i = 1, ..., n$, $j = 1, ..., n$. Then, the input-output technical coefficients updating problem can be posed as:

Find the matrix $\mathbf{A} = (a_{ij}), i, j = 1, ..., n$, *that has the least "discrepancy" with respect to the matrix* $\mathbf{A}^0 = (a_{ij}^0)$, $i, j = 1, ..., n$, *among those such that Conditions* (2.53), (2.54), *and* (2.55) *are satisfied.*

Among the measures of "discrepancy" between $\mathbf{A}$ and $\mathbf{A}^0$ used in the input-output analysis is the following "generalized" cross-entropy [32, 35]:

$$H = \sum_{i=1}^{n} \sum_{j=1}^{n} a_{ij} \ln(a_{ij}/a_{ij}^0). \tag{2.56}$$

When $a_{ij}^0 = 0$, it is assumed that $a_{ij} = 0$ and $\ln(0/0) \equiv 0$. Note that a_{ij}'s and a_{ij}^0's, $i = i, \ldots, n$, $j = 1, \ldots, n$, do not constitute two probability distributions. Therefore, H is not a cross-entropy as defined in Chapter 1, and H being interpreted as a measure of discrepancy between $\mathbf{A}$ and $\mathbf{A}^0$ is not well justified. We will address this issue after first formulating the updating problem. With

H being the measure of discrepancy, the updating problem becomes one that solves the following generalized cross-entropy minimization problem ([32], page 309):

Model 2.7:

$$\min \quad \sum_{i=1}^{n}\sum_{j=1}^{n} a_{ij}\ln(a_{ij}/a_{ij}^{0}) \tag{2.57}$$

$$\text{s.t.} \quad \sum_{j=1}^{n} X_j a_{ij} = U_i, \qquad i = 1, ..., n, \tag{2.58}$$

$$\sum_{i=1}^{n} X_j a_{ij} = V_j, \qquad j = 1, ..., n, \tag{2.59}$$

$$a_{ij} > 0,$$

$$a_{ij} = 0, \text{ only when } a_{ij}^{0} = 0, \ i, j = 1, ..., n.$$

Since both $\sum_{i=1}^{n}\sum_{j=1}^{n} a_{ij}$ and $\sum_{i=1}^{n}\sum_{j=1}^{n} a_{ij}^{0}$ are finite and known, although $\{a_{ij}^{0}\}$ and $\{a_{ij}\}$ are not probability distributions, Model 2.7 is still meaningful, and it can be shown that Model 2.7 is equivalent to a regular cross-entropy minimization problem.

In order to solve Model 2.7, the Lagrangian can be defined as follows:

$$\begin{aligned} L(\mathbf{A}, \boldsymbol{\lambda}, \boldsymbol{\mu}) &= \sum_{i=1}^{n}\sum_{j=1}^{n} a_{ij}\ln(a_{ij}/a_{ij}^{0}) - \sum_{i=1}^{n}\lambda_i(\sum_{j=1}^{n} X_j a_{ij} - U_i) \\ &- \sum_{j=1}^{n}\mu_j(\sum_{i=1}^{n} X_j a_{ij} - V_j), \end{aligned} \tag{2.60}$$

where λ_i and μ_j, $i, j = 1, ..., n$, are the Lagrange multipliers associated with Constraints (2.58) and (2.59), respectively. By setting $\frac{\partial L}{\partial a_{ij}} = 0$, it can be shown that ([32], page 309)

$$\begin{aligned} a_{ij} &= a_{ij}^{0}\exp^{[-1+\lambda_i X_j + \mu_j X_j]} \\ &= \exp^{[\lambda_i X_j - \frac{1}{2}]} a_{ij}^{0} \exp^{[\mu_j X_j - \frac{1}{2}]} \\ &= r_i a_{ij}^{0} s_j, \end{aligned} \tag{2.61}$$

where $r_i = \exp^{[\lambda_i X_j - \frac{1}{2}]}$ and $s_j = \exp^{[\mu_j X_j - \frac{1}{2}]}$, for $i, j = 1, ..., n$. Substituting a_{ij} values into Constraints (2.58) and (2.59) yields

$$r_i = U_i / (\sum_{j=1}^{n} a_{ij}^0 s_j X_j), \quad i = 1, ..., n, \tag{2.62}$$

and

$$s_j = V_j / (\sum_{i=1}^{n} r_i a_{ij}^0 X_j), \quad j = 1, ..., n. \tag{2.63}$$

Values of r_i and s_j can be found by solving Equations (2.62) and (2.63). In fact, a matrix balancing algorithm known as the RAS algorithm actually solves Equations (2.62) and (2.63) by an iterative procedure. The name RAS originated from the matrix notation given to the $n \times n$ diagonal matrices defined as:

$$\mathbf{R} = \text{diag}[r_i],$$

and

$$\mathbf{S} = \text{diag}[s_j].$$

It can be shown that ([32], page 310) $\mathbf{A} = \mathbf{R}\mathbf{A}^0\mathbf{S}$. Also, since $\frac{\partial^2 L}{(\partial a_{ij})^2} = \frac{1}{a_{ij}^0} > 0$, the solution to Equations (2.62) and (2.63) is indeed the unique solution to Model 2.7. More about the RAS algorithm can be found in [35, 30, 3].

The solution to Model 2.7 provides the so-called *point-estimates* of the technical coefficients $a_{ij}, i, j = 1, ..., n$. However, in many cases it may be necessary to replace a_{ij} by a set of discrete points $y = [y_1, ..., y_m]$ with corresponding probabilities p_{ij}^m ([17], page 544). In other words,

$$P[a_{ij} = y_m] = p_{ij}^m, \qquad i, j = 1, ..., n, \text{ and } m = 1, ..., M, \tag{2.64}$$

with

$$\sum_{m=1}^{M} p_{ij}^m = 1, \qquad \forall\, i, j = 1, ..., n.$$

Assuming that $P[a_{ij}^0 = y_m] = q_{ij}^m$, for $i, j = 1, ..., n, m = 1, ..., M$, form a known prior distribution, then Model 2.7 becomes

Model 2.8:

$$\min \qquad \sum_{m=1}^{M} \sum_{i=1}^{n} \sum_{j=1}^{n} p_{ij}^m \ln(p_{ij}^m / q_{ij}^m) \tag{2.65}$$

$$\text{s.t.} \quad \sum_{j=1}^{n}\sum_{m=1}^{M} X_j y_m p_{ij}^m = U_i, \qquad i = 1, ..., n, \tag{2.66}$$

$$\sum_{i=1}^{n}\sum_{m=1}^{M} X_j y_m p_{ij}^m = V_j, \qquad j = 1, ..., n, \tag{2.67}$$

$$\sum_{m=1}^{M} p_{ij}^m = 1, \qquad i, j = 1, ..., n, \tag{2.68}$$

$$p_{ij}^m \geq 0, \qquad i, j = 1, ..., n, \text{ and } m = 1, ..., M. \tag{2.69}$$

Constraints (2.66) and (2.67) are obtained by taking the expected values of the left-hand sides of Constraints (2.58) and (2.59) respectively.

Model 2.8 has more constraints than Model 2.7, but they are very similar in the sense that both have only linearly constrained equalities. Model 2.8 can be generalized when more complex probability distributions for coefficients a_{ij} occur. For example, if it is understood that a_{ij}'s come from a known probability distribution f_{ij}, then:

$$P[a_{ij} = y_m] = f_{ij}(y_m), \tag{2.70}$$

and Model 2.8 becomes

Model 2.9:

$$\min \quad \sum_{m=1}^{M}\sum_{i=1}^{n}\sum_{j=1}^{n} f_{ij}(y_m) \ln(f_{ij}(y_m)/f_{ij}^0(y_m)) \tag{2.71}$$

$$\text{s.t.} \quad \sum_{j=1}^{n}\sum_{m=1}^{M} X_j y_m f_{ij}(y_m) = U_i, \qquad i = 1, ..., n, \tag{2.72}$$

$$\sum_{i=1}^{n}\sum_{m=1}^{M} X_j y_m f_{ij}(y_m) = V_j, \qquad j = 1, ..., n, \tag{2.73}$$

$$\sum_{m=1}^{M} f_{ij}(y_m) = 1, \qquad i, j = 1, ..., n, \tag{2.74}$$

$$f_{ij}(y_m) \geq 0, \qquad i, j = 1, ..., n, \text{ and } m = 1, ..., M, \tag{2.75}$$

where $f_{ij}^0(y_m)$ is an *a priori* distribution. Depending on the functional form of $f_{ij}(y_m)$, Model 2.9 may take various forms of entropy optimization problems including those with nonlinear constraints.

2.4 REGIONAL PLANNING

Regional planning models dealing with locating centers of activities such as shopping areas, residential areas, and work districts have been developed since the 1960's. This section focuses on the estimation and prediction of the distribution of residential activity in urban areas. Based on the theory of land use and economics (e.g., [1, 2]), early models centered around linear programming formulations (e.g., [22]). Wilson's pioneering work in 1970 [48] started using entropy optimization models for regional planning.

Herbert and Stevens [22] proposed a well-known linear programming model to optimally distribute households to residential land. The model was developed for the Penn–Jersey Transportation Study as part of a larger model designed to locate all types of land-using activities. As pointed out by Herbert and Stevens, linear programming model usually is not regarded as a tool for prediction. However, if such a model is formulated to properly reflect house-buyers' economic and behavioral considerations, then its optimal solution may be both prescriptive and predictive.

Since the original linear programming model was developed for practical applications, a certain degree of conceptual elegance was sacrificed for simplicity. We briefly describe the model as follows. First we define

T_i^{kw} $\equiv$ the number of type w households to be located in a type k house in zone i;
b^{kw} $\equiv$ the residential budget, i.e., the budget that a type w household is willing to allocate for the purchase of a type k house;
p_i^{kw} $\equiv$ the cost of having a type k house in zone i for a type w household;
s^{kw} $\equiv$ the area (number of acres) used by a type w household with a type k house;
L_i $\equiv$ the area of available residential land in zone i;
P^w $\equiv$ the number of type w households to be located.

The budget b^{kw} covers the house itself, the size of the lot, the level of amenity and associated trip costs. The cost of housing, p_i^{kw}, does not reflect the variation of cost with respect to different sites within a zone but it includes the cost attributable to all other characteristics such as the cost of all trips made by the household. The quantity $(b^{kw} - p_i^{kw})$ can be interpreted as the bidding power [49, 2]. Building upon Alonso's work [1, 2], Herbert and Stevens [22] proposed the following linear programming model for locating households on residential land.

$$\max \quad \sum_{i,k,w} (b^{kw} - p_i^{kw}) T_i^{kw} \tag{2.76}$$

$$\text{s.t.} \quad \sum_{k,w} s^{kw} T_i^{kw} \leq L_i, \qquad \forall\, i, \tag{2.77}$$

$$\sum_{i,k} T_i^{kw} = P^w, \qquad \forall\, w, \tag{2.78}$$

$$T_i^{kw} \geq 0, \qquad \forall\, i, k, w. \tag{2.79}$$

The model can be interpreted as maximizing the bidding power subject to the constraints of land availability and finding every household a house. The linear objective function can also be interpreted as a utility measure, as described in ([39], page 213).

Herbert and Stevens' model has many variations for realistic considerations. For example, the budget term b^{kw} in the objective function can be refined by using b_i^{kw} to represent the budget that a type w household is willing to allocate for the purchase of a type k house in a zone i. Similarly, the term s^{kw} in the first constraint can be refined by using s_i^{kw} to represent the area used by a type w household with a type k house in a zone i. (See Senior and Wilson [39] for the use of s_i^k, rather than s^{kw} or s_i^{kw}.) Moreover, rather than dealing with vacant land, the model can be generalized to handle completely built-up land by replacing the second constraint with

$$\sum_{w} T_i^{kw} \leq H_i^k, \qquad \forall\, i, k, \tag{2.80}$$

where H_i^k denotes the number of type k houses available in zone i. By introducing a dummy household type that takes care of the "surplus" housing, the above inequality becomes

$$\sum_{w} T_i^{kw} = H_i^k, \qquad \forall\, i, k. \tag{2.81}$$

Note that Herbert and Stevens' model ignored the spatial interaction between home location and work location. To address this issue, let us assume that every household has only one key worker and the worker works in a particular zone j. In this way, T_i^{kw} can be refined by T_{ij}^{kw}, the number of workers from a type w household living in a type k house in zone i and working in zone j. Similarly, both b_i^{kw} and p_i^{kw} can be work-location dependent and refined by b_{ij}^{kw} and p_{ij}^{kw}. In other words, both the residential budget and the cost of a house to a particular household type may depend on the work location of the key worker in the household. Moreover, instead of treating P^w as a given number, we may use an estimate E_j^w to represent the number of jobs in zone j for workers from a type w household.

With all these modifications, Herbert and Stevens' model can be generalized as follows:

$$\max \quad \sum_{ijkw} (b_{ij}^{kw} - p_{ij}^{kw}) T_{ij}^{kw} \tag{2.82}$$

$$\text{s.t.} \quad \sum_{jw} T_{ij}^{kw} = H_i^k, \qquad \forall\, i, k, \tag{2.83}$$

$$\sum_{i,k} T_{ij}^{kw} = E_j^w, \qquad \forall\, j, w, \tag{2.84}$$

$$T_{ij}^{kw} \geq 0, \qquad \forall\, i, j, k, w. \tag{2.85}$$

When the transportation cost is the major concern, instead of the housing cost, Senior and Wilson [39] presented their model using a different objective function:

$$\max \sum_{ijkw} (b_{ij}^{kw} - c_{ij}^{w}) T_{ij}^{kw}, \tag{2.86}$$

where c_{ij}^w is the journey-to-work transportation cost between zone i and zone j for a type w household.

Following Wilson's work [48], Senior and Wilson noted that "imperfect markets, imperfect availability of information to households, and a host of historical accidents and indivisibilities conspire to support the view that an entropy maximizing or statistical averaging view may be the best" ([39], page 217). To reflect this view, they proposed a revised model in the following form of entropy optimization:.

Model 2.10:

$$\max \quad -\sum_{i,j,k,w} T_{ij}^{kw} \ln T_{ij}^{kw} \tag{2.87}$$

$$\text{s.t.} \quad \sum_{j,w} T_{ij}^{kw} = H_i^k, \qquad \forall\, i, k, \tag{2.88}$$

$$\sum_{i,k} T_{ij}^{kw} = E_j^w, \qquad \forall\, j, w, \tag{2.89}$$

$$\sum_{i,j,k,w} (b_{ij}^{kw} - c_{ij}^{w}) T_{ij}^{kw} = Z, \tag{2.90}$$

$$T_{ij}^{kw} \geq 0, \qquad \forall\, i, j, w, k, \tag{2.91}$$

where Z is a value preset by a planner.

The justification for using the entropic objective function is similar to the case of transportation planning described in Section 2.2. The quantity Z is usually chosen to be smaller than the optimal value obtained by solving the linear programming model defined by Equations (2.86), (2.83), (2.84), and (2.85).

Note that Model 2.10 is a linearly constrained entropy optimization problem. In order to find a solution, let us consider its Lagrangian:

$$\begin{aligned} L &= -\sum_{i,j,k,w} T_{ij}^{kw} \ln T_{ij}^{kw} - \sum_{i,k} \lambda_i^{1k} \left(\sum_{j,w} T_{ij}^{kw} - H_i^k \right) \\ &+ \sum_{j,w} \lambda_j^{2w} \left(\sum_{j,w} T_{ij}^{kw} - E_j^w \right) \end{aligned}$$

$$+ \quad \mu \left\{ \sum_{i,j,k,w} T_{ij}^{kw} (b_{ij}^{kw} - c_{ij}^{w}) - Z \right\}, \tag{2.92}$$

where λ_i^{1k}, for each i, k, λ_j^{2w}, for each w, j, and μ are real numbers representing the Lagrange multipliers corresponding to Constraints (2.88), (2.89), and (2.90) respectively.

Setting the partial derivative of $L(\mathbf{p}, \alpha, \beta)$ with respect to T_{ij}^{kw} to zero produces

$$T_{ij}^{kw} = \exp^{-\lambda_i^{1k}} \exp^{+\lambda_j^{2w}} \exp^{\mu(b_{ij}^{kw} - c_{ij}^{w})} \exp^{-1} . \tag{2.93}$$

Define A_i^k and B_j^w by

$$A_i^k H_i^k = \exp^{-\lambda_i^{1k} - 1}, \tag{2.94}$$

and

$$B_j^w E_j^w = \exp^{+\lambda_j^{2w}} . \tag{2.95}$$

Equation (2.93) becomes

$$T_{ij}^{kw} = A_i^k B_j^w H_i^k E_j^w \exp^{\mu(b_{ij}^{kw} - c_{ij}^{w})} . \tag{2.96}$$

Substituting for T_{ij}^{kw} in (2.88) and (2.89) produces

$$A_i^k = \left(\sum_{j,w} B_j^w E_j^w \exp^{\mu(b_{ij}^{kw} - c_{ij}^{w})} \right)^{-1}, \qquad \forall\, i, k, \tag{2.97}$$

and

$$B_j^w = \left(\sum_{j,w} A_i^k H_i^k \exp^{\mu(b_{ij}^{kw} - c_{ij}^{w})} \right)^{-1}, \qquad \forall\, w, j. \tag{2.98}$$

Then, Equations (2.97) and (2.98) can be solved iteratively [39].

While other solution techniques are the subjects for discussion in Chapter 3, it is worth mentioning that a geometric programming approach was proposed by Scott and Jefferson [36] for solving Model 2.10.

Model 2.10 can be made more realistic by incorporating additional constraints. For example, Kapur ([27], page 320), in a different but related context, considered the average rent (paid by an average household for the use of its house) within a zone as a factor for determining population distribution and modeled the average rent r_i of a household in zone i as a function of the number of households residing in that zone as follows:

$$r_i = d_i \left(\sum_{j,k,w} T_{ij}^{kw} \right)^{\alpha}, \qquad \forall\, i, \tag{2.99}$$

where $\alpha > 0$ and $d_i > 0$ are two parameters. To consider the total rent, we may impose one more constraint

$$r = \sum_i d_i \left(\sum_{j,k,w} T_{ij}^{kw} \right)^{\alpha+1} \leq R, \tag{2.100}$$

where R is a given constant and the left-hand-side of the inequality is obtained by first multiplying r_i by the number of households in zone i, i.e., $\left(\sum_{j,k,w} T_{ij}^{kw}\right)$, and then summing over the index i.

Note that

$$\frac{\partial r}{\partial T_{ij}^{kw}} = d_i(\alpha + 1) \left(\sum_{j,k,w} T_{ij}^{kw} \right)^{\alpha},$$

and

$$\frac{\partial^2 r}{\left(\partial T_{ij}^{kw}\right)^2} = d_i(\alpha + 1)\alpha \left(\sum_{j,k,w} T_{ij}^{kw} \right)^{\alpha-1}.$$

Therefore, for $\alpha > 0$, we have

$$\frac{\partial^2 r}{\left(\partial T_{ij}^{kw}\right)^2} > 0.$$

This implies the convexity of r. Thus an expanded version of Model 2.10, formed with the addition of Constraint (2.100), becomes an entropy optimization problem subject to convex constraints, as defined in Subsection 1.2.4.

2.5 PORTFOLIO OPTIMIZATION

The origin of modern portfolio theory can be traced back to the work of Markowitz in 1959 [31], where he applied linear programming theory to analyze the return on an investment.

Assume that there are n assets for an investor to allocate his/her investment. The return on asset j, $j = 1, ..., n$, is given by a random variable r_j. The investor faces a problem of how to allocate a right proportion p_j, $j = 1, ..., n$, of the investment to each asset j. Denote the vector of asset return by $\mathbf{r} \equiv (r_1, ..., r_n)^T$. Let the expected return on asset j be denoted by $m_j = \mathrm{E}\,(r_j)$ and the "variance-covariance" matrix of the random vector $\mathbf{r}$ be denoted by $\mathbf{V}$. Matrix $\mathbf{V}$ is generally known as the *risk matrix* of the investment, and the value of $\mathbf{p}^T\mathbf{V}\mathbf{p}$ is known as the *expected risk* of the portfolio $\mathbf{p} \equiv (p_1, ..., p_n)^T$.

Once a lower bound, denoted by c, of the expected total return of the investor is specified, the Markowitz-Tobin Model [38, 9] considers the following problem to minimize the expected risk involved in the investment :

$$\min \quad \mathbf{p}^T\mathbf{V}\mathbf{p} \tag{2.101}$$

$$\text{s.t.} \quad \sum_{j=1}^{n} m_j p_j \geq c, \tag{2.102}$$

$$\sum_{j=1}^{n} p_j = 1, \tag{2.103}$$

$$p_j \geq 0, \quad j = 1, ..., n. \tag{2.104}$$

This problem is a quadratic programming problem. By varying the lower bound c parametrically, the whole set of optimal or efficient portfolios $\{\mathbf{p}^*(c)\}$, also known as the *efficient frontier*, can be obtained ([9], page 73).

In practice, an efficient portfolio can be defined by minimizing the expected risk, by maximizing the expected return, or by using a mixed criterion. Kapur [27] considered an entropic maximization strategy and defined the following portfolio optimization problem:

Model 2.11 :

$$\max \quad \omega_1 \sum_{j=1}^{n} m_j p_j + \omega_2 \left(- \sum_{j=1}^{n} p_j \ln p_j \right) - \omega_3 \mathbf{p}^T\mathbf{V}\mathbf{p} \tag{2.105}$$

$$\text{s.t.} \quad \sum_{j=1}^{n} p_j = 1, \tag{2.106}$$

$$p_j \geq 0, \quad j = 1, ..., n, \tag{2.107}$$

where ω_1, ω_2 and ω_3 are nonnegative parameters representing the desired weights.

The optimal solution of Model 2.11 depends on the value of ω_i, $i = 1, 2, 3$, which can be specified by a portfolio manager. This allows the practitioners to follow different strategies.

Another strategy is to stay close to an existing portfolio $\{p_1^0, ..., p_n^0\}$ and allocate the investment with considerations such that (i) the expected value of Price/Earning ratio (PER), also known as the earning multiplier, of the portfolio is above a given lower bound b_1; (ii) the expected value of Price Book Value ratio (PBV), which reflects the attractiveness of investment, of the portfolio is above a given lower bond b_2; and (iii) the expected risk of the portfolio remains below a given upper bound b_3.

This results in the following problem for consideration:

Model 2.12:

$$\min \quad \sum_{j=1}^{n} p_j \ln(p_j/p_j^0) \tag{2.108}$$

$$\text{s.t.} \quad \sum_{j=1}^{n} \text{PER}_j p_j \geq b_1, \tag{2.109}$$

$$\sum_{j=1}^{n} \text{PBV}_j p_j \geq b_2, \tag{2.110}$$

$$\mathbf{p}^T \mathbf{V} \mathbf{p} \leq b_3, \tag{2.111}$$

$$\sum_{j=1}^{n} p_j = 1, \tag{2.112}$$

$$p_j \geq 0, \quad j = 1, ..., n, \tag{2.113}$$

where PER_j and PBV_j are the price earning ratio and price book value of asset j, for $j = 1, ..., n$, respectively.

It is clearly seen that Model 2.12 is a minimum cross-entropy problem subject to quadratic constraints. Methods for estimating of the variance-covariance matrix **V** can be found in [9]. A geometric programming solution to Model 2.12 has been provided by Yamada and Rajasekera [52].

Note that the constraint set of Model 2.12 can be supplemented with other requirements. For example, some financial managers may like to select optimum portfolio based on the dividend yield. They may set a target dividend yield D (such as 2%) for the portfolio. Let d_j stand for the dividend yield on asset j, then we can impose an additional constraint to Model 2.12 :

$$\sum_{j=1}^{n} d_j p_j \geq D. \tag{2.114}$$

For diversity and safety consideration, a financial manager may place a cap

$$p_j \leq U_j, \tag{2.115}$$

on some assets. Similarly, upper limits are often placed on the proportion of a portfolio that can be invested in any industrial sector, such as the energy sector or telecommunications sector [9]. In this case, we may add additional constraints to Model 2.12, such as

$$\sum_{j \in I_k} p_j \leq u_k, \quad k = 1, 2, ..., l, \tag{2.116}$$

where l is the total number of industries under consideration, u_k is the upper bound set for industry k, and I_k is the index set identifying those assets considered to be part of industry k, for $k = 1, ..., l$.

2.6 IMAGE RECONSTRUCTION

Images can be used to characterize the underlying distribution of certain physical properties (e.g., density, shape, and brightness) of an object under investigation. In many applications where an image is required, only a finite number of observations and/or indirect measurements can be made. Image reconstruction is a procedure for processing the measurement data to construct an image of the object. Readers are referred to [21, 6, 41] and the references therein for detailed descriptions.

A very important scientific application of image reconstruction is in computerized tomography (CT) for medical diagnosis. Physicians need to know, for example, the location, shape, and size of a suspected tumor inside a patient's brain in order to plan a suitable course of treatment. With computerized tomography, images of cross-sections of a human body can be constructed from the data obtained by measuring the attenuation of X-rays along a large number of straight lines (or strips) through each cross-section.

The image reconstruction problem varies widely with respect to the area of applications. For ease of introduction, we focus on the case of the two-dimensional X-ray CT with the understanding that the discussion can be generalized to higher dimensions. In this case, the distribution is that of the X-ray linear attenuation coefficient of body tissues. The total attenuation of the X-ray beam between a source and a detector is approximately the integral of the linear attenuation coefficient along the line between the source and the detector.

The unknown distribution of a certain physical parameter whose image is to be reconstructed is represented by a function f of two variables. The function is known to be zero-valued outside a squared-shape region, which is referred to as the support of the image. Our objective is to estimate the value of f at n regularly spaced points within its support, where n is set by the user. The values of f at these n sample points are denoted by $\mathbf{f} \equiv (f_1, f_2, \ldots, f_n)$. Assume that the measurement data are obtained by observing m projections. We denote the measurement data by $\mathbf{d} \equiv (d_1, d_2, \ldots, d_m)$.

Usually the relationship between the unknown distribution $\mathbf{f}$ and the observed measurement $\mathbf{d}$ is approximated by a linear relation

$$\mathbf{d} \approx \mathbf{Af}, \tag{2.117}$$

where $\mathbf{A} = [a_{ij}]$ is a projection matrix. The approximation sign in Expression (2.117) reflects the possible errors in modeling and measurement.

A number of models have been developed for solving the image reconstruction problem. In this section, we focus on the method that discretizes the image and partition its support into a finite number of equi-sized squared regions (called pixels or cells) whose centers are the n sample points. To calculate the projection matrix $\mathbf{A}$, we make a fundamental assumption that f is constant in each of the the equi-sized pixels. In other words, the value of f throughout the square region of a pixel j is assumed to be the same as f_j. Under this assumption, a_{ij} is simply the length of the intersection of the line corresponding to the ith projection with the pixel surrounding the jth sample point.

Once the projection matrix $\mathbf{A}$ is defined and the measurement $\mathbf{d}$ is known, the problem is to find an $\mathbf{f}$ satisfying (2.117). To cope with the errors mentioned above, Herman [20] suggested that Expression (2.117) be replaced by an "interval constraint" and a non-negativity constraint be added:

$$\mathbf{d} - \boldsymbol{\epsilon} \leq \mathbf{Af} \leq \mathbf{d} + \boldsymbol{\epsilon}, \tag{2.118}$$

$$\mathbf{f} \geq \mathbf{0}, \tag{2.119}$$

where $\boldsymbol{\epsilon} = (\epsilon_1, \epsilon_2, \ldots, \epsilon_m)$ is an m-vector of user-chosen tolerance levels. When the "interval structure" of Constraint (2.118) is not to be exploited for algorithmic efficiency, Inequality (2.118) can be replaced by an equivalent system

$$\mathbf{A}'\mathbf{f} \leq \mathbf{d}', \tag{2.120}$$

with twice as many one-sided inequalities [6, 20].

There are two basic approaches to solving such an image construction problem. While the feasibility approach tries to find a solution to (2.118) and (2.119), the optimization approach seeks a point that is not only feasible in the above sense but also optimal with respect to a certain criterion. In the literature, at least three different types of optimization models have been proposed, namely, the entropy maximization models, the quadratic minimization models, and the maximum likelihood models.

The entropy optimization models seek to optimize an entropic objective function $-\sum_{j=1}^{n} f_j \ln f_j$, with the convention of $0 \ln 0 = 0$, subject to various constraints. In this way, a solution that is consistent with the measurement data but remains maximally noncommittal can be found by solving the following entropy optimization problem:

Model 2.13:

$$\max \quad -\sum_{j=1}^{n} f_j \ln f_j \tag{2.121}$$

$$\text{s.t.} \quad \mathbf{d} - \epsilon \leq \mathbf{Af} \leq \mathbf{d} + \epsilon, \tag{2.122}$$

$$\mathbf{f} \geq \mathbf{0}. \tag{2.123}$$

Note that this solution can also be interpreted as the most probable solution that is consistent with the measurement data. For the justification of such an interpretation, see Section 2.2 for details.

A different entropy optimization model begins with definition of the error vector $\mathbf{e} = (e_1, ..., e_m)^T$, where

$$e_i \equiv d_i - \sum_{j=1}^{n} a_{ij} f_j, \quad i = 1, ..., m. \tag{2.124}$$

Assume that $e_1, ..., e_m$ exist due to imprecise measurement and are independent noise terms with 0 mean and known variance σ_i^2. Burch et al. [5] observed that the Strong Law of Large Numbers implies that

$$Q(\mathbf{f}) \equiv (1/m) \sum_{i=1}^{m} \left(\sum_{j=1}^{n} a_{ij} f_j - d_i \right)^2 / \sigma_i^2 \rightarrow 1, \quad \text{as } m \rightarrow \infty. \tag{2.125}$$

Thus, if m is sufficiently large, the following entropy optimization problem with quadratic constraints can be useful:

Model 2.14

$$\max \quad -\sum_{j=1}^{n} f_j \ln f_j \tag{2.126}$$

$$\text{s.t.} \quad (1/m)(\mathbf{Af} - \mathbf{d})^T \mathbf{S}^2 (\mathbf{Af} - \mathbf{d}) = 1, \tag{2.127}$$

$$f_j \geq 0, j = 1, ..., n, \tag{2.128}$$

where **S** is a diagonal matrix with $1/\sigma_i$ being its ith diagonal element.

Concerns such as the smoothing effect, nonuniformity, peakness, and exactness [46] of a constructed image can also be addressed by entropy optimization with proper objective functions and constraints.

REFERENCES

[1] Alonso, W., "A Theory of the Urban Land Market," *Papers of the Regional Science Association*, Vol. 6, 1960, pp. 146–158.

[2] Alonso, W., *Location and Land Use*, Harvard University Press, Cambridge, Massachusetts, 1964.

[3] Bregman, L.M., "The Relaxation Method of Finding the Common Point of Convex Sets and Its Application to the Solution of Problems in Convex Programming," *U.S.S.R. Computational Mathematics and Mathematical Physics*, Vol. 7, 1967, pp. 200–217.

[4] Brice, S., "Derivation of Nested Transport Models within a Mathematical Programming Framework," *Transportation Research Part B*, Vol. 23B, 1989, pp. 19–28.

[5] Burch, S.F., Gull, S.F., and Skilling, J.K., "Image Restoration by a Powerful Maximum Entropy Method," *Computer Vision, Graphics, and Image Processing*, Vol. 23, 1983, pp. 113–128.

[6] Censor, Y., and Herman, G.T., "On Some Optimization Techniques in Image Reconstruction," *Applied Numerical Mathematics*, Vol. 3, 1987, pp. 365–391.

[7] Cottle, R.W., "Application of a Block Successive Overrelaxation Method to a Class of Constrained Matrix Problems," *Proceedings of the International Congress on Mathematical Programming*, edited by R.W. Cottle, M.L. Kelmanson, and B. Korte, Rio de Janeiro, Brazil, April, 1981, pp. 89–103.

[8] El-Affendi, M.A., and Kouvatsos, D.D., "A Maximum Entropy Analysis of the M/G/1 and G/M/1 Queueing System at Equilibrium," *Acta Informatica*, Vol. 19, 1983, pp. 339–355.

[9] Elton, E.D., and Gruber, M.J., *Modern Portfolio Theory and Investment Analysis*, 4th Edition, John Wiley, New York, 1991.

[10] Eriksson, J.R., "A Note on Solution of Large Sparse Maximum Entropy Problems with Linear Equality Constraints," *Mathematical Programming*, Vol. 18, 1980, pp. 146–154.

[11] Erlander, S., "Efficient Population Behavior and the Simultaneous Choices of Origins, Destinations and Routes," *Transportation Research Part B*, Vol. 24B, 1990, pp. 363–373.

[12] Erlander, S., and Stewart, N.F., *The Gravity Model in Transportation Analysis*, VSP, The Netherlands, 1990.

[13] Erlang, A.K., "Theory of Probabilities and Telephone Conversations," *Nyt Tidsskrift Matematik*, Vol. B-20, 1909, pp. 33–39.

[14] Fang, S.-C., and Tsao, H.-S.J., "An Unconstrained Convex Programming Approach to Solving Convex Quadratic Programming Problems," *Optimization*, Vol. 27, 1993, pp. 235–243.

[15] Fang, S.-C., and Tsao, H.-S.J., "Linear-Constrained Entropy Maximization Problem with Quadratic Cost and Its Applications to Transportation Planning Problems," *Transportation Science*, Vol. 29, 1995, pp. 353–365.

[16] Ferdinand, A.E., "A Statistical Mechanics Approach to System Analysis," *IBM Journal of Research and Development*, Vol. 14, 1970, pp. 539–547.

[17] Golan, A., Judge, G., and Robinson, S., "Recovering Information from Incomplete or Partial Multi-sectoral Economic Data," *Review of Economics and Statistics*, Vol. LXXVI, 1994, pp. 541–549.

[18] Gross, D., and Harris, C.M., *Fundamentals of Queueing Theory*, 2nd Edition, John Wiley, New York, 1985.

[19] Guiasu, S., "Maximum Entropy Condition in Queueing Theory," *Journal of Operational Research Society*, Vol. 37, 1986, pp. 293–301.

[20] Herman, G.T., "A Relaxation Method for Reconstructing Objects from Noisy X-Rays," *Mathematical Programming*, Vol. 8, 1975, pp. 1–19.

[21] Herman, G.T., editor, *Image Reconstruction from Projections: Implementation and Applications*, Springer-Verlag, Berlin, 1979.

[22] Herbert, D.J., and Stevens, B.H., "A Model for the Distribution of Residential Activity in Urban Areas," *Journal of Regional Science*, Vol. 2, 1960, pp. 21–36.

[23] Jaynes, E.T., "Information Theory and Statistical Mechanics," *Physics Review*, Vol. 106, 1957, pp. 620–630.

[24] Johnston, R.J., and Pattie, C.J., "Using Entropy-Maximizing Procedure to Estimate Territorial Social Indicators: An Introduction and Illustration," *Social Indicators Research*, Vol. 27, 1992, pp. 235–256.

[25] Jornsten, K.O., "An Algorithm for the Combined Distribution and Assignment Problem," *Transportation Research Part B*, Vol. 15B, 1981, pp. 21–33.

[26] Jornsten, K.O., and Lundgren, J.T., "An Entropy-Based Model Split Model," *Transportation Research Part B*, Vol. 23B, 1989, pp. 345–359.

[27] Kapur, J.N., *Maximum-Entropy Models in Science and Engineering*, Wiley Eastern Ltd., New Delhi, 1989.

[28] Leontief, W., "Quantitative Input-Output Relations in the Economic System of the United States," *Review of Economic and Statistics*, Vol. 18, 1936, pp. 105–125.

[29] Leontief, W., *Input-Output Economics*, 2nd Edition, Oxford University Press, New York, 1986.

[30] Macgill, S.M., "Theoretical Properties of Bi-proportional Matrix Adjustments," *Environment and Planning A*, Vol. 9, 1977, pp. 687–701.

[31] Markowitz, H.M., *Portfolio Selection: Efficient Diversification of Investments*, John Wiley, New York, 1959.

[32] Miller, R.E., and Blair, P.D., *Input-Output Analysis: Foundations and Extensions*, Prentice Hall, New Jersey, 1985.

[33] Mishra, S., and Fang, S.-C., "A Maximum Entropy Optimization Approach to Tandem Queues with Generalized Blocking," *Performance Evaluation*, Vol. 702, 1997, pp. 1–25.

[34] Safwat, K.N.A., and Magnanti, T.L., "A Combined Trip Generation, Trip Distribution, Modal Split, and Trip Assignment Model," *Transportation Science*, Vol. 22, 1988, pp. 14–30.

[35] Schneider, M.H., and Zenios, S.A., "Comparative Study of Algorithms for Matrix Balancing," *Operations Research*, Vol. 38, 1990, pp. 439–455.

[36] Scott, C.H., and Jefferson, T.R., "Entropy Maximizing Models of Residential Location via Geometric Programming," *Geographical Analysis*, Vol. 9, 1977, pp. 181–187.

[37] Sen, A., and Smith, T.E., *Gravity Models of Spatial Interaction Behavior*, Springer-Verlag, Berlin, 1995.

[38] Sengupta, J.K., "Mixed Strategy and Information Theory in Optimal Portfolio Choice," *International Journal of Systems Science*, Vol. 20, 1989, pp. 215–227.

[39] Senior, M.L., and Wilson, A.G., "Exploration and Synthesis of Linear Programming and Spatial Interaction Models of Residential Location," *Geographical Analysis*, Vol. 6, 1974, pp. 209–238.

[40] Shore, J.E., "Information Theoretic Approximations for M/G/1 and G/G/1 Queueing Systems," *Acta Informatica*, Vol. 17, 1982, pp. 47–61.

[41] Stark, H., editor, *Image Recovery: Theory and Application*, Academic Press, Orlando, Florida, 1987.

[42] Takahashi, Y., and Takami, Y., "A Numerical Method for the Steady-State Probabilities of a G1/G/C Queue in a General Class," *Journal of Operations Research Society of Japan*, Vol. 19, 1976, pp. 147–157.

[43] Tomlin, J.A., "A Mathematical Programming Model for the Combined Distribution-Assignment of Traffic," *Transportation Science*, Vol. 5, 1971, pp. 122–140.

[44] Tomlin, J.A., and Tomlin, S.G., "Traffic Distribution and Entropy," *Nature*, Vol. 220, 1968, pp. 974–976.

[45] Viel, V.Q., "Practices in Input-Output Table Compilation," *Regional Science and Urban Economics*, Vol. 24, 1994, pp. 27–54.

[46] Wang, Y., and Lu, W., "Multicriterion Maximum Entropy Image Reconstruction from Projections," *IEEE Transactions on Medical Imaging*, Vol. 11, 1992, pp. 70–75.

[47] Wilson, A.G., "A Statistical Theory of Spatial Distribution Models," *Transportation Research*, Vol. 1, 1967, pp. 253–269.

[48] Wilson, A.G., *Entropy in Urban and Regional Planning*, Pion, London, 1970.

[49] Wilson, A.G., *Urban and Regional Models in Geography and Planning*, John Wiley, London, 1974.

[50] Wu, J.-S., and Chan, W.C., "Maximum Entropy Analysis of Multiple-Server Queueing Systems," *Journal of Operational Research Society*, Vol. 40, 1989, pp. 815–826.

[51] Wu, J-S., "Maximum Entropy Analysis of Open Queueing Networks with Group Arrivals," *Journal of Operational Research Society*, Vol. 43, 1992, pp. 1063–1078.

[52] Yamada, M., and Rajasekera, J.R., "Portfolio Re-balancing with the Entropy Criteria," Report No. 310, QUICK Research Institute Corp., Tokyo, 1993.

[53] Zhuang, X., Haralick, R.M., and Zhao, Y., "Maximum Entropy Image Reconstruction," *IEEE Transactions on Signal Processing*, Vol. 39, 1991, pp. 1478–1480.

3

ENTROPY OPTIMIZATION METHODS: LINEAR CASE

Let $\mathbf{x} \equiv (x_1, \ldots, x_n)^T \geq \mathbf{0}$ be a nonnegative n-dimensional column vector and $\mathbf{p} \equiv (p_1, \ldots, p_n)^T > \mathbf{0}$ be a positive n-dimensional column vector. With the convention of $0 \ln 0 = 0$, we define the quantity $\sum_{j=1}^{n} x_j \ln(x_j/p_j)$ to be the cross-entropy of $\mathbf{x}$ with respect to $\mathbf{p}$, in a general sense. Note that when $\mathbf{x}$ and $\mathbf{p}$ are both probability distributions, i.e., $\sum_{j=1}^{n} x_j = \sum_{j=1}^{n} p_j = 1$, this quantity becomes the commonly defined cross-entropy between the two probability distributions (see Chapter 1).

Given $\mathbf{p} > \mathbf{0}$, an $m \times n$ matrix $\mathbf{A} \equiv [a_{ij}]$ and an m-dimensional column vector $\mathbf{b} = (b_1, \ldots, b_m)^T$, we are interested in solving the following linearly constrained minimum cross-entropy problem:

Program EL:

$$\min \quad \sum_{j=1}^{n} x_j \ln(\frac{x_j}{p_j}) \tag{3.1}$$

$$\text{s.t.} \quad \sum_{j=1}^{n} a_{ij} x_j = b_i, \quad i = 1, 2, \ldots, m, \tag{3.2}$$

$$x_j \geq 0, \quad j = 1, 2, \ldots, n. \tag{3.3}$$

Equation (3.2) represents a set of linear equality constraints. Theory and algorithms for linearly constrained minimum cross-entropy problems with inequality constraints will also be discussed in this chapter.

This entropy optimization model has been successfully applied to various fields including information theory [36, 46], statistical mechanics [40], pattern recognition [60], image reconstruction [50], actuarial science [69], queueing theory [37, 68], etc. Some of the applications have been explained in Chapter 2. More applications can be found in the *Proceedings* of the annual conferences on Maximum Entropy and Bayesian Methods, e.g., [35, 60]. In the area of transportation and regional planning, Wilson [67] pioneered the use of entropy models in the study of spatial interaction. These models have since become an integral part of both the theory and practice of many branches of transportation planning. They play a pivotal role in trip distribution [63, 21], combined trip distribution and trip assignment [62, 24], modal choice [42, 2], and many other branches [57]. In the area of matrix theory and linear algebra, the matrix scaling and matrix balancing problems have also been shown to be special cases of Program EL [13, 22].

In the past two decades, many algorithms have been proposed for solving the linearly constrained minimum cross-entropy problem. Better known algorithms include Bregman's method [11, 22], MART (Multiplicative Algebraic Reconstruction Technique) [34, 49, 22], Newton's method [25, 23, 1], the Generalized Iterative Scaling Method (GISM) [8, 17] and interior-point methods [61, 55]. Basically, Bregman's method can be viewed as a dual convex programming algorithm of row-action type. Many balancing methods have been identified as special cases of Bregman's method by Lamond and Stewart [48]. MART can be considered as a particular approximation strategy to the Bregman's method. Some computational results were reported in [22] and [23]. Newton's method has been applied to solving a system of nonlinear equations derived from the Karush-Kuhn-Tucker (K.K.T.) conditions of Program EL. The Generalized Iterative Scaling Method (GISM) deals with a special case of Program EL in which $\sum_{j=1}^{n} x_j = 1$ is also assumed. The special problem is first transformed into one with nonnegative $\mathbf{A}$ and positive $\mathbf{b}$ before an initial solution is iteratively scaled. It turns out that GISM is closely related to MART. After the advent of the Karmarkar algorithm for solving linear programs, interior-point methods have also been proposed for solving Program EL [61, 55].

Recently Fang and Tsao [30, 64] proposed an unconstrained dual method for optimizing entropy subject to linear constraints. We will derive the theory in detail and provide numerical results. Also discussed in this chapter, as a direct consequence of the finite dimensional study, is a semi-infinite entropy optimization problem with a finite number of variables but an infinite number of linear inequality constraints. An efficient cutting-plane algorithm will be introduced with convergence proofs and numerical results.

The K.K.T. conditions have played an important role in developing solution methods for the linearly constrained entropy optimization problems. Both the primal and dual approaches known in the existing literature can be best explained and compared from a unified perspective. To help accentuate the common thread, we start with the derivation of a dual program of Program EL and its K.K.T. conditions. Denote the feasible region of Program EL by

$$F \equiv \{\mathbf{x} \in R^n | \mathbf{A}\mathbf{x} = \mathbf{b}; \mathbf{x} \geq \mathbf{0}\} \tag{3.4}$$

and the relative interior of F by

$$F^0 \equiv \{\mathbf{x} \in R^n | \mathbf{A}\mathbf{x} = \mathbf{b}; \mathbf{x} > \mathbf{0}\}. \tag{3.5}$$

An n-vector $\mathbf{x}$ is called an interior solution of Program EL if $\mathbf{x} \in F^0$. For convenience, we will refer to F^0 as the interior of F. With these definitions, we have the following verifiable result:

Lemma 3.1. If F is nonempty and $\mathbf{p} > \mathbf{0}$, then Program EL has a unique optimal solution. Moreover, if F has a nonempty interior, then the unique optimal solution is strictly positive.

Throughout this chapter, we assume that F has a nonempty interior, i.e., $F^0 \neq \emptyset$. Under this assumption, we can capitalize on the regular Kuhn-Tucker theorem. Define the following Lagrangian function

$$L(\mathbf{x}, \mathbf{w}) \equiv \sum_{j=1}^{n} x_j \ln(x_j/p_j) - \sum_{i=1}^{m} w_i (\sum_{j=1}^{n} a_{ij} x_j - b_i), \tag{3.6}$$

where $\mathbf{0} \leq \mathbf{x} \in R^n$ and $\mathbf{w} \in R^m$. This leads to the K.K.T. conditions:

$$x_j = p_j \exp(\sum_{i=1}^{m} a_{ij} w_i - 1), \quad j = 1, 2, ..., n, \tag{3.7}$$

$$\sum_{j=1}^{n} a_{ij} x_j = b_i, \; i = 1, 2, ..., m, \tag{3.8}$$

$$x_j \geq 0, \; j = 1, 2, ..., n. \tag{3.9}$$

Note that since $\mathbf{p} > \mathbf{0}$, Condition (3.7) implies $\mathbf{x} > \mathbf{0}$. Consequently, Condition (3.9) becomes redundant. Therefore, only Conditions (3.7) and (3.8) remain.

Under the assumption that F has a nonempty interior, Equations (3.7) and (3.8) provide a necessary and sufficient condition for guaranteeing that Program EL has a unique and positive optimal solution $\mathbf{x}^*$. However, if the matrix $\mathbf{A}$ does not have full row-rank, then the corresponding dual vector may not be unique. In fact, if $\mathbf{w}^*$ is a solution to the K.K.T. conditions, then it is easy to see that $\mathbf{w}^* + \mathbf{u}$ is also a solution as long as $\mathbf{u}$ is in the null space of $\mathbf{A}^T$, i.e., $\mathbf{A}^T \mathbf{u} = \mathbf{0}$.

Actually, we can integrate Condition (3.7) into Condition (3.8) to generate the following system of equations:

$$h_i(\mathbf{w}) \equiv \sum_{j=1}^{n} a_{ij} p_j \exp(\sum_{k=1}^{m} a_{kj} w_k - 1) - b_i = 0, \quad i = 1, 2, \ldots, m. \tag{3.10}$$

This is a key expression for the entire chapter. Under the assumption that F has a nonempty interior, it embodies the entirety of the K.K.T. conditions. By solving this system of m equations in m variables and by using Condition (3.7) for conversion, we can find an optimal solution $\mathbf{x}^*$ of Program EL.

Since Equation (3.10) involves only the dual variables, the Lagrangian dual of Program EL can be shown to be independent of $\mathbf{x}$. First, we define

$$g(\mathbf{w}) \equiv \inf_{\mathbf{x} \geq \mathbf{0}} L(\mathbf{x}, \mathbf{w}), \tag{3.11}$$

where $L(\mathbf{x}, \mathbf{w})$ is defined by (3.6). Then, substituting the necessary and sufficient first-order optimality condition (3.7) in (3.6) results in

$$g(\mathbf{w}) = -\sum_{j=1}^{n} p_j \exp(\sum_{i=1}^{m} a_{ij} w_i - 1) + \sum_{i=1}^{m} b_i w_i. \tag{3.12}$$

The Lagrangian dual can be defined as:

Program DEL:

$$\min_{\mathbf{w} \in R^m} \{d(\mathbf{w}) \equiv \sum_{j=1}^{n} p_j \exp(\sum_{i=1}^{m} a_{ij} w_i - 1) - \sum_{i=1}^{m} b_i w_i\}. \tag{3.13}$$

This is an unconstrained convex program. This dual program can also be derived via the theory of generalized geometric programming and the use of a simple inequality. Such derivation, in the context of optimizing the entropy function with additional linear functions of $\mathbf{x}$ in the objective function, will be detailed in Section 5.2. As will be shown there, if F has a nonempty interior, then there exists no duality gap between Program EL and Program DEL. Since this dual program is derived directly from the K.K.T. conditions, Equation (3.10) remains a necessary and sufficient condition for the existence of an optimal dual solution to Program DEL. Moreover, Equation (3.7) converts a dual optimal solution into a primal optimal solution. More precisely, if F has a nonempty interior, then Program DEL has an optimal solution $\mathbf{w}^*$ and

$$x_j^* = p_j \exp(\sum_{i=1}^{m} a_{ij} w_i^* - 1), \quad j = 1, 2, ..., n, \tag{3.14}$$

is an optimal solution to Program EL. If the matrix $\mathbf{A}$ of Program EL has full row-rank, then the dual objective $d(\mathbf{w})$ is strictly convex.

For the special case in which Program EL contains an additional constraint

$$\sum_{j=1}^{n} x_j = 1, \tag{3.15}$$

the following dual program has been derived, based on a geometric programming approach [65]:

$$\min_{\mathbf{w}\in R^m} \ln\left\{\sum_{j=1}^{n} p_j \exp\left(\sum_{i=1}^{m} a_{ij}w_i\right)\right\} - \sum_{i=1}^{m} b_i w_i. \tag{3.16}$$

To obtain a primal optimal solution $\mathbf{x}^*$ after having obtained a dual optimal solution $\mathbf{w}^*$, simply use the following conversion formula:

$$x_j^* = \frac{p_j \exp\{\sum_{i=1}^{m} a_{ij} w_i^*\}}{\sum_{k=1}^{n} p_k \exp\{\sum_{i=1}^{m} a_{ik} w_i^*\}}, \quad j = 1, \ldots, n. \tag{3.17}$$

We now provide an introduction to the following entropy optimization problems with inequality constraints:

Program ELI:

$$\min \quad \sum_{j=1}^{n} x_j \ln(\frac{x_j}{p_j}) \tag{3.18}$$

$$\text{s.t.} \quad \sum_{j=1}^{n} a_{ij} x_j \leq b_i, \quad i = 1, 2, \ldots, m, \tag{3.19}$$

$$x_j \geq 0, \quad j = 1, 2, \ldots, n. \tag{3.20}$$

Define the feasibility region of Program ELI as

$$FI \equiv \{\mathbf{x} \in R^n \mid \mathbf{Ax} \leq \mathbf{b};\ \mathbf{x} \geq \mathbf{0}\}. \tag{3.21}$$

The K.K.T. conditions of Program ELI are

$$x_j = p_j \exp(\sum_{i=1}^{m} a_{ij} w_i - 1), \ j = 1, 2, ..., n, \tag{3.22}$$

$$\sum_{j=1}^{n} a_{ij} x_j \leq b_i, \ i = 1, 2, ..., m, \tag{3.23}$$

$$\sum_{i=1}^{m} w_i (\sum_{j=1}^{n} a_{ij} x_j - b_i) = 0, \tag{3.24}$$

$$w_i \leq 0, \ i = 1, 2, ..., m, \tag{3.25}$$

$$x_j \geq 0, \ j = 1, 2, ..., n. \tag{3.26}$$

Similar to the equality case, Constraint (3.26) is redundant. Note that, with Constraints (3.23) and (3.25), Equation (3.24) specifies the complementary slackness condition. Moreover, the corresponding Lagrangian dual of Program ELI can be derived as:

Program DELI :

$$\min_{\mathbf{w}\leq\mathbf{0}} \{d(\mathbf{w}) \equiv \sum_{j=1}^{n} p_j \exp(\sum_{i=1}^{m} a_{ij}w_i - 1) - \sum_{i=1}^{m} b_i w_i\}. \tag{3.27}$$

As will become clear in the rest of this chapter, Programs DEL and DELI play a central role in unifying the existing solution methods. Note that Program DELI is identical to Program DEL except that the dual variables in Program DELI are required to be nonpositive.

In the existing literature of entropy optimization, many variations exist. Some dealt with the maximization of entropy, while others dealt with the minimization of cross-entropy. Some explicitly specified the probability constraint, i.e., $\sum_{j=1}^{n} x_j = 1$ and $\mathbf{x} \geq \mathbf{0}$, while others did not. Yet, some considered the minimization of $\sum_{j=1}^{n} x_j(\ln(x_j/p_j) - 1)$ [48] in order to eliminate the term of -1 in the exponent of Equation (3.7). Some treated the constraint $\mathbf{Ax} \geq \mathbf{b}$, rather than $\mathbf{Ax} \leq \mathbf{b}$. Instead of Program DEL, the following equivalent form, obtained by changing the sign of variables, is often seen in the literature.

Program DEL$'$:

$$\min_{\mathbf{w}\in R^m} \{d'(\mathbf{w}) \equiv \sum_{j=1}^{n} p_j \exp(-\sum_{i=1}^{m} a_{ij}w_i - 1) + \sum_{i=1}^{m} b_i w_i\}. \tag{3.28}$$

Similarly, Program DELI is sometimes rewritten as:

Program DELI$'$:

$$\min_{\mathbf{w}\geq\mathbf{0}} \{d'(\mathbf{w}) \equiv \sum_{j=1}^{n} p_j \exp(-\sum_{i=1}^{m} a_{ij}w_i - 1) + \sum_{i=1}^{m} b_i w_i\}. \tag{3.29}$$

Correspondingly, Conditions (3.7), (3.8), and (3.10) become

$$x_j = p_j \exp(-\sum_{i=1}^{m} a_{ij}w_i - 1), \quad j = 1, 2, \ldots, n,$$

$$\sum_{j=1}^{n} a_{ij}x_j = b_i, \ i = 1, 2, ..., m,$$

$$h_i'(\mathbf{w}) \equiv \sum_{j=1}^{n} a_{ij}p_j \exp(-\sum_{k=1}^{m} a_{kj}w_k - 1) - b_i = 0, \ i = 1, \ldots, m.$$

For consistency purposes and to avoid confusion, we will treat Programs DEL and DELI, not Programs DEL$'$ and DELI$'$, throughout the rest of this book. Other variations will be mentioned in what follows for clarification.

The rest of this chapter is organized as follows. In Section 3.1, we review some well-known existing solution methods in an integrated fashion. Then a complete study of the unconstrained dual approach is discussed in Section 3.2. An extension of the unconstrained dual approach to treating entropy optimization problems with infinitely many constraints via semi-infinite programming theory is the subject of Section 3.3.

3.1 EXISTING METHODS

Five major existing methods, namely, Bregman's balancing method, Multiplicative Algebraic Reconstruction Technique (MART), Newton's method, Generalized Iterative Scaling Method (GISM), and interior-point method, for solving linearly constrained entropy optimization problems are included in this section.

3.1.1 Bregman's Balancing Method

Bregman's balancing method was developed for solving a general class of constrained optimization problems. Here we focus on the special case of linearly constrained entropy optimization problems. The method is capable of handling problems with either equality or inequality constraints. The basic concept becomes particularly simple when it is viewed from a dual perspective. To explicitly explain Bregman's balancing method in a dual framework, we use Programs DEL and DELI as the basis for discussion.

Observe that by solving Equation (3.10) for a dual optimal solution $\mathbf{w}^*$ a corresponding primal optimal solution $\mathbf{x}^*$ can be obtained by using Equation (3.14). Bregman's method is an iterative one. In each iteration, the basic idea of Bregman's balancing method is to minimize the dual objective function $d(\mathbf{w})$ with respect to only one dual variable. To be more precise, for those with equality constraints, the method starts with an arbitrarily chosen $\mathbf{w}^0 \in R^m$ and then iterates. In the kth iteration, it seeks to satisfy *one* of the m equations in (3.10) by adjusting only *one* component of the current solution $\mathbf{w}^k$. The iterative process is performed on each of the m equations of (3.10) as well as on each of the m dual variables in a cyclic manner. In this way, the algorithm moves in the dual space from a current dual solution to the minimal solution along a straight line defined by allowing only one single variable to vary while all other variables are fixed at their current values.

A step-by-step procedure of Bregman's balancing method for solving linearly constrained entropy optimization problems with equality constraints can be described as follows:

Algorithm 3.1: Bregman's Balancing Method – Equality Constraints

Step 1: Choose any $\mathbf{w}^0 \in R^m$ and a sufficiently small tolerance level $\epsilon > 0$. Set $k = 0$ and

$$x_j^0 = p_j \exp(\sum_{l=1}^{m} a_{lj} w_l^0 - 1), \quad j = 1, 2, \ldots, n. \tag{3.30}$$

Step 2: Set $i = (k \bmod m) + 1$. Find the unique solution of λ^k such that

$$\phi^k(\lambda^k) = 0, \tag{3.31}$$

where

$$\phi^k(\lambda^k) \equiv \sum_{j=1}^{n} a_{ij} p_j \exp(\sum_{l=1}^{m} a_{lj} w_l^k - 1 + a_{ij}\lambda^k) - b_i \tag{3.32}$$

$$= \sum_{j=1}^{n} a_{ij} x_j^k \exp(a_{ij}\lambda^k) - b_i. \tag{3.33}$$

Step 3: For $l = 1, 2, ..., m$, set

$$w_l^{k+1} = \begin{cases} w_l^k + \lambda^k, & \text{if } l = i, \\ w_l^k, & \text{if } l \neq i. \end{cases} \tag{3.34}$$

Step 4: If Equation (3.10) is satisfied within the tolerance level ϵ, then stop with $\mathbf{w}^* \equiv \mathbf{w^k}$ and obtain the primal solution $\mathbf{x}^*$ according to Equation (3.14). Otherwise, calculate

$$x_j^{k+1} = p_j \exp(\sum_{l=1}^{m} a_{lj} w_l^{k+1} - 1) \tag{3.35}$$

$$= x_j^k \exp(a_{ij}\lambda^k), \quad j = 1, 2, \ldots, n. \tag{3.36}$$

Step 5: Set $k \leftarrow k + 1$ and go to Step 2.

Different versions of Algorithm 3.1 exist in the literature. A minus sign has also been used in the updating Equation (3.34) of Step 3 (e.g., [48]), in addition to the many variations mentioned earlier in this chapter. As a result, Equations (3.32) and (3.33) of Step 2 appear slightly different from their counterparts in [48]. In addition to the use of this variation, some other papers (e.g., [16, 14]) stated the Bregman's algorithm based on Program DEL$'$, rather than Program DEL. This further variation actually leads to expressions that are identical to Equations (3.32) and (3.33) of Step 2.

Note that the updates performed in Step 3 and Step 4, namely, Equation (3.34) and Equation (3.36), have been referred to as the "dual iteration" and "primal iteration," respectively. Moreover, since Equation (3.31) can be expressed in terms of either dual variables or primal variables, Algorithm 3.1 can be written exclusively in either dual variables or primal variables once the algorithm is appropriately initialized in Step 1. In fact, many papers on this subject have discussed only the primal iterations without addressing any dual connection. Lamond and Stewart [48] is a good example. Although the dual variables did appear and play an important role in their version of the algorithm, they were not acknowledged as being the dual variables, i.e., the decision variables of Program DEL.

However, Bregman's balancing method can be viewed as a dual algorithm because the dual feasibility is maintained throughout the iterative process. But for an implementation based only on dual iterations, the calculations of the corresponding primal iterates $\mathbf{x}^k$ are still implicitly carried out.

From time to time, Bregman's balancing method is referred to as a primal-dual approach in the sense that it is based on the solution of the K.K.T. conditions. We now state the conditions under which Algorithm 3.1 converges. For a detailed proof, the reader is referred to [11].

Theorem 3.1. Suppose that $\mathbf{p} > \mathbf{0}$, $\mathbf{A}$ has no zero rows, F has a nonempty interior, and ϵ is set to 0. Then, Algorithm 3.1 generates a sequence of $\mathbf{x}^k$ that converges to the unique optimal solution of Program EL.

The condition that $\mathbf{A}$ has no zero rows ensures that Equation (3.31) not only has a solution but also has a unique solution. Elfving [22] showed that Algorithm 3.1 has a linear rate of local convergence. Lamond and Stewart [48] showed that many independently discovered balancing methods, used in transportation planning and other fields, are in fact special cases of Bregman's balancing method. Many papers noted that if $a_{ij} = 0$ or 1, for each i and j, then Equation (3.31) can be solved easily with a closed-form solution of λ^k. Also noted was that when a_{ij} = -1, 0, or 1, for each i and j, the evaluation of the exponential function for solving Equation (3.31) is not necessary and the solution can be found very efficiently.

Bregman's balancing method also has been introduced in terms of "entropy projection" [15]. In this setting, let $\boldsymbol{\alpha}$ be an n-dimensional column vector, β a real number, H a hyperplane defined by $H \equiv \{\mathbf{x} \in R^n | \boldsymbol{\alpha}^T\mathbf{x} = \beta\}$, and $\mathbf{y}$ a nonnegative n-dimensional column vector. Then consider the following system of $n+1$ equations

$$y_j^{'} = y_j \exp(\gamma\alpha_j), \quad j = 1, 2, ..., n, \tag{3.37}$$

$$\boldsymbol{\alpha}^T\mathbf{y}^{'} = \beta. \tag{3.38}$$

This system determines $n+1$ unknowns $y_j^{'} \geq 0$, $j = 1, 2, ..., n$, and $\gamma \in R$. $\mathbf{y}^{'}$ has been referred to as the *entropy projection* of $\mathbf{y}$ *onto* H and γ the *entropy projection coefficient* associated with the projection. We denote $P_H(\cdot)$ as the entropy projection operator, then $\mathbf{y}^{'}$ and $\mathbf{y}$ are related by

$$\mathbf{y}' = P_H(\mathbf{y}). \tag{3.39}$$

It is interesting to point out that the λ^k in Step 2 of Bregman's balancing method is exactly the entropy projection coefficient associated with the entropy projection of $\mathbf{x}^k$ onto the ith hyperplane, $i = (k \bmod m) + 1$, of the system $\mathbf{Ax} = \mathbf{b}$, i.e., $H_i \equiv \{\mathbf{x} \in R^n \mid \mathbf{A}_i\mathbf{x} = b_i\}$, where $\mathbf{A}_i$ is the ith row of matrix $\mathbf{A}$. With Equations (3.31) and (3.33) in Step 2 and Equation (3.36) in Step 4, Bregman's balancing method can be viewed as an iterative process of Entropy Projection. In other words, for each $i = (k \bmod m) + 1$, Bregman's algorithm actually finds

$$\mathbf{x}^{k+1} \equiv P_{H_i}(\mathbf{x}^k). \tag{3.40}$$

When Program EL contains the probability constraint, treat $\sum_{j=1}^{n} x_j = 1$ as any other equality constraint. Extensions of Bregman's balancing method have been developed. One version, which includes an "iteration parameter" ω in the updating scheme of (3.34) so as to gain an additional degree of freedom [22], turned out to be very useful in handling problems with noisy data.

In other words, Equation (3.34) in Step 3 of Algorithm 3.1 is replaced by the following updating scheme:

$$w_l^{k+1} = \begin{cases} w_l^k + \omega\lambda^k, & \text{if } l = i, \\ w_l^k, & \text{if } l \neq i. \end{cases} \tag{3.41}$$

This extended version was shown to exhibit local convergence with a linear rate under certain conditions [22].

Bregman's method can be modified to solve entropy optimization problems with linear inequality constraints, i.e., Program ELI. The fundamental change is to cope with the nonpositivity constraints of Program DELI. Therefore, instead of moving to the minimal solution on the straight line in the setting of the unconstrained optimization, the modified algorithm moves on the feasible portion of the straight line to the minimal possible solution. With the first partial derivative of the dual objective

$$\frac{\partial d(\mathbf{w})}{\partial w_i} = \sum_{j=1}^{n} p_j \exp(\sum_{l=1}^{m} a_{lj} w_l - 1) a_{ij} - b_i \tag{3.42}$$

in mind, we outline the modified algorithm as follows.

Algorithm 3.2: Bregman's Method – Inequality Constraints

Step 1: Choose any $\mathbf{w}^0 \leq \mathbf{0}$ and a sufficiently small tolerance level $\epsilon \geq 0$. Set $k = 0$ and

$$x_j^0 = p_j \exp(\sum_{l=1}^{m} a_{lj} w_l^0 - 1), \quad j = 1, 2, \ldots, n. \tag{3.43}$$

Step 2: Set $i = (k \bmod m) + 1$. Calculate

$$r_i = b_i - \sum_{j=1}^{n} a_{ij} x_j^k. \tag{3.44}$$

If $r_i = 0$ or ($r_i > 0$ and $w_i^k = 0$), then set

$$w_l^{k+1} = w_l^k, \quad l = 1, 2, ..., m, \tag{3.45}$$

$$x_j^{k+1} = x_j^k, \quad j = 1, 2, ..., n, \tag{3.46}$$

and go to Step 5. Otherwise, solve

$$\phi^k(\lambda^k) = \sum_{j=1}^{n} a_{ij} x_j^k \exp(a_{ij}\lambda^k) - b_i = 0 \tag{3.47}$$

for the unique solution of λ^k.

Step 3:
If $r_i < 0$ or ($r_i > 0$ and $\lambda^k + w_i^k \leq 0$), then set

$$w_i^{k+1} = w_i^k + \lambda^k, \tag{3.48}$$

$$w_l^{k+1} = w_l^k, \ \forall\, l \neq i, \tag{3.49}$$

$$x_j^{k+1} = p_j \exp(\sum_{l=1}^{m} a_{lj} w_l^{k+1} - 1) \tag{3.50}$$

$$= x_j^k \exp(a_{ij}\lambda^k), \quad j = 1, 2, \ldots, n, \tag{3.51}$$

and go to Step 5. Otherwise, continue.

Step 4:

$$\begin{aligned} w_i^{k+1} &= 0, && (3.52) \\ w_l^{k+1} &= w_l^k, \ \forall\, l \neq i, && (3.53) \\ x_j^{k+1} &= p_j \exp(\sum_{l=1}^{m} a_{lj} w_l^{k+1} - 1) && (3.54) \\ &= x_j^k \exp(-a_{ij} w_i^k), \quad j = 1, 2, \ldots, n. && (3.55) \end{aligned}$$

Step 5: If Equations (3.22) through (3.25) are all satisfied within the tolerance level ϵ, then stop with $\mathbf{w}^* = \mathbf{w}^k$ and $\mathbf{x}^* = \mathbf{x}^k$. Otherwise, set $k \leftarrow k + 1$, and go to Step 2.

We now briefly explain the steps of Algorithm 3.2. First of all, unlike Algorithm 3.1, Algorithm 3.2 requires explicit inclusion of the dual variables. Step 2 ensures that if (i) the current dual solution is already optimal along the straight line (i.e., the ith primal constraint is satisfied by $\mathbf{x}^k$), or (ii) the current dual solution is on the boundary of the dual feasible region and no further improvement is possible along the straight line, then one should not move at all from the current solution. Step 3 indicates a movement along the straight line to the minimal solution, which is also dual feasible. However, the movement in Step 4 is limited by the boundary of the dual feasible region.

Note that, Lamond and Stewart [48] dealt with the linearly constrained entropy optimization problem with the constraint $\mathbf{Ax} \geq \mathbf{b}$, rather than $\mathbf{Ax} \leq \mathbf{b}$, and used $w_i^{k+1} = w_i^k - \lambda^k$ (in a different notation though), rather than Equation (3.48). Therefore, their version of the Bregman's Algorithm appears slightly different from Algorithm 3.2. Although Censor et al. [16] and Censor et al. [14] treated the case with $\mathbf{Ax} \leq \mathbf{b}$, their version of the Bregman's algorithm for solving Program ELI is based on Program DELI$'$. Their use of $w_i^{k+1} = w_i^k - \lambda^k$ for updating the current dual solution further contributes to the difference in appearance between Algorithm 3.2 and their version.

In terms of Entropy Projection, Algorithm 3.2, once appropriately initialized in Step 1, can be captured by

$$x_j^{k+1} = x_j^k \exp(a_{ij}\pi^k), \quad j = 1, 2, \ldots, n, \tag{3.56}$$

$$\mathbf{w}^{k+1} = \mathbf{w}^k + \pi^k \mathbf{e}^i, \tag{3.57}$$

where $\mathbf{e}^i$ is the ith unit vector of R^m,

$$\pi^k \equiv \min(-w_i^k, \lambda^k), \tag{3.58}$$

and λ^k is the entropy projection coefficient associated with the entropy projection of $\mathbf{x}^k$ onto the hyperplane $H_i \equiv \{\mathbf{x} \in R^n | \mathbf{A}_i\mathbf{x} = b_i\}$, which in turn is the binding hyperplane associated with the half-space $L_i \equiv \{\mathbf{x} \in R^n | \mathbf{A}_i\mathbf{x} \leq b_i\}$ defined by the ith inequality constraint of Program ELI, where $i = (k \bmod m)+1$.

The following convergence proof of Algorithm 3.2 follows from the general theory developed by Bregman [11].

Theorem 3.2. Suppose that $\mathbf{p} > \mathbf{0}$, $\mathbf{A}$ has no zero rows, FI has a nonempty interior and ϵ is set to 0. Then, Algorithm 3.2 generates a sequence of $\mathbf{x}^k$ that converges to the unique optimal solution of Program ELI.

It is worth mentioning that some numerical results on using Algorithm 3.1 to solve entropy optimization problems with equality constraints can be found in [22, 31]. But little is known about using Algorithm 3.2 to solve those with inequality constraints. When the probability constraint is added to Program ELI, $\sum_{j=1}^n x_j = 1$ can be represented by two inequality constraints: $\sum_{j=1}^n x_j \leq 1$ and $-\sum_{j=1}^n x_j \leq -1$. Therefore, Algorithm 3.2 is readily applicable.

Bregman's method has been extended in many directions. One direction leads to the so-called "underrelaxed" version by replacing the requirement of solving Equation (3.31) or (3.47) with that of solving

$$\sum_{j=1}^n a_{ij} x_j^k \exp(a_{ij}\lambda^k) = \xi^k b_i + (1 - \xi^k) \sum_{j=1}^n a_{ij} x_j^k, \tag{3.59}$$

where $\delta \leq \xi^k \leq 1$ for some positive δ. Details of this extension and related convergence results can be found in [54], for the equality-constrained case, and in [16], for the inequality-constrained case.

Another extension is to solve entropy optimization problems with "interval constraints" in the form of

$$b_i^1 \leq \sum_{j=1}^{n} a_{ij} x_j \leq b_i^2, \quad i = 1, 2, \ldots, m, \tag{3.60}$$

instead of the form in (3.19). More details together with related convergence results and the advantage of using this extension over Algorithm 3.2 can be referred to [15, 14].

In summary, Bregman's method or any of its extensions is best viewed as one that minimizes the dual objective with respect to one variable at a time in a cyclic manner. The one-variable minimization problem is unconstrained for those entropy optimization problems with equality constraints but is constrained by the nonpositivity requirement for those with inequality constraints.

3.1.2 MART

The Multiplicative Algebraic Reconstruction Technique (MART) was first proposed by Gordon et al. [34] for image construction. It is actually quite similar to Bregman's balancing method. Although MART can be stated completely without any reference to the dual Programs DEL or DELI, to better motivate the method and to point out its similarities to other methods, we present the method from a dual perspective.

Similar to Bregman's balancing method, MART also tries to find a dual optimal solution $\mathbf{w}^*$ of Programs DEL or DELI. After that, an optimal primal solution $\mathbf{x}^*$ to Program EL can be obtained by using Equation (3.14). Starting from an initial m-vector $\mathbf{w}^0$, MART, in iteration k, seeks to satisfy *one* of the m equations in Equation (3.10) by updating only one component of $\mathbf{w}^k$. The iterative process is performed on each of the m equations in system (3.10) as well as on each of the m dual variables in a cyclic manner.

The major difference between MART and Bregman's method occurs in Step 2 of the Bregman's algorithms (i.e., Algorithms 3.1 and 3.2). Instead of solving Equation (3.31) of Algorithm 3.1 or Equation (3.47) of Algorithm 3.2 to obtain the exact change of one of the m dual variables (in order to achieve the minimum solution along a straight line), MART specifies an explicit closed-form formula:

$$\lambda^k \equiv \ln \frac{b_i}{\mathbf{A}_i \mathbf{x}^k}, \tag{3.61}$$

for the change, where $\mathbf{A}_i$ denotes the ith row of $\mathbf{A}$.

This formula certainly eliminates the need for solving the one-dimensional search problem called for in Step 2 of Algorithms 3.1 and 3.2. However, the convergence results hold only for a smaller class of linearly constrained entropy optimization problems where, for every $i = 1, \ldots, m$, either

$$b_i > 0 \text{ and } 0 \leq a_{ij} \leq 1, \quad j = 1, 2, ..., n, \tag{3.62}$$

or

$$b_i < 0 \text{ and } 0 \geq a_{ij} \geq -1, \quad j = 1, 2, ..., n. \tag{3.63}$$

Note that Inequalities (3.62) and (3.63) are equivalent for the equality-constrained case. Also note that, for the equality-constrained case where the probability constraint is present, a systematic way of converting an arbitrary feasible problem into a form in which Inequality (3.62) holds exists and will be discussed in Subsection 3.1.4.

Note that the close-form formula (3.61) provides an approximate solution to the root of Equation (3.31) (for the equality case) and Equation (3.47) (for the inequality case). Bregman's method and MART were developed independently of each other. This approximation connection was established in [16]. To motivate the approximation method, consider the case (3.62). For $i = (k \bmod m) + 1$, define

$$\psi^k(u^k) \equiv \sum_{j=1}^{n} a_{ij} x_j^k (u^k)^{a_{ij}} - b_i. \tag{3.64}$$

Then ψ^k is related to ϕ^k through a simple change of variable: $u^k = \exp(\lambda^k)$. Therefore, one can solve Equations (3.31) or (3.47) by first solving Equation (3.64). Consider the graph of $\psi^k(u^k)$ versus u^k and the straight line passing through $(0, -b)$ and $(1, \psi^k(1))$ on the graph. Approximating ψ^k by the straight line gives an approximate solution of Equations (3.31) and (3.47). Since the secant to the graph intersects the u^k axis at

$$u^{k*} \equiv \frac{b_i}{\sum_{j=1}^{n} a_{ij} x_j^k} = \frac{b_i}{\mathbf{A}_i \mathbf{x}^k}, \tag{3.65}$$

this leads directly to formula (3.61).

The approximation for the case of (3.63) can be obtained similarly, and the approximate root is found to be

$$\lambda^k \equiv -\ln \frac{b_i}{\mathbf{A}_i \mathbf{x}^k}. \tag{3.66}$$

With this understanding, MART for the equality-constrained case can be described as follows:

Algorithm 3.3: MART – Equality Constraints

Step 1: Choose any $\mathbf{w}^0 \in R^m$ and a sufficiently small tolerance level $\epsilon > 0$. Set $k = 0$ and

$$x_j^0 = p_j \exp(\sum_{l=1}^{m} a_{lj} w_l^0 - 1), \quad j = 1, 2, \ldots, n. \tag{3.67}$$

Step 2: Set $i = (k \bmod m) + 1$ and

$$w_l^{k+1} = \begin{cases} w_l^k + \lambda^k, & \text{if } l = i, \\ w_l^k, & \text{if } l \neq i, \end{cases} \tag{3.68}$$

where

$$\lambda^k \equiv \ln \frac{b_i}{\mathbf{A}_i \mathbf{x}^k}. \tag{3.69}$$

Step 3: If system (3.10) is satisfied within the tolerance level, then stop with $\mathbf{w}^* = \mathbf{w}^k$ and obtain an optimal primal solution $\mathbf{x}^*$ by using Equation (3.14).

Otherwise, calculate

$$x_j^{k+1} = p_j \exp(\sum_{l=1}^{m} a_{lj} w_l^{k+1} - 1) \tag{3.70}$$

$$= x_j^k \exp(a_{ij}\lambda^k), \quad j = 1, 2, \ldots, n. \tag{3.71}$$

Step 4: Set $k \leftarrow k+1$ and go to Step 2.

A convergence theorem can be found in [50].

Theorem 3.3. Assume that the interior of F is nonempty and, for every $i = 1, \ldots, m$, $b_i > 0$ and $0 \le a_{ij} \le 1$, $j = 1, 2, ..., n$. If Algorithm **3.3** is initiated with $\mathbf{w}^0 = \mathbf{0}$ (or equivalently $x_j^0 = p_j \exp(-1)$, $j = 1, 2, ..., n$) and ϵ is set to 0, then Algorithm **3.3** generates a sequence of $\mathbf{x}^k$ that converges to the unique optimal solution of Program EL.

Elfving [22] further proved that Algorithm **3.3** exhibits a linear rate of local convergence. Since MART is an approximation strategy of Bregman's method, it shares many of the attributes of Bregman's method. For example, MART can also be extended to include an "iteration parameter" (ω) so as to gain an additional degree of freedom. Instead of using Equation (3.68) to update $\mathbf{w}^{k+1}$, the extended version uses

$$w_l^{k+1} = \begin{cases} w_l^k + \omega\lambda^k, & l = i \equiv (k \bmod m) + 1, \\ w_l^k, & l \neq i. \end{cases} \tag{3.72}$$

The related convergence results of this extended version of MART can be found in [22].

It is interesting to point out that if the constraint matrix $\mathbf{A}$ consists of only 0's and 1's and if Algorithms **3.1** and **3.3** start with the same initial solution, then they generate an identical sequence of $\{\mathbf{x}^k\}$. Observing the similarity between MART and Bregman's method, Censor et al. [16] proposed the following algorithm for solving entropy optimization problems with linear inequality constraints. Again, the algorithm is identical to its Bregman's counterpart except that a closed-form formula is used to approximate the root of Equation (3.47), instead of solving it for the exact solution of λ^k.

Algorithm 3.4: MART – Inequality Constraints

Step 1: Choose any initial solution $\mathbf{w}^0 \leq \mathbf{0}$ and a sufficiently small tolerance level $\epsilon > 0$. Set $k = 0$, and

$$x_j^0 = p_j \exp(\sum_{l=1}^{m} a_{lj} w_l^0 - 1), \quad j = 1, 2, \ldots, n. \tag{3.73}$$

Step 2: Set $i = (k \bmod m) + 1$. Calculate

$$r_i = b_i - \sum_{j=1}^{n} a_{ij} x_j^k. \tag{3.74}$$

If $r_i = 0$ or ($r_i > 0$ and $w_i^k = 0$), then set

$$w_l^{k+1} = w_l^k, \quad l = 1, 2, ..., m, \tag{3.75}$$

$$x_j^{k+1} = x_j^k, \quad j = 1, 2, ..., n, \tag{3.76}$$

and go to Step 5. Otherwise, set

$$\lambda^k = \begin{cases} \ln \frac{b_i}{\mathbf{A}_i \mathbf{x}^k}, & \text{if } b_i > 0, \\ -\ln \frac{b_i}{\mathbf{A}_i \mathbf{x}^k}, & \text{if } b_i < 0. \end{cases} \tag{3.77}$$

Step 3:
If $r_i < 0$ or ($r_i > 0$ and $\lambda^k + w_i^k \leq 0$), then set

$$w_i^{k+1} = w_i^k + \lambda^k, \tag{3.78}$$

$$w_l^{k+1} = w_l^k, \ \forall \ l \neq i, \tag{3.79}$$

$$x_j^{k+1} = p_j \exp(\sum_{l=1}^{m} a_{lj} w_l^{k+1} - 1) \tag{3.80}$$

$$= x_j^k \exp(a_{ij} \lambda^k), \quad j = 1, 2, \ldots, n, \tag{3.81}$$

and go to Step 5. Otherwise, go to the next step.

Step 4: Set

$$w_i^{k+1} = 0, \tag{3.82}$$

$$w_l^{k+1} = w_l^k, \ \forall \ l \neq i, \tag{3.83}$$

$$x_j^{k+1} = p_j \exp(\sum_{l=1}^{m} a_{lj} w_l^{k+1} - 1) \tag{3.84}$$

$$= x_j^k \exp(-a_{ij} w_i^k), \quad j = 1, 2, \ldots, n. \tag{3.85}$$

Step 5: If Equations (3.22) through (3.25) are all satisfied within the tolerance level ϵ, then stop with $\mathbf{w}^* = \mathbf{w}^k$ and $\mathbf{x}^* = \mathbf{x}^k$. Otherwise, set $k \leftarrow k + 1$ and go to Step 2.

The convergence proof of the above algorithm can be found in Censor et al. [16]. We state it as the following theorem:

Theorem 3.4. Assume that (i) the interior of FI is nonempty and, (ii) for every $i = 1, \ldots, m$, either $b_i > 0$ and $0 \leq a_{ij} \leq 1$, $j = 1, 2, ..., n$, or $b_i < 0$ and $0 \geq a_{ij} \geq -1$, $j = 1, 2, ..., n$. Let ϵ be set to 0. Then, Algorithm 3.4 generates a sequence of $\mathbf{x}^k$ that converges to the unique optimal solution of Program ELI.

Like Bregman's method for the inequality case, MART can be captured by the following iterative procedure:

$$x_j^{k+1} = x_j^k \exp(a_{ij} \pi^k), \quad j = 1, 2, \ldots, n, \tag{3.86}$$

$$\mathbf{w}^{k+1} = \mathbf{w}^k + \pi^k \mathbf{e}^i, \tag{3.87}$$

where $\mathbf{e}^i$ is the ith unit vector of R^m,

$$\pi^k \equiv \min(-w_i^k, \lambda^k), \tag{3.88}$$

(3.89)

and

$$\lambda^k = \begin{cases} \ln \frac{b_i}{\mathbf{A}_i \mathbf{x}^k}, & \text{if } b_i > 0, \\ -\ln \frac{b_i}{\mathbf{A}_i \mathbf{x}^k}, & \text{if } b_i < 0. \end{cases} \tag{3.90}$$

When the probability constraint is added to Program ELI, the equality constraint $\sum_{j=1}^{n} x_j = 1$ can be represented by two inequality constraints: $\sum_{j=1}^{n} x_j \leq 1$ and $-\sum_{j=1}^{n} x_j \leq -1$. Since the former inequality constraint satisfies Condition (3.62) and the latter satisfies Condition (3.63), Algorithm 3.4 is readily applicable. Since the dual feasibility is maintained throughout the iterative process, MART can be viewed as a dual algorithm. Similar to

Bregman's method, MART can also be generalized by using the concept of "underrelaxation." More details about this concept, including the convergence theorems, can be found in [16].

Both Bregman's method and MART have been referred to as "row-action" type algorithms because, in each iteration, only the knowledge of one constraint is required. This alleviates greatly the requirement of computer storage and hence enables us to solve large-scale problems. An excellent review on row-action methods can be found in Censor [12]. Parallel computation for row-action methods is a worthy topic for future research.

3.1.3 Newton's Method

Instead of updating only one component of the current dual vector $\mathbf{w}$ in each iteation, Erlander [25] proposed using Newton's method to solve the system of nonlinear equations (3.10). This is equivalent to solving the optimality conditions of the unconstrained dual program DEL of the linearly constrained entropy optimization problem defined by Program EL.

Newton's method requires the information embedded in the Jacobian matrix $J(\mathbf{w}) = \left(\frac{\partial h_i}{\partial w_k}\right)$. Since

$$\frac{\partial h_i}{\partial w_k} = \sum_{j=1}^{n} a_{ij} p_j \exp(\sum_{l=1}^{m} a_{lj} w_l - 1) a_{kj} = \sum_{j=1}^{n} a_{ij} x_j a_{kj}, \quad 1 \le i, k \le m, \tag{3.91}$$

the Jacobian matrix can be expressed as

$$J(\mathbf{w}) = \mathbf{AXA}^T, \tag{3.92}$$

where $\mathbf{X} \equiv \operatorname{diag}(x_1, x_2, ..., x_n)$ is a diagonal matrix formed by the components of $\mathbf{x}$. Note that when the matrix $\mathbf{A}$ has full row-rank and $\mathbf{x} > \mathbf{0}$, then the Jacobian matrix $\mathbf{AXA}^T$ is positive-definite and hence invertible. The inverse matrix decides the direction of movement in the following version of Newton's method:

Algorithm 3.5: Newton's Method – Equality Constraints

Step 1: Choose an initial point $\mathbf{w}^0 \in R^m$ and a sufficiently small tolerance level $\epsilon \geq 0$. Set $k = 0$ and

$$x_j^k = p_j \exp(\sum_{i=1}^{m} a_{ij} w_i^k - 1), \quad j = 1, 2, \ldots, n. \tag{3.93}$$

Step 2: Find the Newton direction $\boldsymbol{\lambda}^k$ by solving

$$\mathbf{A}\mathbf{X}^k\mathbf{A}^T\boldsymbol{\lambda}^k = -(\mathbf{A}\mathbf{x}^k - \mathbf{b}). \tag{3.94}$$

Step 3: Set

$$\mathbf{w}^{k+1} = \mathbf{w}^k + \boldsymbol{\lambda}^k. \tag{3.95}$$

Step 4: If system (3.10) is satisfied within the tolerance level ϵ, then stop with $\mathbf{w}^* = \mathbf{w}^k$ and obtain the corresponding primal optimal solution $\mathbf{x}^*$ according to Equation (3.14). Otherwise, calculate

$$x_j^{k+1} = p_j \exp(\sum_{i=1}^{m} a_{ij} w_i^{k+1} - 1) \tag{3.96}$$

$$= x_j^k \exp(\sum_{i=1}^{m} a_{ij} \lambda_i^k), \quad j = 1, 2, \ldots, n. \tag{3.97}$$

Step 5: Set $k \leftarrow k + 1$. Then, go to Step 2.

It is well known that Newton's method enjoys a quadratic rate of local convergence but does not necessarily possess the property of global convergence. This is in contrast with Bregman's method and MART, which exhibit global convergence property but have a linear rate of convergence. The convergence behavior of Newton's method can be stated as follows.

Theorem 3.5. Suppose that $\mathbf{p} > \mathbf{0}$, $\mathbf{F}$ has a nonempty interior and the system $\mathbf{APx} = \mathbf{b}$, $\mathbf{x} \geq \mathbf{0}$, is consistent, where $\mathbf{P}$ denotes the diagonal matrix formed by the components of $\mathbf{p}$. If the initial point $\mathbf{w}^0$ is sufficiently close to $\mathbf{w}^*$ (the solution of Equation (3.10)), the sequence $\{\mathbf{w}^k\}$ generated by Algorithm 3.5 converges to $\mathbf{w}^*$ with an order of convergence at least two.

For those linearly constrained entropy optimization problems in which the row-rank of matrix **AP** is less than m or those with inequality constraints, modified Newton's method can be used. The reader is referred to [23] for some reported results.

3.1.4 Generalized Iterative Scaling Method

As will be shown, GISM is closely related to MART and can be unified with many other methods within a dual framework. Given $\mathbf{p} \equiv (p_1, \ldots, p_n)^T > \mathbf{0}$ such that $\sum_{j=1}^{n} p_j = 1$, the Generalized Iterative Scaling Method (GISM), deals with a special case of Program EL in the following form:

Program ELS:

$$\min \quad \sum_{j=1}^{n} x_j \ln(\frac{x_j}{p_j}) \tag{3.98}$$

$$\text{s.t.} \quad \sum_{j=1}^{n} a_{ij} x_j = b_i, \quad i = 1, 2, \ldots, m, \tag{3.99}$$

$$\sum_{j=1}^{n} x_j = 1, \tag{3.100}$$

$$x_j \geq 0, \quad j = 1, 2, \ldots, n. \tag{3.101}$$

Again, the last two constraints are usually referred to as the "probability constraint" or "simplex constraint" in the literature. A solution vector $\mathbf{x} > \mathbf{0}$ satisfying Constraints (3.99) and (3.100) is called an "interior solution."

Since Program ELS has a bounded feasible domain and the objective function is strictly convex, it has a unique minimum solution as long as the feasible domain is not empty. Moreover, as pointed out in [8], if Program ELS has an interior feasible solution, then its optimal solution $\mathbf{x}^*$ is an interior solution that can be expressed in the following form:

$$x_j^* = p_j \mu \prod_{i=1}^{m} \mu_i^{a_{ij}}, \quad j = 1, 2, ..., n, \tag{3.102}$$

where $\mu > 0$ and $\mu_i > 0$, $i = 1, 2, ..., m$, are parameters to be determined from Equations (3.99) and (3.100).

This special form is called a "log-linear" model, because, for $j = 1, \ldots, n$, $\ln(x_j^*/p_j)$ is linear in a_{ij}, $i = 1, \ldots, m$. The log-linear model was originally used to solve a special case of Program ELS with $a_{ij} = 0$ or 1, $\forall\ i, j$ [8]. The solution method is known as an iterative scaling method because, when the method iterates, a new solution is generated by multiplying the current solution with a scaling factor.

Darroch and Ratcliff [17] generalized the iterative scaling method to allow a_{ij}'s to be real numbers. This generalization is now known as the Generalized Iterative Scaling Method or GISM.

The crux of the theory behind GISM consists of: (i) if Program ELS has an interior solution $\mathbf{x} > 0$ in the form of Equation (3.102), then $\mathbf{x}$ is the unique optimal solution to program ELS; (ii) under certain conditions, such an interior solution $\mathbf{x}$ can be constructed via an iterative scaling procedure. After establishing the first claim, we motivate the second claim and describe the associated algorithm and convergence results.

Given a probability distribution $\mathbf{p} > \mathbf{0}$, we denote the objective function of Program ELS by $D(\mathbf{x}, \mathbf{p})$. From [46] and page 22 of [43], we have the following result:

Lemma 3.2. Given $p_j > 0$, for $j = 1, ..., n$, and $\sum_{j=1}^{n} p_j = 1$, if $\mathbf{x}$ satisfies the probability constraint, then $D(\mathbf{x}, \mathbf{p}) \geq 0$. Moreover, $D(\mathbf{x}, \mathbf{p}) = 0$ if and only if $\mathbf{x} = \mathbf{p}$.

With this lemma, we can establish the first claim for GISM.

Theorem 3.6. If Program ELS has an interior solution $\mathbf{x}^* > 0$ in the form of Equation (3.102), then $\mathbf{x}^*$ is the unique optimal solution of Program ELS.

Proof: Since $\mathbf{x}^*$ satisfies Equation (3.102), we know that $\mathbf{x}^* > \mathbf{0}$ and

$$
\begin{aligned}
D(\mathbf{x}^*, \mathbf{p}) &= \sum_{j=1}^{n} x_j^* \ln(\frac{x_j{}^*}{p_j}) \\
&= \sum_{j=1}^{n} x_j^* (\ln \mu + \sum_{i=1}^{m} a_{ij} \ln \mu_i)
\end{aligned}
$$

$$= (\ln \mu) \sum_{j=1}^{n} x_j^* + \sum_{i=1}^{m} (\ln \mu_i)(\sum_{j=1}^{n} a_{ij} x_j^*). \tag{3.103}$$

Let $\mathbf{x}$ be any feasible solution of Program ELS. We now show that $D(\mathbf{x}, \mathbf{p}) \geq D(\mathbf{x}^*, \mathbf{p})$. Note that $\sum_{j=1}^{n} a_{ij} x_j^* = \sum_{j=1}^{n} a_{ij} x_j = b_i$, for $i = 1, 2, ..., m$, and $\sum_{j=1}^{n} x_j^* = \sum_{j=1}^{n} x_j = 1$. Also recall that $\ln(x_j^*/p_j) = \ln \mu + \sum_{i=1}^{m} a_{ij} \ln \mu_i$. Therefore, Equation (3.103) becomes

$$\begin{aligned} \mathrm{D}(\mathbf{x}^*, \mathbf{p}) &= (\ln \mu) \sum_{j=1}^{n} x_j + \sum_{i=1}^{m} (\ln \mu_i)(\sum_{j=1}^{n} a_{ij} x_j) \\ &= \sum_{j=1}^{n} x_j (\ln \mu + \sum_{i=1}^{m} a_{ij} \ln \mu_i) \\ &= \sum_{j=1}^{n} x_j \ln(\frac{x_j^*}{p_j}) \\ &= \sum_{j=1}^{n} x_j \ln(\frac{x_j}{p_j}) - \sum_{j=1}^{n} x_j \ln(\frac{x_j}{x_j^*}) \\ &= D(\mathbf{x}, \mathbf{p}) - D(\mathbf{x}, \mathbf{x}^*), \end{aligned} \tag{3.104}$$

where $D(\mathbf{x}, \mathbf{p})$ and $D(\mathbf{x}, \mathbf{x}^*)$ are well defined because both $\mathbf{p}$ and $\mathbf{x}^*$ are positive probability distributions.

Consequently, $D(\mathbf{x}, \mathbf{p}) - D(\mathbf{x}^*, \mathbf{p}) = D(\mathbf{x}, \mathbf{x}^*)$. Lemma 3.2 implies that $D(\mathbf{x}, \mathbf{x}^*) \geq 0$. Hence, $D(\mathbf{x}, \mathbf{p}) \geq D(\mathbf{x}^*, \mathbf{p})$ and $\mathbf{x}^*$ is an optimal solution to Program ELS. Moreover, if $D(\mathbf{x}, \mathbf{p}) = D(\mathbf{x}^*, \mathbf{p})$, then $D(\mathbf{x}, \mathbf{x}^*) = 0$. Consequently, by Lemma 3.2, $\mathbf{x} = \mathbf{x}^*$. Hence, $\mathbf{x}^*$ is the unique optimal solution of Program ELS.

□

The key implication of this theorem is that it allows us to focus on the interior solutions in the form of Equation (3.102). Once such a solution is found, it is assured to be the unique optimal solution of Program ELS. Our next goal is to see how such a solution can be found by iterative rescaling.

To achieve this goal, Darroch and Ratcliff [17] first proposed the idea of transforming Program ELS, without actually specifying the transformation, into an equivalent problem in which all program parameters are nonnegative and then proposed the GISM algorithm for solving the equivalent problem. In this section, we first describe a procedure that transforms Program ELS into an equivalent problem with strictly positive problem parameters and then introduce GISM. More precisely, the equivalent problem takes the following form:

Program ELS$'$:

$$\min \quad \sum_{j=1}^{n} x_j \ln(\frac{x_j}{p_j}) \tag{3.105}$$

$$\text{s.t.} \quad \sum_{j=1}^{n} a'_{ij} x_j = b'_i, \quad i = 1, 2, \ldots, m, \tag{3.106}$$

$$\sum_{j=1}^{n} x_j = 1, \tag{3.107}$$

$$x_j \geq 0, \quad j = 1, 2, \ldots, n, \tag{3.108}$$

where $1 > a'_{ij} > 0$ and $1 > b'_i > 0$, $i = 1, \ldots, m$ and $j = 1, \ldots, n$. Note that Program ELS$'$ is a special case of Program ELS and Theorem 3.6 applies. Also note that if the original program is already in the form of Program ELS$'$, then the transformation is not necessary.

We now describe the transformation. For $i = 1, \ldots, m$, let us define

$$u_i \equiv [\max_j(-a_{ij})] + 1. \tag{3.109}$$

Hence, for $i = 1, ..., m$,

$$(u_i + a_{ij}) > 0, j = 1, 2, ..., n. \tag{3.110}$$

For $i = 1, \ldots, m$, define

$$M_i \equiv \max_j(u_i + a_{ij}), \tag{3.111}$$

and

$$t_i = \frac{1}{m(M_i + 1)}. \tag{3.112}$$

It is easily seen that, for $i = 1, ..., m$, $M_i > 0$, $t_i > 0$, $(M_i + 1)t_i = (1/m)$, $t_i M_i < (1/m)$ and

$$t_i(u_i + a_{ij}) \leq t_i M_i < (1/m). \tag{3.113}$$

Now, for $i = 1, \ldots, m$ and $j = 1, \ldots, n$, define

$$a'_{ij} \equiv t_i(u_i + a_{ij}). \tag{3.114}$$

Apparently,

$$1/m > a'_{ij} > 0, \tag{3.115}$$

for $i = 1, ..., m$ and $j = 1, ..., n$. Moreover, for $j = 1, ..., n$,

$$0 < \sum_{i=1}^{m} a'_{ij} = \sum_{i=1}^{m} t_i(u_i + a_{ij}) < 1. \tag{3.116}$$

Note that, for any i, $i = 1, 2, ..., m$, if $\mathbf{x}$ is a feasible solution to Program ELS, then

$$\begin{aligned} \sum_{j=1}^{n} a'_{ij} x_j &= \sum_{j=1}^{n} t_i(u_i + a_{ij}) x_j \\ &= t_i u_i \sum_{j=1}^{n} x_j + t_i \sum_{j=1}^{n} a_{ij} x_j \\ &= t_i u_i + t_i b_i \\ &= t_i(u_i + b_i). \end{aligned} \tag{3.117}$$

Hence, for $i = 1, \ldots, m$, if we define

$$b'_i \equiv t_i(u_i + b_i), \tag{3.118}$$

then

$$\sum_{j=1}^{n} a'_{ij} x_j = b'_i. \tag{3.119}$$

If Program ELS has a feasible solution, then, by Equation (3.116) and the probability constraint, we have

$$1/m > b_i^{'} > 0 \ , \quad i = 1, \ldots, m. \tag{3.120}$$

It can be verified that if $\mathbf{x}$ satisfies Equations (3.119) and the probability constraint, it must also be a feasible solution to Program ELS. Thus, Programs ELS and ELS$^{'}$ are equivalent. Note that since the probability constraint is taken advantage of in the derivation of Equation (3.117), Equation (3.99) is equivalent to Equations (3.106) or (3.119) only in the presence of the probability constraint. Also note that the problem parameters associated with Program ELS$^{'}$ satisfy Inequality (3.62) and therefore this transformation can be used as a preprocessor not only for GISM but also for MART, for solving Program ELS.

Again, since Program ELS$^{'}$ is a special case of Program ELS, Theorem 3.6 applies. Particularly, the optimal solution to Program ELS$^{'}$, when it exists and is an interior solution, should be in the following form:

$$x_j^* = p_j \mu^{'} \prod_{i=1}^{m} \mu_i^{'a_{ij}^{'}}, j = 1, 2, ..., n, \tag{3.121}$$

where $\mu^{'} > 0$ and $\mu_i^{'} > 0$, $i = 1, 2, ..., m$, are parameters to be determined from Equations (3.106) and (3.107).

To pave the way for the GISM algorithm, we derive an equivalent expression for Equation (3.121) as follows. For $j = 1, ..., n$, by Inequality (3.116), we can define, for $j = 1, 2, \ldots, n$,

$$1 > a_{m+1,j}^{'} \equiv 1 - \sum_{i=1}^{m} a_{ij}^{'} \ > 0. \tag{3.122}$$

If Program ELS has a feasible solution, then, by Inequality (3.120),

$$1 > b_{m+1}^{'} \equiv 1 - \sum_{i=1}^{m} b_i^{'} > 0. \tag{3.123}$$

More interestingly, we have, for $i = 1, 2, \ldots, m + 1$,

$$\sum_{j=1}^{n} a_{ij}^{'} x_j = b_i^{'}. \tag{3.124}$$

Note that Equation (3.124) is also true for $i = m+1$. The corresponding equation is obtained by first summing the first m constraints of Program ELS$'$ and then subtracting the resulting equation from Equation (3.107). Therefore, Program ELS$'$ is equivalent to

Program ELS$''$:

$$\min \quad \sum_{j=1}^{n} x_j \ln(\frac{x_j}{p_j}) \tag{3.125}$$

$$\text{s.t.} \quad \sum_{j=1}^{n} a'_{ij} x_j = b'_i, \quad i = 1, 2, \ldots, m, m+1, \tag{3.126}$$

$$x_j \geq 0, \quad j = 1, 2, \ldots, n. \tag{3.127}$$

This program will serve in what follows as a key to recognizing an interesting relationship between GISM and MART.

Equation (3.114) implies $a_{ij} = (1/t_i)a'_{ij} - u_i$, for $i = 1, 2, ..., m$ and $j = 1, 2, ..., n$. It can be verified that Equation (3.102) can be transformed to the following form:

$$\begin{aligned} x_j &= p_j \mu \prod_{i=1}^{m} \mu_i^{[(1/t_i)a'_{ij} - u_i]} \\ &= p_j \mu \prod_{i=1}^{m} (\mu_i^{(1/t_i)})^{a'_{ij}} \prod_{i=1}^{m} \mu_i^{-u_i}, \end{aligned} \tag{3.128}$$

for $j = 1, ...n$. Since, $\sum_{i=1}^{m+1} a'_{ij} = 1$, the above equation becomes

$$x_j = p_j \prod_{i=1}^{m} (\mu \mu_i^{(1/t_i)})^{a'_{ij}} \mu^{a'_{m+1,j}} \{\prod_{i=1}^{m} \mu_i^{-u_i}\}^{\sum_{i=1}^{m+1} a'_{ij}}, \tag{3.129}$$

for $j = 1, ..., n$.

Defining $\upsilon \equiv \mu \prod_{i=1}^{m} \mu_i^{-u_i}$, the above equation reduces to

$$x_j = p_j \{\prod_{i=1}^{m} (\upsilon \mu_i^{(1/t_i)})^{a'_{ij}}\} \upsilon^{a'_{m+1,j}}. \tag{3.130}$$

Now, if we further define

$$\lambda_i \equiv (\upsilon \mu_i^{(1/t_i)}) > 0, \tag{3.131}$$

for $i = 1, ..., m$, and

$$\lambda_{m+1} \equiv v > 0, \tag{3.132}$$

then Equation (3.102) becomes

$$x_j = p_j \prod_{i=1}^{m+1} \lambda_i^{a'_{ij}}, \quad j = 1, \ldots, n. \tag{3.133}$$

The generalized iterative scaling method outlined below searches for an interior solution of Program ELS in the form of Equation (3.133).

Algorithm 3.6: GISM Algorithm

Step 1: (Transformation)

Let $\epsilon \geq 0$ be a sufficiently small tolerance level. Set $k \leftarrow 0$, and

$$x_j^{(0)} \equiv p_j > 0, j = 1, 2, \ldots, n. \tag{3.134}$$

For $i = 1, 2, ..., m$, compute

$$u_i \equiv [\max_j(-a_{ij})] + 1, \tag{3.135}$$

$$M_i \equiv \max_j(u_i + a_{ij}), \tag{3.136}$$

$$t_i \equiv \frac{1}{m(M_i + 1)}. \tag{3.137}$$

For $i = 1, 2, ..., m$ and $j = 1, 2, ..., n$, compute

$$a'_{ij} \equiv t_i(u_i + a_{ij}) \tag{3.138}$$

and

$$a'_{m+1,j} \equiv 1 - \sum_{i=1}^{m} a'_{ij}. \tag{3.139}$$

For $i = 1, 2, ..., m$, compute

$$b'_i \equiv t_i(u_i + b_i) \tag{3.140}$$

and

$$b'_{m+1} \equiv 1 - \sum_{i=1}^{m} b'_i. \tag{3.141}$$

If $b'_i \leq 0$ or $b'_i \geq 1/m$, for some $i = 1, 2, \ldots, m$, then stop, and Program ELS is infeasible.

Step 2: Compute

$$b_i^{'(k)} \equiv \sum_{j=1}^{n} a'_{ij} x_j^{(k)}, \; i = 1, ..., m+1. \tag{3.142}$$

Step 3: If Equations (3.106) and (3.107) are satisfied within the tolerance level ϵ, then stop with $\mathbf{x}^* \equiv \mathbf{x}^{(k)}$ as the optimal solution. Otherwise, compute

$$x_j^{(k+1)} \equiv x_j^{(k)} \prod_{i=1}^{m+1} \left(\frac{b'_i}{b_i^{'(k)}} \right)^{a'_{ij}}, \; j = 1, 2, ..., n. \tag{3.143}$$

Step 4: Set $k \leftarrow k + 1$; go to Step 2.

It can be shown [17] that, in Step 2, $b_i^{'(k)} > 0$ and $\sum_{i=1}^{m+1} b_i^{'(k)} = 1$. From Step 3, we know that $\mathbf{x}^k$ is in the form of Equation (3.133). Intuitively, we can see that a converging solution should have the same form. Note that $\mathbf{x}^{k+1}$ is obtained by multiplying $\mathbf{x}^k$ with a "scaling factor." This occasioned the name of this method.

The key convergence result for Algorithm 3.6 is stated as follows.

Theorem 3.7. If Program ELS has an interior solution and ϵ is set to 0, then the sequence $\{\mathbf{x}^{(k)}\}$ generated by GISM converges to the unique optimum solution of Program ELS in the form of Equation (3.133).

A detailed proof of this theorem can be found in [17]. It was noted by Kumar et al. [45] that the GISM outlined above may take many iterations before reaching an acceptable converging point. Revised versions with improvement can be found in Hoshino et al. [38].

We now show that GISM can be viewed as "MART with simultaneous updates." When MART is applied to Program ELS$''$, the updating step defined by Equation (3.71), in the notation of this subsection, becomes

$$x_j^{(k+1)} = x_j^{(k)} \left(\frac{b'_i}{b_i'^{(k)}} \right)^{a'_{ij}}. \tag{3.144}$$

By comparing this equation with Equation (3.143), we observe that, in each iteration, GISM takes $m+1$ steps simultaneously along each of the $m+1$ dual variables using identical step sizes as specified in MART.

3.1.5 Interior-Point Method

The existing literature on the interior-point approach actually deals with the following linearly constrained entropy optimization problem, which are more general than Program EL.

Program ELL:

$$\begin{array}{ll} \min & f(\mathbf{x}) \equiv \mathbf{c}^T\mathbf{x} + \sum_{j=1}^{n} w_j x_j \ln x_j \\ \text{s.t.} & \mathbf{x} \in F_p \equiv \{\mathbf{x} \in R^n \mid A\mathbf{x} = \mathbf{b}, \mathbf{x} \geq 0\}, \end{array}$$

where $\mathbf{c}, \mathbf{w} \in R^n$, $\mathbf{w} \geq \mathbf{0}$, $\mathbf{A} \in R^{m \times n}$, and $\mathbf{b} \in R^m$ are given problem parameters. The interior-point approach assumes that F_p has a nonempty interior, i.e., $F_p^0 \equiv \{\mathbf{x} \in R^n \mid \mathbf{Ax} = \mathbf{b}, \mathbf{x} > 0\} \neq \emptyset$. It is obvious to see that, when $\mathbf{c} = \mathbf{0}$, the objective function $f(\mathbf{x})$ becomes a weighted entropy function. Its dual problem can be written as:

Program DELL:

$$\begin{array}{ll} \max & d(\mathbf{x}, \mathbf{y}) \equiv \mathbf{b}^T\mathbf{y} - \mathbf{w}^T\mathbf{x} \\ \text{s.t.} & (\mathbf{x}, \mathbf{y}) \in F_d \equiv \{(\mathbf{x}, \mathbf{y}) \mid \nabla f(\mathbf{x}) - \mathbf{A}^T\mathbf{y} \geq 0, \mathbf{x} \in F_p\}. \end{array}$$

We denote the excess vector $\nabla f(\mathbf{x}) - \mathbf{A}^T\mathbf{y}$ by $\mathbf{s}$ and further assume that F_d also has a nonempty interior.

From the K.K.T. conditions, we know that $\mathbf{x}^*$ is an optimal solution of Program ELL if and only if

(i) (Primal feasibility): $\mathbf{x}^* \in F_p$,
(ii) (Dual feasibility): $\exists\ \mathbf{y}^*$ such that $(\mathbf{x}^*, \mathbf{y}^*) \in F_d$,
(iii) (Complementary slackness): $\mathbf{X}^*(\nabla f(\mathbf{x}^*) - \mathbf{A}^T\mathbf{y}^*) = \mathbf{0}$,

where $\mathbf{X}^*$ denotes the diagonal matrix formed by the components of the vector $\mathbf{x}^*$.

A primal-dual interior point method iterates from a pair of interior solutions $(\mathbf{x}^k, \mathbf{y}^k)$, where $\mathbf{A}\mathbf{x}^k = \mathbf{b}, \mathbf{x}^k > 0$ and $\mathbf{s}^k = \nabla f(\mathbf{x}^k) - \mathbf{A}^T\mathbf{y}^k > 0$, to a new interior solution pair $(\mathbf{x}^{k+1}, \mathbf{y}^{k+1})$ such that the complementary slackness is reduced from $(\mathbf{x}^k)^T\mathbf{s}^k$ to $(\mathbf{x}^{k+1})^T\mathbf{s}^{k+1}$. In other words, the primal feasibility and dual feasibility conditions are maintained and the complementary slackness is reduced in each iteration until the complementary slackness vanishes and, in the meantime, the algorithm reaches an optimal solution.

To describe the algorithm, we use the boldface upper-case letters $\mathbf{X}$, $\mathbf{S}$, and $\mathbf{W}$ to denote the diagonal matrices formed by the components of vectors $\mathbf{x}$, $\mathbf{s}$, and $\mathbf{w}$, respectively. We also denote the vectors of all ones of appropriate dimensions by $\mathbf{e}$, the l_2 norm by $\|\cdot\|$, and the vector whose components are $\ln(x_j)$'s, $j = 1, 2, \ldots, n$ by $\ln \mathbf{x}$.

Consider the primal-dual "potential function" proposed by Todd and Ye [61]

$$\psi(\mathbf{x}, \mathbf{s}) = \rho \ln(\mathbf{x}^T\mathbf{s}) - \sum_{j=1}^{n} \ln(x_j s_j), \tag{3.145}$$

where $\rho \geq n + \sqrt{n}$. This potential function can be written as

$$\psi(\mathbf{x}, \mathbf{s}) = (\rho - n)\ln(\mathbf{x}^T\mathbf{s}) - \sum_{j=1}^{n} \ln(\frac{x_j s_j}{\mathbf{x}^T\mathbf{s}}).$$

The inequality of the geometric and arithmetic means yields

$$-\sum_{j=1}^{n} \ln(\frac{x_j s_j}{\mathbf{x}^T\mathbf{s}}) \geq n \ln n.$$

Therefore,

$$(\rho - n)\ln(\mathbf{x}^T\mathbf{s}) \leq \psi(\mathbf{x}, \mathbf{s}) - n\ln n. \tag{3.146}$$

Inequality (3.146) implies the exact amount, $-(\rho - n) \mid \ln \epsilon \mid$, by which ψ should be reduced to reach

$$\mathbf{x}^T\mathbf{s} \leq \epsilon.$$

Given $0 < \mathbf{x}^k \in F_p$, $\mathbf{s}^k = \nabla f(\mathbf{x}^k) - \mathbf{A}^T\mathbf{y}^k > 0$, and $\delta^k = (\mathbf{x}^k)^T\mathbf{s}^k$, the algorithm proposed in [55, 44] solves the following system of nonlinear equations for $\triangle\mathbf{x}$ and $\triangle\mathbf{y}$:

$$\mathbf{X}^k\triangle\mathbf{s} + \mathbf{S}^k\triangle\mathbf{x} = \theta\mathbf{p}^k, \tag{3.147}$$

$$\mathbf{A}\triangle\mathbf{x} = \mathbf{0}, \tag{3.148}$$

$$\triangle\mathbf{s} = \nabla f(\mathbf{x}^k + \triangle\mathbf{x}) - \nabla f(\mathbf{x}^k) - \mathbf{A}^T\triangle\mathbf{y}, \tag{3.149}$$

where $\theta > 0$ is a constant to be specified later and

$$\mathbf{p}^k \equiv \frac{\delta^k}{\rho}\mathbf{e} - \mathbf{X}^k\mathbf{S}^k\mathbf{e}. \tag{3.150}$$

Let

$$\mathbf{x}^{k+1} \equiv \mathbf{x}^k + \triangle\mathbf{x}, \tag{3.151}$$

$$\mathbf{y}^{k+1} \equiv \mathbf{y}^k + \triangle\mathbf{y}, \tag{3.152}$$

$$\mathbf{s}^{k+1} \equiv \nabla f(\mathbf{x}^{k+1}) - \mathbf{A}^T\mathbf{y}^{k+1}. \tag{3.153}$$

Then, by choosing

$$\theta = \frac{\alpha \min_j(\sqrt{x_j^k s_j^k})}{\| (\mathbf{X}^k\mathbf{S}^k)^{-0.5}\mathbf{p}^k \|} \tag{3.154}$$

for some $0 < \alpha < 1$, we obtain

$$\psi(\mathbf{x}^{k+1}, \mathbf{s}^{k+1}) \leq \psi(\mathbf{x}^k, \mathbf{s}^k) - \gamma$$

for a constant $\gamma > 0$.

Actually, the system of Equations (3.147), (3.148), and (3.149) does not need to be solved exactly. Potra and Ye [55] showed that it suffices to solve the following system of linear equations:

$$\mathbf{X}^k(\nabla^2 f(\mathbf{x}^k)\triangle\mathbf{x} - \mathbf{A}^T\triangle\mathbf{y}) + \mathbf{S}^k\triangle x = \theta\mathbf{p}^k, \tag{3.155}$$

$$\mathbf{A}\triangle\mathbf{x} = \mathbf{0}. \tag{3.156}$$

Note that this system can be written as

$$(\mathbf{X}^k\nabla^2 f(\mathbf{x}^k) + \mathbf{S}^k)\Delta\mathbf{x} = \mathbf{X}^k\mathbf{A}^T\Delta\mathbf{y} + \theta\mathbf{p}^k, \tag{3.157}$$

$$\mathbf{A}(\mathbf{X}^k\nabla^2 f(\mathbf{x}^k) + \mathbf{S}^k)^{-1}\mathbf{X}^k\mathbf{A}^T\Delta\mathbf{y} + \mathbf{A}(\mathbf{X}^k\nabla^2 f(\mathbf{x}^k) + \mathbf{S}^k)^{-1}\theta\mathbf{p}^k = \mathbf{0}, \tag{3.158}$$

where

$$\nabla^2 f(\mathbf{x}^k) = (\mathbf{X}^k)^{-1}\mathbf{W}. \tag{3.159}$$

Given an initial interior-point solution, the interior-point algorithm can be stated as follows.

Algorithm 3.7: Interior-Point Method

Initialization:
Given an initial primal interior-point solution $\mathbf{x}^0$ and an initial dual solution $\mathbf{y}^0$ such that $\mathbf{A}\mathbf{x}^0 = \mathbf{b}$, $\mathbf{x}^0 > 0$, and $\mathbf{s}^0 = \nabla f(\mathbf{x}^0) - \mathbf{A}^T\mathbf{y}^0 > 0$, calculate $\delta^0 = (\mathbf{x}^0)^T\mathbf{s}^0$; select $0 < \alpha < 1$; set $k \leftarrow 0$.

Iteration:

If $\delta^k < \epsilon$, then stop. Otherwise,
solve Equations (3.155) and (3.156) for $\Delta\mathbf{x}$ and $\Delta\mathbf{y}$; set
$\mathbf{x}^{k+1} \equiv \mathbf{x}^k + \Delta\mathbf{x}$;
$\mathbf{y}^{k+1} \equiv \mathbf{y}^k + \Delta\mathbf{y}$;
$\mathbf{s}^{k+1} \equiv \nabla f(\mathbf{x}^{k+1}) - \mathbf{A}^T\mathbf{y}^{k+1}$;
$\delta^{k+1} \equiv (\mathbf{x}^{k+1})^T\mathbf{s}^{k+1}$;
reset $k \leftarrow k+1$ for the next iteration.

Potra and Ye [55] proved the following convergence theorem.

Theorem 3.8. Suppose that $\epsilon > 0$, $2n \geq \rho \geq n + \sqrt{n}$ and $\psi(\mathbf{x}^0, \mathbf{s}^0) \leq O((\rho - n)|\ln\epsilon|)$. Then, Algorithm 3.7 terminates in $O((\rho - n)|\ln\epsilon|)$ iterations.

Potra and Ye [55] suggested that, in practical implementation, the step size can be set to $\bar{\eta}$ based on a line search as follows:

$$\bar{\eta} \equiv \arg \min_{\eta \geq 0} \psi(\mathbf{x}^k + \eta\triangle\mathbf{x}, \mathbf{s}^k + \eta\triangle\mathbf{s}). \tag{3.160}$$

With this step size, one can set

$$\mathbf{x}^{k+1} \equiv \mathbf{x}^k + \bar{\eta}\triangle\mathbf{x}, \quad \mathbf{y}^{k+1} \equiv \mathbf{y}^k + \bar{\eta}\triangle\mathbf{y}. \tag{3.161}$$

We now provide a standard procedure for obtaining an initial interior-point solution [28]. In order to find a primal interior-point solution, we solve the following problem in the form of Program ELL:

$$\min \quad \sum_{j=1}^{n} x_j \ln x_j + Mz \tag{3.162}$$

$$\text{s.t.} \quad \mathbf{A}\mathbf{x} + nz(\mathbf{b} - \mathbf{A}\mathbf{e}/n) = \mathbf{b}, \tag{3.163}$$

$$\mathbf{e}^T\mathbf{x} = 1, \tag{3.164}$$

$$\mathbf{x} \geq 0, \mathbf{z} \geq 0, \tag{3.165}$$

where $M = n^2\text{max}_{i,j}(a_{i,j})$. Note that $\mathbf{x}^0 = \mathbf{e}/n$ and $z^0 = 1/n$ is a feasible solution of this problem. Hence, we have the following problem

$$\min \quad f(\mathbf{x}) = \mathbf{c}'^T\mathbf{x} + \sum_{j=1}^{n+1} w_j' x_j \ln x_j \tag{3.166}$$

$$\text{s.t.} \quad \mathbf{A}'\mathbf{x} = \mathbf{b}', \tag{3.167}$$

$$\mathbf{x} \geq 0, \tag{3.168}$$

where

$$\mathbf{c}' = (0, \cdots, 0, M)^T \in R^{n+1}, \mathbf{w}' = (1, \cdots, 1, 0)^T \in R^{n+1}, \mathbf{b}' = (\mathbf{b}^T, 1)^T,$$

$$\mathbf{A}' = \begin{pmatrix} \mathbf{A} & \mathbf{d} \\ \mathbf{e}^T & \mathbf{0} \end{pmatrix} \text{ and } d_i = nb_i - \sum_{j=1}^{n} a_{ij}, \; i = 1, \cdots, m.$$

It is clear that

$$y_i^0 = \begin{cases} 0, & i = 1, \cdots, m, \\ -\ln n, & i = m+1, \end{cases}$$

is a feasible dual solution $\mathbf{y}^0$ such that $\mathbf{s}^0 = \nabla f(\mathbf{x}^0) - \mathbf{A}'^T \mathbf{y}^0 > 0$. For details about how to generate an initial feasible solution $(\mathbf{x}^0, \mathbf{s}^0)$ such that $\psi(\mathbf{x}^0, \mathbf{s}^0)$ is bounded by $(\rho - n)|\ln(\mathbf{x}^{0^T}\mathbf{s}^0)|$, see [55].

In addition to the "potential reduction" interior-point method described above, the "path-following" interior-point method, which follows an ideal interior trajectory to reach an optimal solution, was proposed in [70, 59]. The convergence of the path-following interior-point method has been established. However, to the best of our knowledge, the polynomial-time convergence behavior remains an open issue.

3.2 AN UNCONSTRAINED CONVEX PROGRAMMING APPROACH

It is important to note that none of the methods discussed in the previous section, as originally proposed, works on the dual side of the problem directly. Also, none of them possesses the property of being globally convergent with a quadratic rate of convergence. In fact, for some of those methods [22], even the number of iterations required to reach the optimum could be very sensitive to the individuality and size of the problem. The objective of this section is to design an efficient and robust dual algorithm for solving Program EL.

The dual program of Program EL can be obtained from either Lagrangian dual [5] or geometric dual [27, 29, 65, 19, 20]. Both approaches result in Program DEL. Since an optimal solution to Program EL can be easily constructed from an optimal solution to Program DEL, we can solve Program EL by solving Program DEL first. Notice that Program DEL is an unconstrained minimization problem with a convex objective function. Therefore, one can explore different unconstrained optimization techniques [52, 18], including the descent methods, conjugate direction methods, and quasi-Newton methods, for its solution.

After having conducted careful experiments, we found that the curved-search method proposed by Ben-Tal et al. [6] looks promising for achieving the objective. In particular, they provided certain sufficient conditions under which the curved-search method is globally convergent with a quadratic rate of convergence. We will customize the curved-search method for solving Program EL with the desired efficiency and robustness.

Section 3.2 is organized as follows. A family of curved-search algorithms for solving Program EL are introduced in Subsection 3.2.1. Their global convergence and quadratic rate of local convergence will be proved in Subsection 3.2.2. Subsection 3.2.3 is devoted to the implementation issues and the analysis of computational effort. Related computational experience is reported in Subsection 3.2.4, and concluding remarks are made in Subsection 3.2.5.

3.2.1 A Class of Curved-Search Algorithms

Our approach to solving Program EL is to solve its dual first by the curved-search method [6] and then construct a corresponding primal optimal solution from the dual optimal solution. We briefly describe the curved-search method and then tailor it for solving Program DEL.

Most classical iterative methods improve a current solution by moving along a straight line. The basic idea of the curved-search method is to improve a current solution by moving along a quadratic curve which is determined by minimizing a certain model of the objective function subject to suitable constraints. Ben-Tal et al. [6] showed that the quadratic curve used in their method turns out to be a nonlinear combination of the "signed" Newton direction and the steepest descent direction.

To be more precise, for an unconstrained convex minimization problem with a twice continuously differentiable objective function f, the curved-search method moves from one solution $\mathbf{w}_k$ to the next solution $\mathbf{w}_{k+1}$ along the quadratic curve (in variable $t \geq 0$)

$$\mathbf{q}_k(t) \equiv \mathbf{w}_k + t\mathbf{d}_k + \frac{1}{2}t^2\mathbf{z}_k, \tag{3.169}$$

where

$$\begin{aligned} \mathbf{d}_k &\equiv -\beta_k \frac{\|\nabla f(\mathbf{w}_k)\|^2}{\nabla f^T(\mathbf{w}_k)[\nabla^2 f(\mathbf{w}_k)]^{-1}\nabla f(\mathbf{w}_k)}[\nabla^2 f(\mathbf{w}_k)]^{-1}\nabla f(\mathbf{w}_k), \\ \mathbf{z}_k &\equiv -\alpha_k\|\nabla f(\mathbf{w}_k)\|\nabla f(\mathbf{w}_k). \end{aligned}$$

By searching for an appropriate step size t_k such that

$$t_k \in \arg\min_{t>0} f(\mathbf{w}_k + t\mathbf{d}_k + \frac{1}{2}t^2\mathbf{z}_k), \tag{3.170}$$

a new solution $\mathbf{w}_{k+1}$ is defined by

$$\mathbf{w}_{k+1} \equiv \mathbf{w}_k + t_k\mathbf{d}_k + \frac{1}{2}t_k^2\mathbf{z}_k. \tag{3.171}$$

Notice that α_k and β_k are adjustable positive parameters associated with the constraints with which the above-mentioned model of the objective function is minimized. Therefore, Equations (3.170) and (3.171) actually define a family of curved-search algorithms. The performance of this family of algorithms can be fine-tuned by choosing appropriate values of α_k and β_k. Also note that when $\alpha_k = 0$, the curved-search algorithm becomes the (signed) Newton method. When $\beta_k = 0$, it becomes the steepest descent algorithm.

Consider Program DEL again.

Program DEL:

$$\min_{\mathbf{w}\in R^m} \{f(\mathbf{w}) \equiv \sum_{j=1}^{n} p_j \exp\{\sum_{i=1}^{m} a_{ij}w_i - 1\} - \sum_{i=1}^{m} b_i w_i\}. \tag{3.172}$$

The strong duality theorem [5, 27] shows that if Program EL has an interior feasible solution, i.e., if there exists a feasible solution $\mathbf{x}$ with $x_j > 0$, $j = 1, \ldots, n$, then the duality gap between Programs EL and DEL vanishes. In other words, we have min(EL) = − min(DEL). Moreover, a primal optimal solution $\mathbf{x}^*$ to Program EL is given by

$$x_j^* = p_j \exp\{\sum_{i=1}^{m} a_{ij}w_i^* - 1\}, \text{ for } j = 1, \ldots, n, \tag{3.173}$$

where $\mathbf{w}^*$ is an optimal solution to Program DEL. For a detailed development for the duality theory, see Section 5.2, where a more general problem is studied.

We now focus on Program DEL and study the gradient vector and the Hessian matrix of the dual objective function f. Denote $\mathbf{g} \equiv \nabla f(\mathbf{w})$ and $\mathbf{H} \equiv \nabla^2 f(\mathbf{w})$. The k_1th element of the gradient is given by

$$\frac{\partial f}{\partial w_{k_1}} = \sum_{j=1}^{n} p_j \exp\{\sum_{i=1}^{m} a_{ij} w_i - 1\} a_{k_1 j} - b_{k_1}, \tag{3.174}$$

and the (k_1,k_2)th element of the Hessian matrix is given by

$$\frac{\partial^2 f}{\partial w_{k_1} \partial w_{k_2}} = \sum_{j=1}^{n} p_j \exp\{\sum_{i=1}^{m} a_{ij} w_i - 1\} a_{k_1 j} a_{k_2 j}. \tag{3.175}$$

Equations (3.174) and (3.175) clearly show that the dual objective function f is twice continuously differentiable. In addition, we have the following lemma:

Lemma 3.3. If the constraint matrix $\mathbf{A}$ in Program EL has full row-rank, then the Hessian matrix $\mathbf{H}$ of the dual objective function f in Program DEL is nonsingular and positive-definite.

Proof: Equation (3.175) indicates that the Hessian matrix $\mathbf{H}$ of f can be written as

$$\mathbf{H} = \mathbf{A}\mathbf{D_s}(\mathbf{w})\mathbf{A}^T, \tag{3.176}$$

where $\mathbf{D_s}(\mathbf{w})$ is an $n \times n$ diagonal matrix with $s_j(\mathbf{w})$ as its jth diagonal element and

$$s_j(\mathbf{w}) = p_j \exp\{\sum_{i=1}^{m} a_{ij} w_i - 1\} > 0, \text{ for } j = 1, 2, \ldots, n. \tag{3.177}$$

Therefore, by matrix theory, $\mathbf{H}$ must be positive-definite and, hence, nonsingular as long as $\mathbf{A}$ has full row-rank.

□

By convexity theory, Lemma 3.3 implies that function f is strictly convex when $\mathbf{A}$ has full row-rank. Moreover, from the theory of convex programming, we know that, under this condition, Program DEL has a unique optimal solution if it attains its minimum.

Based on the curved-search method discussed earlier [6], we propose the following algorithm for solving Program DEL and refer to it as the CS-ENT algorithm.

Algorithm 3.8: The CS-ENT Algorithm

Initialization:

Choose small real numbers $\epsilon \geq 0$, $\delta \geq 0$ and $\gamma > 0$.
Choose a sequence α_k such that $\overline{\alpha} \geq \alpha_k \geq 0$ with $\overline{\alpha} > 0$.
Choose a sequence β_k such that $\overline{\beta} \geq \beta_k \geq \underline{\beta}$ with $\overline{\beta} \geq \underline{\beta} > 0$.
Set $k = 0$ and the starting solution $\mathbf{w}_0 \equiv \mathbf{0}$.

Iteration:

Compute $\mathbf{g}_k \equiv \nabla f(\mathbf{w}_k)$ with its k_1th element defined by Equation (3.174).
Compute $\mathbf{H}_k \equiv \nabla^2 f(\mathbf{w}_k)$ with its (k_1,k_2)th element defined by Equation (3.175).
If $||\mathbf{g}_k|| \leq \epsilon$, stop with $\mathbf{w}^* \equiv \mathbf{w}_k$. Otherwise, continue.
Solve $\mathbf{H}_k \mathbf{v}_k = \mathbf{g}_k$ for $\mathbf{v}_k$.
Compute $G_k \equiv \mathbf{g}_k^T \mathbf{v}_k$, $\gamma_k \equiv \frac{|G_k|}{||\mathbf{v}_k||\,||\mathbf{g}_k||^2}$ and $\delta_k \equiv (\det \mathbf{H}_k)^2$.
If $\delta_k > \delta$ and $\gamma_k > \gamma$, then
compute $\mathbf{d}_k \equiv -\beta_k \frac{||\mathbf{g}_k||^2}{G_k} \mathbf{v}_k$ and $\mathbf{z}_k \equiv -\alpha_k ||\mathbf{g}_k|| \mathbf{g}_k$.
Otherwise, set $\mathbf{d}_k \equiv -\mathbf{g}_k$ and $\mathbf{z}_k \equiv \mathbf{0}$.
Compute a step size $t_k \in \arg\min_{t>0} f(\mathbf{w}_k + t\mathbf{d}_k + \frac{1}{2}t^2 \mathbf{z}_k)$.
Set $\mathbf{w}_{k+1} = \mathbf{w}_k + t_k \mathbf{d}_k + \frac{1}{2} t_k^2 \mathbf{z}_k$.
Reset $k \leftarrow k + 1$ for the next iteration.

After having obtained a dual optimal solution $\mathbf{w}^*$, we can easily compute the primal optimal solution $\mathbf{x}^*$ according to Equation (3.173).

For the special case in which Program EL contains a probability constraint, Tsao et al. [65] obtained the following dual program using a geometric programming approach:

$$\min_{\mathbf{w}\in R^m} \left\{ h(\mathbf{w}) \equiv \ln \left[\sum_{j=1}^{n} p_j \exp\left(\sum_{i=1}^{m} a_{ij} w_i \right) \right] - \sum_{i=1}^{m} b_i w_i \right\}. \tag{3.178}$$

Similar curved-search algorithms can be applied to this program. To obtain a primal optimal solution $\mathbf{x}^*$ after having obtained a dual optimal solution $\mathbf{w}^*$, simply use Equation (3.17).

Up to now, the feasibility of Program EL has been assumed. Ben-Tal and Charnes [5] showed that Program EL is infeasible if and only if Program DEL is unbounded. Since Program DEL is unconstrained, it is not difficult to verify the unboundedness of Program DEL numerically and hence the infeasibility of Program EL.

3.2.2 Convergence Properties

As discussed in Subsection 3.2.1, when α_k and β_k are set to be positive, the curved-search algorithm is a combination of the (signed) Newton and steepest descent methods. As a result, we expect that, under some regularity conditions, the CS-ENT algorithm converges globally (like the steepest descent method) with a quadratic rate of convergence (like Newton method). In this subsection, we show that, for a Program EL with an interior feasible solution and a full row-rank constraint matrix $\mathbf{A}$, the CS-ENT algorithm is indeed globally convergent with a quadratic convergence rate. The next two theorems state these facts.

Theorem 3.9. If the constraint matrix $\mathbf{A}$ has full row-rank and Program EL has an interior feasible solution, then the CS-ENT algorithm either stops after a finite number of steps or generates an infinite sequence $\{\mathbf{w}_k \in R^m | k = 1, 2, \ldots\}$ such that (i) $\mathrm{f}(\mathbf{w}_{k+1}) < f(\mathbf{w}_k)$ for all k; (ii) $\{\mathbf{w}_k \in R^m | k \in N\}$ has at least a cluster point $\mathbf{w}^c$ in the level set $L_0 \equiv \{\mathbf{w} \in R^m | f(\mathbf{w}) \leq f(\mathbf{0})\}$; (iii) for each such cluster point $\mathbf{w}^c$, $|\nabla f(\mathbf{w}^c)| \leq \epsilon$.

Proof: Recall that the dual objective function f is twice continuously differentiable and $\mathbf{w}_0 \equiv \mathbf{0}$ is the initial solution. If we can further show that the level set L_0 is compact, then Theorem 3.9 is a direct consequence of Theorem 3.1 of [6]. By the assumption that Program EL has an interior feasible solution, there is no duality gap between Programs EL and DEL. Since $x_j \ln(\frac{x_j}{p_j}) \geq -1$ for $x_j \geq 0$ and $p_j > 0$, Program EL has a finite infimum. Consequently, the strong duality theorem [3] implies that the dual Program DEL achieves its finite optimum at a point $\mathbf{w}^* \in R^m$. In addition, since the constraint matrix $\mathbf{A}$ is assumed to be of full row-rank, f is strictly convex. Therefore, f has a unique minimizer $\mathbf{w}^* \in R^m$ with a finite minimum value and hence each level set of f, including L_0, is compact. This concludes the proof.

□

Note that, under the assumptions of Theorem 3.9, since f has a unique optimal solution $\mathbf{w}^*$ satisfying the first-order optimality conditions, the infinite sequence generated by the CS-ENT algorithm, with $\epsilon = 0$, converges to the unique cluster point $\mathbf{w}^c = \mathbf{w}^*$. Moreover, this infinite sequence has a quadratic rate of convergence.

Theorem 3.10. Under the assumptions of Theorem 3.9, the infinite sequence $\{\mathbf{w}_k \in R^m | k \in N\}$ generated by the CS-ENT algorithm, with ϵ set to 0, converges to $\mathbf{w}^*$ at an order of at least 2.

Proof: Theorem 4.4 of [6] establishes the quadratic rate of convergence for the CS-ENT algorithm, if we can show that the following four conditions are satisfied: (i) $\{\mathbf{w}_k \in R^m | k \in N\}$ converges to $\mathbf{w}^*$; (ii) $\nabla \mathrm{f}(\mathbf{w}^*) = 0$; (iii) $\nabla^2 f(\mathbf{w}^*)$ is positive-definite; (iv) $\mathbf{H}(\cdot) \equiv \nabla^2 f(\cdot)$ is locally Lipschitz-continuous in a neighborhood of $\mathbf{w}^*$.

From earlier discussion, the first three conditions are clearly met. As to Condition (iv), recall that $\mathbf{H} = \mathbf{A}\mathbf{D}_\mathbf{s}(\mathbf{w})\mathbf{A}^T$ and $\mathbf{D}_\mathbf{s}(\mathbf{w})$ is a diagonal matrix with $s_j(\mathbf{w})$ as its jth diagonal element. From Equation (3.177), we know that $s_j(\mathbf{w})$, $j = 1, 2, \ldots, n$, belongs to C^∞, i.e., $s_j(\mathbf{w})$ has derivatives of every order. Therefore, every element of $\mathbf{H}$ is a finite combination of C^∞ functions. Since differentiability implies Lipschitz continuity, we know that Condition (iv) is also satisfied and the proof is complete.

□

3.2.3 Implementation and Complexity

Before discussing the complexity issues, let us focus on the implementation first. In the CS-ENT algorithm, the key computational steps include the calculation of $\mathbf{v}_k$ by solving $\mathbf{H}_k\mathbf{v}_k = \mathbf{g}_k$ and finding t_k by minimizing $f(\mathbf{w}_k + t\mathbf{d}_k + \frac{1}{2}t^2\mathbf{z}_k)$ over $t > 0$.

To calculate $\mathbf{v}_k$, in our implementation, we chose the Gaussian Elimination method. In fact, since $\mathbf{H}_k$ is positive-definite, an implementation by using Cholesky Factorization method can certainly improve the computational performance. However, our purpose here is not to find the best implementation but to introduce the CS-ENT algorithm. Hence we use the commonly applied Gaussian method and leave other sophisticated implementations for future studies. Following the same philosophy, we chose a commonly used line-search procedure consisting of the search for a three-point pattern and the Golden-section search [52] to determine the step size t_k. A class of curved-search algorithms without line search can be found in [6].

Consider Program DEL again. Whether one is calculating the objective function f, its gradient ∇f, or the Hessian $\nabla^2 f$, the evaluation of exponential functions is always required. Hence there are two potential problems related to the implementation of the CS-ENT algorithm: (i) arithmetic overflow problem, (ii) evaluation of the exponential functions.

For the first problem, in order to prevent arithmetic overflow in evaluating the exponential functions, first note that Program DEL is a minimization problem. Therefore, if the overflow problem does not occur at the initial solution $\mathbf{w}_0$, this potential problem could be avoided throughout the whole optimization process. Consequently, in the implementation, we chose $\mathbf{w}_0 \equiv \mathbf{0}$. In this way, each of the n exponential functions in the objective function is reduced to a common number e^{-1} at the initial solution. Also, in search of the three-point pattern, we need to make sure that the first step away from the current solution should be sufficiently small in order to avoid overflow. With these two precautionary steps, the overflow problem never occurred in our tests.

For the second problem, the evaluation of the exponential functions: The evaluation of an exponential function, like any other transcendental function, requires significantly more computational effort than an elementary arithmetic operation. In the CS-ENT algorithm, exponential functions appear in the objective function, its gradient, and its Hessian. Realizing the fact that there are n exponential functions involved in the gradient vector and m^2n exponential functions involved in the Hessian matrix, one might think the CS-ENT algorithm would require an excessive amount of computation. Fortunately, this is not the case. In our implementation, only n "seed" exponential functions need to be evaluated in each iteration. These seeds are the diagonal elements $s_j(\mathbf{w})$'s of the diagonal matrix $\mathbf{D_s}(\mathbf{w})$ given by Equation (3.177). Once they are computed and stored, both the gradient vector and Hessian matrix can be computed via simple arithmetic operations. Therefore, no excessive computational effort is required. Also, because the total number of these seeds grows linearly in n, the number of primal variables, the computational time spent on these calculations becomes less significant in solving large problems.

Having addressed some of the implementation issues, let us focus on the complexity analysis. Basically, three steps in the CS-ENT algorithm require a significant amount of computational effort: (i) the calculation of Hessian, (ii) the inversion of Hessian, and (iii) the line search. For small problems, the step of line search is usually the computational bottleneck. But even for large problems, (i) and (ii) really dominate the computational complexity. To be more precise, once the n seeds of exponential functions are calculated and stored, it takes $O(m^2n)$ elementary operations to compute the Hessian. This is because the Hessian matrix contains m^2 elements and each element requires $O(n)$ operations for the evaluation. After having obtained the Hessian, the Gaussian Elimination method in the implementation requires $O(m^3)$ elementary operations to invert $\mathbf{H}_k$ for obtaining $\mathbf{v}_k$.

Note that since every element of the Hessian matrix $\mathbf{H}_k$ can be independently calculated, an implementation with parallel computation can greatly improve the performance of the CS-ENT algorithm for solving large problems. The use of sparse matrix techniques and the Cholesky factorization method in matrix inversion can also lead to more efficient implementation. A study of CPU breakdown based on a number of experiments, which confirm this complexity analysis, will be summarized in the next subsection.

3.2.4 Computational Experience

In this subsection, we report our computational experience with the CS-ENT algorithm in two scenarios. In Scenario 1, we solve two matrix scaling problems which were solved and reported by Elfving [22] and make simple comparisons. In Scenario 2, Programs EL of different sizes, ranging from 100 variables with 10 constraints to 1,000 variables with 500 constraints, are randomly generated and tested. The computational results confirm our earlier analysis and discussion. For purposes of simplicity, in all applications, we set $\alpha_k = \beta_k = 1$ for the CS-ENT algorithm.

Scenario 1: The matrix scaling problem is to find two diagonal matrices, $\mathbf{D}_1$ and $\mathbf{D}_2$, for a given square matrix $\mathbf{\Phi}$ such that the scaled matrix $\mathbf{D}_1\mathbf{\Phi}\mathbf{D}_2$ has prescribed row and column sums. This problem is equivalent to a special case of Program EL and has applications in transportation planning, economics, and statistics [48].

Two matrix scaling problems with

$$\mathbf{M}_1 = \begin{bmatrix} 10^4 & 10^2 & 10^2 \\ 10^2 & 1 & 1 \\ 10^2 & 1 & 1 \end{bmatrix} \text{ and } \mathbf{M}_2 = \begin{bmatrix} 10^2 & 10^2 & 0 \\ 10^2 & 10^4 & 1 \\ 0 & 1 & 10^2 \end{bmatrix} \tag{3.179}$$

were studied by Elfving [22]. The prescribed row sums and column sums are identically 1. Note that scaling $\mathbf{M}_1$ is equivalent to an incidence of Program EL with the prior $\mathbf{p}_1$, constraint matrix $\mathbf{A}_1$, and right-hand-side vector $\mathbf{b}_1$ defined as follows:

$$\mathbf{p}_1 = \begin{bmatrix} 10^4 \\ 10^2 \\ 10^2 \\ 10^2 \\ 1 \\ 1 \\ 10^2 \\ 1 \\ 1 \end{bmatrix}, \mathbf{A}_1 = \begin{bmatrix} 1 & 1 & 1 & 0 & 0 & 0 & 0 & 0 & 0 \\ 0 & 0 & 0 & 1 & 1 & 1 & 0 & 0 & 0 \\ 0 & 0 & 0 & 0 & 0 & 0 & 1 & 1 & 1 \\ 1 & 0 & 0 & 1 & 0 & 0 & 1 & 0 & 0 \\ 0 & 1 & 0 & 0 & 1 & 0 & 0 & 1 & 0 \\ 0 & 0 & 1 & 0 & 0 & 1 & 0 & 0 & 1 \end{bmatrix} \text{ and } \mathbf{b}_1 = \begin{bmatrix} 1 \\ 1 \\ 1 \\ 1 \\ 1 \\ 1 \end{bmatrix}. \tag{3.180}$$

Table 3.1. CS-ENT Algorithm for Matrix Scaling (Time in Seconds; % in parenthesis)

Program	it.	cpu	cpu-Hessian	cpu-inversion	cpu-other
3.1	7	0.336	0.029(9%)	0.021(6%)	0.286(85%)
3.2	4	0.197	0.016(8%)	0.012(6%)	0.169(86%)

Since the row-rank of matrix $\mathbf{A}_1$ is 5 (one of the scaling constraints is redundant), we can eliminate the last row to obtain a full-rank matrix $\mathbf{A}_1'$ with a corresponding $\mathbf{b}_1'$. In this way, we have an equivalent incidence of Program EL (Program 3.1) with

$$\mathbf{p}_1 = \begin{bmatrix} 10^4 \\ 10^2 \\ 10^2 \\ 10^2 \\ 1 \\ 1 \\ 10^2 \\ 1 \\ 1 \end{bmatrix}, \mathbf{A}_1' = \begin{bmatrix} 1 & 1 & 1 & 0 & 0 & 0 & 0 & 0 & 0 \\ 0 & 0 & 0 & 1 & 1 & 1 & 0 & 0 & 0 \\ 0 & 0 & 0 & 0 & 0 & 0 & 1 & 1 & 1 \\ 1 & 0 & 0 & 1 & 0 & 0 & 1 & 0 & 0 \\ 0 & 1 & 0 & 0 & 1 & 0 & 0 & 1 & 0 \end{bmatrix} \text{ and } \mathbf{b}_1' = \begin{bmatrix} 1 \\ 1 \\ 1 \\ 1 \\ 1 \end{bmatrix}. \tag{3.181}$$

Similarly, we have another incidence of Program EL (Program 3.2) for the second matrix scaling problem with $\mathbf{p}_2 = [10^2, 10^2, 0, 10^2, 10^4, 1, 0, 1, 10^2]^T$, $\mathbf{A}_2 = \mathbf{A}_1$, and $\mathbf{b}_2 = \mathbf{b}_1$. Since $\mathbf{p}_2$ contains two zero elements, Program 3.2 is not well-defined. One can handle this situation as Elfving [22] did by first setting x_3 and x_7 to 0 and then solving a reduced problem with a matrix $\bar{\mathbf{A}}_2'$ obtained by deleting the third and the seventh columns from $\mathbf{A}_2'$. But this situation does not affect our dual approach at all.

In solving Programs 3.1 and 3.2 on an Alliant FX8 using the CS-ENT algorithm, we set $\alpha_k = \beta_k = 1$ and $\epsilon = 10^{-5}$. Also, the line search stops when the difference between the objective values of two consecutive test points is within 10^{-10}. Although Alliant FX8 has parallel/vector processing capability, we suppressed this capability for easy comparison with other reports. The computational results are provided in Table 3.1.

Table 3.1 lists the CPU time spent on the whole algorithm (cpu), computation of Hessian and gradient (cpu-Hessian), the inversion of the Hessian (cpu-inversion), and the rest of the computation (cpu-other). Within the category of cpu-other, the repeated evaluation of the objective function in line search accounts for most of the time consumption. For small problems like these two, most of the computational time is indeed spent on the line search.

Compared to Elfving's result, the CS-ENT algorithm seems to be much more robust since Elfving's Gauss-Seidel algorithm takes 33 iterations for solving Program 3.1 and 7833 iterations for Program 3.2 with the "iteration parameter" $\omega = 1$. Even when ω is set to its optimal value ω_{opt}, Elfving's algorithm still takes 17 iterations for Program 3.1 and 294 iterations for Program 3.2. The low number of iterations required by the CS-ENT algorithm also reflects its advantage of having a quadratic rate of convergence.

It is worth noting that our comparison is not completely fair. To begin with, these two algorithms use different initial solutions because there exists no initial solution for the CS-ENT algorithm that is equivalent to the one used in the Gauss-Seidel algorithm. Similarly, these two algorithms use different stopping criteria. In addition, the CS-ENT algorithm can be fine-tuned with a better choice of α_k and β_k.

Scenario 2: We test the performance of the CS-ENT algorithm with randomly generated incidences of Program EL of various sizes. In our experiment, the feasible region $\{\mathbf{x}|\mathbf{A}\mathbf{x} = \mathbf{b}, \mathbf{x} \geq \mathbf{0}\}$ is randomly generated by a subroutine created in [39] and the prior p_j, for $j = 1, \ldots, n$, are uniformly distributed between 0 and 1. The CS-ENT algorithm stops when $||\mathbf{g}_k|| \leq \epsilon = 10^{-5}$, and the line search stops when the difference between the objective values of two consecutive test points is within 10^{-10}. The size of the test problems ranges from 10 (constraints) $\times$ 100 (variables) to $500 \times 1,000$. The results are tabulated in Table 3.2.

In our experiment, we observed the following facts:

(i) The CS-ENT algorithm always converges.

Table 3.2. CS-ENT Algorithm for Randomly Generated Problems (Time in Seconds)

Size	it.	cpu	cpu-Hessian	cpu-inversion	cpu-other
10×100	5	1.8	0.4(22%)	0.1(4%)	1.3(74%)
100×200	7	193.1	78.4(41%)	82.5(43%)	32.3(16%)
100×500	8	388.9	225.9(58%)	90.9(23%)	72.1(19%)
250×500	11	4237.9	1882.0(44%)	2036.4(48%)	319.6(8%)
500×1000	14	43887.0	19089.5(43%)	23155.3(53%)	1642.3(4%)

(ii) In order to verify the quadratic convergence rate, we calculated $r_k \equiv \|\mathbf{w}_k - \mathbf{w}^*\| / \|\mathbf{w}_{k-1} - \mathbf{w}^*\|^2$. In our experiment, r_k indeed converges up until the current solution is in an "immediate" neighborhood of the optimal solution $\mathbf{w}^*$. Then the ratio becomes sensitive to numerical accuracy. This observation together with the low number of iterations could confirm the quadratic rate of convergence of the CS-ENT algorithm.

(iii) By setting the initial solution at $\mathbf{0}$ and choosing the first step size in the three-point search pattern to be $\frac{1}{n}$, no arithmetic overflow problem ever occurred in our experiment.

(iv) The percentage of cpu-other decreases as problem size increases. This indicates that the computational effort spent in evaluating the exponential functions is relatively small in solving large problems.

(v) Table 3.2 also confirmed that the calculation of the Hessian matrix $\mathbf{H}$ and its inversion really dominate the computational complexity in solving large problems.

(vi) The number of total iterations appears to be very insensitive to the size of the problem. It grows very slowly as the problem size increases. This shows that the CS-ENT algorithm is very robust and promising for solving large problems.

(vii) Since the complexity of the calculation of the Hessian matrix and its inversion can be analyzed and the total number of iterations is quite insensitive to the problem size, the CPU time of the CS-ENT algorithm is more predictable than many other algorithms.

3.2.5 Concluding Remarks

In this section, we proposed a dual-based CS-ENT algorithm for solving Program EL. This algorithm exhibits not only the property of global convergence but also a quadratic rate of convergence. In addition to these theoretical advantages, our computational experience has demonstrated its efficiency and robustness. Coupled with the effortless dual-to-primal conversion, this algorithm allows us to manipulate both primal and dual information freely. In particular, the post-optimality analysis, including sensitivity analysis and parametric programming, is made much easier owing to the unconstrained nature of the dual. Extensions of this approach for solving linear programming problems will be studied in Sections 5.1 and 5.2.

3.3 ENTROPY OPTIMIZATION PROBLEMS WITH INFINITELY MANY LINEAR CONSTRAINTS

In this section, we consider the following entropy optimization problem with n variables and infinitely many linear constraints:

Program SIE:

$$\inf \quad f(\mathbf{x}) = \sum_{j=1}^{n} x_j \ln x_j \tag{3.182}$$

$$\text{s.t.} \quad \sum_{j=1}^{n} g_j(t) x_j \geq h(t), \ \forall\, t \in T, \tag{3.183}$$

$$x_j \geq 0, \quad j = 1, \ldots, n, \tag{3.184}$$

where T is a compact metric space and g_j, $j = 1, 2, \ldots, n$, and h are real-valued continuous functions on T. Throughout this section, we denote the feasible region of Program SIE as FSI. Note that adding any finite number of linear constraints, e.g., $\sum_{j=1}^{n} x_j = 1$, to an SIE problem results in another SIE problem. Also note that when T is instead a finite set, the problem is equivalent to Program ELI.

The SIE formulation has many potential applications. One example is in the area of dosage distribution in pharmacokinetics [4]. Suppose that n different drugs contain a common effective ingredient for a particular treatment. However, they are absorbed and disposed of by different compartments of the human body at different time-dependent rates. A single drug may not be sufficient to provide a sustainable concentration of the effective ingredient to the target compartment. Together with the important fact that each drug has some potential adverse effects or side effects, a *mixture* of these drugs is often more desirable than one drug alone.

Suppose that one dose of drug j, $j = 1, ..., n$, induces a concentration $g_j(t)$, a function of time, of the ingredient in the desired compartment at time t. To ensure the effectiveness of treatment, the total concentration of the ingredient released for a target compartment by the mixture must be kept above a minimum level $h(t)$ at time t throughout a given time period T.

By denoting the proportion of drug j used in the mixture by a decision variable x_j, $j = 1, ..., n$, we clearly see the physical meanings of the constraints in Program SIE. Moreover, to minimize the potential adverse effects or side effects caused by any one of these drugs, mixing of these drugs should be as even as possible. This leads to the objective function of our entropy maximization model.

Actually, any finite entropy optimization model with linear constraints incurred in a continuous temporal or spatial domain can be conceptually extended to Program SIE. Such entropy optimization models have been successfully applied to various engineering disciplines.

The difficulty of solving Program SIE lies in how to deal with the infinite number of constraints effectively. One simple approach, namely, the "Discretization Method," is to discretize the set T as a finite collection of points $\{t_1, ..., t_k\}$ and solve the following finite entropy optimization problem with n variables and k inequality constraints :

Program ELI^k:

$$\min \quad f(x) = \sum_{j=1}^{n} x_j \ln x_j \tag{3.185}$$

$$\text{s.t.} \quad \sum_{j=1}^{n} g_j(t_i)x_j \geq h(t_i), \quad i = 1, \ldots, k, \tag{3.186}$$

$$x_j \geq 0, \quad j = 1, \ldots, n. \tag{3.187}$$

The solution of Program ELI^k provides an approximation to the optimal solution of the original SIE problem. It is easy to see that, in general, a finer discretization of T results in a better approximation. Moreover, unless the functions $g_j(t)$ and $h(t)$ are well-behaved, a good approximation may require solving a very large scale entropy optimization problem. Hence a more efficient algorithm for solving Program SIE is desirable.

In this section, we incorporate the basic concept of "adding constraints" used in solving linear and quadratic semi-infinite programming problems [32, 33] into a "cutting-plane method" for solving Program SIE. Under a mild condition, we show, in Subsection 3.3.1, that the algorithm converges. Then the implementation issues are discussed in Subsection 3.3.2, and computational experience is reported in Subsection 3.3.3. Some concluding remarks are made in Subsection 3.3.4.

3.3.1 A Cutting-Plane Algorithm

The basic concept of the proposed "cutting-plane method" is to add one constraint at a time until an optimal solution is identified. Given a subset $T_k = \{t_1, t_2, \ldots, t_k\}$ of T, $k \geq 1$, we consider a corresponding Program ELI^k as defined by (3.185) - (3.187). Let FSI^k be the feasible region of Program ELI^k. Supposing that $\mathbf{x}^k = (x_1^k, \ldots, x_n^k)^T$ is an optimal solution of ELI^k, we define a "constraint violation function" as follows:

$$v_{k+1}(t) = h(t) - \sum_{j=1}^{n} g_j(t)\, x_j^k, \ \forall\, t \in T. \tag{3.188}$$

Since $g_j(t)$ and $h(t)$ are continuous over T, and T is compact, the function $v_{k+1}(t)$ achieves its maximum over T. Let t_{k+1} be such a maximizer and consider the value of $v_{k+1}(t_{k+1})$. If this value is less than or equal to zero, then we know $\mathbf{x}^k$ must be an optimal solution of Program SIE (because the feasible region FSI^k of Program ELI^k is no smaller than the feasible region FSI of Program SIE). Otherwise, $\mathbf{x}^k$ is not optimal and t_{k+1} must not be an element of T_k. We then augment T_k to a larger subset $T_{k+1} = \{t_1, \ldots, t_k, t_{k+1}\}$ of T. By repeating this process (i.e., increasing one additional constraint a time), $\{\mathbf{x}^k\}$ will converge to the optimal solution of SIE. We now state the cutting-plane algorithm as follows:

Algorithm 3.9: CPSIE ALGORITHM:

Initialization:

Set $k = 1$; select any $t_1 \in T$; set $T_1 = \{t_1\}$.

Iteration:

Step 1: Solve Program ELI^k and obtain an optimal solution $\mathbf{x}^k$.

Step 2: Find a maximizer t_{k+1} of $v_{k+1}(t)$ over T.

Step 3: If $v_{k+1}(t_{k+1}) \leq 0$, then stop, and $\mathbf{x}^k$ is an optimal solution of Program SIE. Otherwise, set $T_{k+1} = T_k \cup \{t_{k+1}\}$; reset $k \leftarrow k+1$; go to Step 1.

Under the assumption that Program SIE has at least one feasible solution, i.e., $FSI \neq \emptyset$, we know that Program ELI^k has a nonempty feasible region FSI^k, for $k \geq 1$. Moreover, if the CPSIE algorithm terminates in a finite number of iterations, denoted by m, then $\mathbf{x}^m$ must be an optimal solution of Program SIE. Now, if the algorithm does not terminate in a finite number of iterations, we obtain an infinite sequence of points $\{\mathbf{x}^1, \mathbf{x}^2, \ldots, \mathbf{x}^k, \ldots\}$. Our objective for the remaining part of this subsection is to show that this sequence converges to a point $\mathbf{x}^*$ that indeed solves Program SIE. For a detailed treatment of cutting-plane methods for solving general convex programs, the reader is referred to [51, 52].

Some simple results are needed for the convergence proof of the CPSIE algorithm:

Lemma 3.4. If $FSI \neq \emptyset$, then Program SIE achieves a finite minimum at a unique point $\mathbf{x}^*$ in FSI.

Proof: With the convention of $0 \ln 0 = 0$, $f(\mathbf{x})$ is continuous on the set $\{\mathbf{x} | \mathbf{x} \in R^n, \mathbf{x} \geq 0\}$. Denoting $f_j(x_j) \equiv x_j \ln x_j$, we see that (i) $f_j(0) = 0$, (ii) $f_j(x_j) \to \infty$ as $x_j \to \infty$, and (iii) $f_j^{'}(x_j) \to -\infty$ as $x_j \to 0$. These imply that, as x_j increases from 0 to ∞, $f_j(x_j)$ decreases from 0 and then increases after reaching the minimum at $1/e$. Therefore, all level sets of $f_j(x_j)$ and hence those of $f(\mathbf{x})$ are bounded. The continuity of $f(\mathbf{x})$ implies the closedness of all these level sets. Therefore, all level sets of $f(\mathbf{x})$ are compact. Since the intersection of any collection of closed sets is closed, the feasible region FSI of SIE is closed. This implies that the intersection of any level set and FSI is also compact. Since $FSI \neq \emptyset$, there exists a $\mathbf{y} \in FSI$ such that $f(\mathbf{y}) < \infty$. By denoting the level set associated with $f(\mathbf{y})$ as L_0, we have $L_0 \cap FSI \neq \emptyset$. Recall that this intersection set is compact and $f(\mathbf{x})$ is continuous over it. The Bolzano-Weierstrass Theorem assures that Program SIE attains a finite minimum. Moreover, the strict convexity of $f(\mathbf{x})$ and the convexity of FSI imply that Program SIE achieves a unique finite minimum $\mathbf{x}^*$ in FSI.

□

In a similar manner, we can prove the following result.

Corollary 3.1. If $FSI \neq \emptyset$, then Program ELI^k achieves a finite minimum at a unique point $\mathbf{x}^k$ in FSI^k, for $k = 1, 2, \ldots$.

If the CPSIE Algorithm does not terminate in finite iterations, then it generates an infinite sequence $\{\mathbf{x}^k | k = 1, 2, \ldots\}$. Here are some properties possessed by the sequence:

Lemma 3.5. If $FSI \neq \emptyset$, then the sequence $\{\mathbf{x}^k | k = 1, 2, \ldots\}$ is bounded.

Proof: Since $FSI \subseteq FSI^k$, the point $\mathbf{y}$ defined in the proof of Lemma 3.4 belongs to FSI^k, for $k = 1, 2, \ldots$. Therefore, $f(\mathbf{x}^k) \leq f(\mathbf{y})$ and $\mathbf{x}^k \in L_0$, for $k = 1, 2, \ldots$. Finally, since L_0 is compact, the sequence $\{\mathbf{x}^k | k = 1, 2, \ldots\}$ is bounded.

□

Lemma 3.6. Let $\{\mathbf{x}^{k_i} | i = 1, 2, \ldots\}$ be a convergent subsequence of $\{\mathbf{x}^k | k = 1, 2, \ldots\}$ with $\mathbf{x}^{k_i} \to \mathbf{z}$. Then, $\mathbf{z} \in FSI$.

Proof: Note that $\mathbf{x}^{k_i}$ satisfies

$$h(t_m) - \sum_{j=1}^{n} g_j(t_m) x_j^{k_i} \leq 0, \quad m = 1, 2, \ldots, k_i. \tag{3.189}$$

With the continuity of g_j, $j = 1, 2, \ldots, n$, and the compactness of T, the convergence of $\mathbf{x}^{k_i}$ to $\mathbf{z}$ implies that $v(t_k) \leq 0$, $k = 1, 2, \ldots$, where

$$v(t) \equiv h(t) - \sum_{j=1}^{n} g_j(t) z_j, \ \forall \ t \in T. \tag{3.190}$$

Due to the compactness of T and the continuity of g_j and h, $v(t)$ attains its maximum at a point in T, denoted by $\bar{t}$. It suffices to show that $v(\bar{t}) \leq 0$. Let $\{\mathbf{x}^{m_i}\}$ be any subsequence of $\{\mathbf{x}^{k_i}\}$ such that $\{t_{m_i+1}\}$ converges. Such a sequence must exist because T is compact. Denote the limit of $\{t_{m_i+1}\}$ by t^*. Clearly, $v(t_{m_i+1}) \leq 0$, $i = 1, 2, \ldots$. Since $v(t)$ is continuous, we further have

$$v(t^*) \leq 0. \tag{3.191}$$

Since t_{m_i+1} maximizes $v_{m_i+1}(t)$ over T, we know

$$v_{m_i+1}(t_{m_i+1}) \geq v_{m_i+1}(\bar{t}), \quad i = 1, 2, \ldots, \tag{3.192}$$

or, equivalently,

$$h(t_{m_i+1}) - \sum_{j=1}^{n} g_j(t_{m_i+1}) x_j^{m_i} \geq h(\bar{t}) - \sum_{j=1}^{n} g_j(\bar{t}) x_j^{m_i}, \quad i = 1, 2, \ldots \tag{3.193}$$

As $i \to \infty$, this implies

$$h(t^*) - \sum_{j=1}^{n} g_j(t^*) z_j \geq h(\bar{t}) - \sum_{j=1}^{n} g_j(\bar{t}) z_j \tag{3.194}$$

or, equivalently,

$$v(t^*) \geq v(\bar{t}). \tag{3.195}$$

With Equation (3.191), we have $v(\bar{t}) \leq 0$ and hence $\mathbf{z} \in FSI$.

□

Theorem 3.11. If $FSI \neq \emptyset$, then $\{\mathbf{x}^k | k = 1, 2, \ldots\}$ converges to the unique optimal solution $\mathbf{x}^*$ of Program SIE.

Proof: By Lemma 3.4, Program SIE has a unique optimal solution $\mathbf{x}^*$. Since $FSI^k \supseteq FSI$, we have

$$f(\mathbf{x}^1) \leq f(\mathbf{x}^2) \leq \ldots \leq f(\mathbf{x}^*). \tag{3.196}$$

Let $\{\mathbf{x}^{k_i}\}$ be a subsequence of $\{\mathbf{x}^k\}$ which converges to a limit $\mathbf{z}$. Because f is continuous, $f(\mathbf{z}) \leq f(\mathbf{x}^*)$. By Lemma 3.6, we have $\mathbf{z} \in FSI$. Therefore, $f(\mathbf{z}) \geq f(\mathbf{x}^*)$ and hence $f(\mathbf{z}) = f(\mathbf{x}^*)$. By Lemma 3.4, $\mathbf{z} = \mathbf{x}^*$. In other words, every convergent subsequence of $\{\mathbf{x}^k | k = 1, 2, \ldots\}$ converges to $\mathbf{x}^*$. Finally, if $\{\mathbf{x}^k\}$ does not converge to $\mathbf{x}^*$, then there exists an $\epsilon > 0$ such that there exists a subsequence of $\{\mathbf{x}^k | k = 1, 2, \ldots\}$ lying completely outside of the open sphere (neighborhood) centered at $\mathbf{x}^*$ with radius ϵ. By Lemma 3.5, this infinite sequence lies completely in a compact set, which could be the intersection of the complement of the open sphere and any compact set bounding the sequence $\{\mathbf{x}^k | k = 1, 2, \ldots\}$. From Bolzano-Weierstrass Theorem, there exists an accumulation point other than $\mathbf{x}^*$. But, this contradicts the fact that every convergent subsequence of $\{\mathbf{x}^k | k = 1, 2, \ldots\}$ converges to $\mathbf{x}^*$. Therefore, $\mathbf{x}^k$ must converge to $\mathbf{x}^*$.

□

3.3.2 Implementation Issues on the CPSIE Algorithm

The CPSIE Algorithm proposed in Subsection 3.3.1 is a "conceptual" algorithm. In this subsection, we focus on the implementation issues.

Solving Program ELI^k

In Step 1 of the CPSIE Algorithm, we face the challenge of solving Program ELI^k, for $k \geq 1$. There are several different solution methods available. In this study, we follow the dual approach developed in [30, 64, 31, 5, 7, 9, 10]. Before stating the fundamental theorem of this approach, let us define

$$FSI^o \equiv \left\{ \mathbf{x} | \sum_{j=1}^{n} g_j(t) x_j > h(t), \ \forall \ t \in T, \ \text{and} \ x_j > 0, j = 1, 2, \ldots, n \right\} \tag{3.197}$$

as the set of "interior solutions" to Program SIE. The fundamental theorem of the dual approach is as follows:

Theorem 3.12. If $FSI^o \neq \emptyset$, then Program ELI^k has an optimal solution $\mathbf{x}^k$ with

$$x_j^k \equiv \exp(\sum_{i=1}^{k} g_j(t_i) w_{k,i}^* - 1), \ j = 1, 2, \ldots, n, \tag{3.198}$$

where $\mathbf{w}_k^* \equiv (w_{k,1}^*, w_{k,2}^*, \ldots, w_{k,k}^*)^T$ solves

Program DELI^k:

$$\max_{\mathbf{w} \geq \mathbf{0}} \{ d^k(\mathbf{w}) \equiv \sum_{i=1}^{k} h(t_i) w_i - \sum_{j=1}^{n} \exp(\sum_{i=1}^{k} g_j(t_i) w_i - 1) \}. \tag{3.199}$$

The proof is similar to that associated with the case where Inequality (3.186) is replaced by an equality. A detailed proof for the equality case will be provided in Section 5.2, which studies a more general problem. A detailed derivation of the dual problem DELI^k and the dual-to-primal conversion formula (3.198) can be found in [64].

Note that Program ELI^k has a unique optimal solution as long as Program SIE is feasible. Since the dual program DELI^k has only sign constraints, we may apply the "projected gradient method" to solve it. In this case, finding a gradient vector is simple and the projection operation on the gradient is trivial.

As to finding an initial solution, in the absence of any particular insight, $\mathbf{w}_k^0 \equiv \mathbf{0}$ is a natural choice for solving DELI^k. We will refer to it as a *cold start.* Another natural choice is $\mathbf{w}_{k-1}^*$, the optimal solution obtained for DELI^{k-1}, augmented with 0 for the additional variable. This will be referred to as a *warm start.*

Maximizing $v_{k+1}(t)$ over T

In Step 2 of the CPSIE Algorithm, it is necessary to evaluate a continuous function v_{k+1} and find a maximizer over T. When $g_j(t)$ and $h(t)$ exhibit certain special properties so that v_{k+1} is concave, the concavity can be exploited for efficient maximization. However, we consider, for generality, arbitrary continuous $g_j(t)$ and $h(t)$ in this section. This leads to an unconstrained nonconcave maximization problem in Step 2. In our implementation, we take a straightforward approach by discretizing the compact domain T into a reduced set of finite discrete points, denoted by $T^{'}$. Then, we evaluate and maximize the constraint violation function $v_{k+1}(t)$ over $T^{'}$. This of course raises some potential problems.

First, in general, a bigger $T^{'}$ may result in a better approximate maximizer of $v_{k+1}(t)$ over T. But it increases the computational requirement. On the other hand, a coarse approximation combined with an inexact solution of an DELI^k may cause problems in generating a valid new point t_{k+1}. This could stall the algorithm numerically. Therefore, we have to control the accuracy in finding an approximate maximizer of $v_{t+1}(t)$ and an inexact optimal solution of Program DELI^k very carefully. These control parameters will be given in the next subsection.

Second, without carefully examining the special structure of a constraint violation function $v_{k+1}(t)$, an exhaustive search may be required. However, if the functions $g_j(t)$ and $h(t)$ are smooth, finer approximation should be done only near those t's which represent binding constraints. Note that the approximation can be arbitrarily refined, if necessary. To trade memory space for speed, the values of $g_j(t_i)$ and $h(t_i)$ could be calculated and stored once and for all.

3.3.3 Computational Experience

In this subsection, we illustrate the potential of the proposed CPSIE Algorithm for solving Program SIE. Two sets of problems were tested, and all tests were run on a standard SUN SPARCstation 2.

Test Problems

Program 3.3:

$$\inf \quad f(\mathbf{x}) = \sum_{j=1}^{n} x_j \ln x_j \tag{3.200}$$

$$\text{s.t.} \quad \sum_{j=1}^{n} |sin(jt)| x_j \geq nt, \ \forall\, t \in T = [0,1], \tag{3.201}$$

$$x_j \geq 0, \quad j = 1, 2, ..., n. \tag{3.202}$$

Program 3.4:

$$\inf \quad f(\mathbf{x}) = \sum_{j=1}^{n} x_j \ln x_j \tag{3.203}$$

$$\text{s.t.} \quad \sum_{j=1}^{n} t^{j-1} x_j \geq n^{1/2} sin(t), \ \forall\, t \in T = [0,1], \tag{3.204}$$

$$x_j \geq 0, \quad j = 1, 2, \ldots, n. \tag{3.205}$$

Note that by controlling the dimensionality n, we actually have two sets of testing problems. In the first set (Program 3.3), all variables play comparable roles in defining the constraints. In the second set (Program 3.4), variables with smaller indices contribute more to the left-hand-side values of the constraints.

Establishing Basis for Testing

We first apply the simple Discretization Method to solve Programs 3.3 and 3.4 as the basis for testing the performance of the CPSIE Algorithm. In our baseline study, dimensionalities were set to n = 10, 30, 100, 300, and 1,000. For each problem, we discretized $T = [0,1]$ into m equal parts and evaluated the Constraints (3.201) and (3.204) at $t = i/m$, $i = 1, \ldots, m$. For each case, this mechanism resulted in a finite entropy optimization problem with n variables and m explicit inequality constraints. We then applied the projected gradient algorithm to solve its dual problem DELI^m and converted the dual optimal solution obtained to the optimal solution of ELI^m according to Equation (3.198) of Theorem 3.12.

More details are given as follows. To solve Program DELI^m, we always initiated the projected gradient algorithm from a *cold start* $\mathbf{w} = \mathbf{0}$ and, in each iteration, a line search (golden-section search) was performed to improve the objective value along the direction of the projected gradient. The line search stopped when the improvement was less than a given parameter ϵ_l. Initially, ϵ_l was set to 0.001 for a rough control without consuming too much CPU time. When such an ϵ_l produced no better solution in the line search than the current one, it was switched to 0.0000001 for a precise validation. Using two different ϵ_l's at two different stages of the algorithm in general saved CPU time. At the end of each iteration, an estimate of the primal solution was obtained through Equation (3.198). The algorithm terminated when the estimate satisfied the K.K.T. conditions. In our implementation, the K.K.T. conditions were considered satisfied when the maximum constraint violation was less than a parameter $\epsilon_{cv} = 0.01$. Since the final $\epsilon_l = 0.0000001$ was small enough, the algorithm, in all our tests, stopped due to the satisfaction of the K.K.T. conditions.

The testing results for Programs 3.3 and 3.4 are tabulated in Table 3.3 and Table 3.4, respectively. The results show that, for a fixed problem size (a fixed n), a finer discretization (with larger m) usually results in a better approximate optimal solution but with longer CPU time. However, this is not always true because, according to our discretization scheme, the constraint set associated with a larger m may not be a superset of that associated with a smaller m. This also indicates a potential problem of how to set a "large enough" m in applying the Discretization Method. The results also show that, to achieve the same accuracy level, larger problems (larger n) require finer discretization. This further indicates the potential problem of requiring heavy computation in solving large SIE problems by using the Discretization Method.

Performance of the CPSIE Algorithm

For a fair comparison, when solving Program ELI^k in Step 1 of the CPSIE Algorithm, we used the same dual approach with exactly the same parameter values as we did when using the Discretization Method. A *cold start* was also employed in all cases. In Step 2, the reduced domain was set to be $T' \equiv \{0.001 \times i | i = 1, \ldots, 1000\}$, and the maximum allowable constraint violation was 0.01. For simplicity, an exhaustive search was performed for finding a maximizer of the constraint violation functions.

Table 3.3. Performance of Discretization Method for Program 3.3

n	m	optimal obj. value	cpu (seconds)	no. iterations
10	10	4.720	0.27	28
10	30	4.721	0.87	63
10	100	4.715	9.79	238
10	300	4.727	44.59	487
10	1000	4.708	530.57	1644
30	10	13.950	0.80	34
30	30	15.280	2.98	77
30	100	15.393	35.68	309
30	300	15.454	151.05	654
30	1000	15.467	1479.55	1739
100	10	48.310	1.12	22
100	30	53.352	6.82	58
100	100	58.453	94.28	233
100	300	58.544	751.17	684
100	1000	58.696	6607.19	2228
300	10	147.075	9.30	46
300	30	162.661	10.33	39
300	100	183.486	207.82	172
300	300	189.281	3378.04	754
300	1000	189.208	26898.20	2317
1000	10	492.176	19.20	30
1000	30	549.064	43.44	40
1000	100	601.569	877.24	174
1000	300	635.354	6085.59	392
1000	1000	663.548	58205.40	1450

Table 3.4. Performance of Discretization Method for Program 3.4

n	m	optimal obj. value	cpu (seconds)	no. iterations
10	10	-3.194	0.37	31
10	30	-3.203	0.63	40
10	100	-3.201	3.94	122
10	300	-3.199	12.14	216
10	1000	-3.202	141.62	668
30	10	-8.780	0.91	42
30	30	-8.750	3.02	71
30	100	-8.746	25.11	222
30	300	-8.755	64.41	328
30	1000	-8.751	625.77	856
100	10	-28.749	3.44	58
100	30	-28.758	51.01	306
100	100	-28.769	150.24	381
100	300	-28.770	465.13	555
100	1000	-28.766	3608.28	1253
300	10	-89.552	38.64	182
300	30	-89.437	334.47	600
300	100	-89.454	409.88	320
300	300	-89.452	3297.50	938
300	1000	-89.464	15372.70	1803
1000	10	-315.003	367.75	366
1000	30	-314.561	1449.26	603
1000	100	-314.584	6182.36	872
1000	300	-314.583	20829.60	1177
1000	1000	-314.588	125122.00	2725

Table 3.5. Performance of CPSIE for Program 3.3: n varied (cpu in seconds)

n	k	no. a.c.	opt. obj.	f.o. cpu	c.v. cpu	total cpu
10	2	2	4.717	0.15	0.33	0.48
30	3	3	15.481	0.40	1.24	1.64
100	11	9	58.702	30.05	11.18	41.23
300	26	20	189.215	3223.58	75.49	3299.04
1000	33	32	663.538	3168.96	307.83	3476.79

Table 3.6. Performance of CPSIE for Program 3.4: n varied (cpu in seconds)

n	k	no. a.c.	opt. obj.	f.o. cpu	c.v. cpu	total cpu
10	2	1	-3.196	0.09	0.23	0.32
30	4	2	-8.754	0.61	0.94	1.55
100	5	2	-28.751	14.75	3.80	18.55
300	6	3	-89.460	196.99	18.78	215.77
1000	9	5	-314.588	227.86	215.29	443.15

The testing results are tabulated in Table 3.5 and Table 3.6. The second column in both Tables 3.5 and 3.6 indicates that the algorithm terminated with k constraints. The third column provides the number of constraints, among those k constraints generated, that were "active" when the algorithm terminated. The fourth column gives the objective value of the approximate optimal solution obtained by CPSIE. The remaining three columns show the CPU time spent on solving the finite entropy optimization problems, the CPU time spent on finding a maximizer of the constraint violation functions, and the total CPU time consumed.

Note that the CPSIE Algorithm has the capability of providing an approximate optimal solution with the same accuracy level as the Discretization method but is much faster for large problems – *one order of magnitude* for Program 3.3 and *two orders of magnitude* for Program 3.4. Also note that, in solving large problems, the CPU time spent on solving finite entropy optimization problems dominates that spent on maximizing the constraint violation functions.

Behavior of the CPSIE Algorithm

Table 3.7. Performance of CPSIE for Program 3.3: ϵ_{cv} varied (cpu in seconds)

ϵ_{cv}	k	no. a.c.	opt. obj.	f.o. cpu	c.v. cpu	total cpu
10.0	2	2	56.295	0.19	3.41	3.60
1.0	6	6	58.633	1.53	7.08	8.61
0.1	9	7	58.695	48.98	9.55	58.53
0.01	11	9	58.702	30.05	11.18	41.23
0.001	12	10	58.700	278.18	12.22	290.40
0.0001	13	11	58.700	320.65	12.88	333.53

Table 3.8. Performance of CPSIE for Program 3.4: ϵ_{cv} varied (cpu in seconds)

ϵ_{cv}	k	no. a.c.	opt. obj.	f.o. cpu	c.v. cpu	total cpu
1.0	2	1	-29.294	0.16	2.17	2.33
0.1	3	1	-28.930	0.71	2.68	3.39
0.01	5	2	-28.751	14.75	3.80	18.55
0.001	8	6	-28.745	60.70	5.34	66.04
0.0001	12	12	-28.730	124.98	54.73	179.72

Having demonstrated the potential superiority of the CPSIE Algorithm, we focus on studying its behavior by varying some key control parameters.

In the first experiment, we fixed $n = 100$ and varied the control parameter ϵ_{cv}. The results are shown in Tables 3.7 and 3.8. Note how sensitive the CPU time is and how insensitive the approximate optimal objective value is to the change in the parameter ϵ_{cv}. It suggests that, when limited by the amount of computing resources, one may consider using a large ϵ_{cv} to save CPU time while not sacrificing too much in solution quality. This is particularly appealing for problems with "soft" constraints.

In the second experiment, we tested the performance of the CPSIE Algorithm with different discretization of the domain T to get a maximizer t_{k+1}. In our test, we focused on Program 3.3 only. The interval T was partitioned into 10,000 subintervals, as opposed to 1,000 subintervals. Actually, we let $T' \equiv \{0.0001 \times i | i = 1, 2, ..., 10000\}$ and varied the number of primal variables n from 5 to 5000. We also varied (i) the tolerance for constraint violation ϵ_{cv} and (ii) the line-search accuracy ϵ_l to study the impact of a finer discretization

Table 3.9. Performance of CPSIE for Program 3.3: 10,000 subintervals

n	ϵ_{cv}	ϵ_l	k	no. a.c.	f.o. cpu	c.v. cpu	total cpu
5	0.01	0.00001	2	2	0.17	1.31	1.48
10	0.01	0.00001	2	2	0.12	2.48	2.60
50	0.1	0.0001	5	5	3.90	25.85	29.75
100	0.1	0.0001	9	9	75.57	87.05	162.62
500	1.0	0.001	18	18	364.98	849.91	1214.89
1000	1.0	0.001	23	23	1521.08	2134.29	3655.37
5000	10.0	0.01	36	36	11327.30	16308.20	27635.50

of T. A single value for ϵ_l, instead of two, was used throughout the search process. These ϵ's reflect possible accuracy requirements that depend on the problem size. The results are shown in Table 3.9, with the CPU time measured in seconds.

Note that, due to the refined approximation of domain T, the algorithm spent more CPU time on finding the maximizer of the constraint violation functions than on solving the finite entropy optimization problems. Also note that all constraints are binding at the end when the algorithm terminates. Therefore, we may want to use a finer partition only when it is needed.

3.3.4 Concluding Remarks

1. We proposed a cutting-plane type algorithm for solving entropy optimizations with a finite number of variables but an infinite number of linear constraints. One obvious advantage of the proposed cutting-plane algorithm, as opposed to the simple Discretization Method, is that only those constraints which tend to be binding are generated. Our computational experience confirmed its efficiency in terms of both CPU and memory requirements, especially for solving large-scale problems.

2. Although a dual approach combined with the projected gradient method was used in our implementation, it may not be the best way. Actually any efficient algorithm, whether it is based on a primal or dual approach, can be incorporated into the scheme of the CPSIE Algorithm.

3. According to our experience, although it is not reported in this section, a *warm start* in many cases indeed helps. Since there is no theoretic basis for this behavior, we only mention it here.

4. In some cases and especially for problems with a small number of variables, since the constraint violation function v_{k+1} is continuous, sometimes its maximizer t_{k+1} lies very close to those points that are in T_k already. In this case, we may have some nearly identical constraints which intersect with one another in a small neighborhood near the optimal solution. This could cause some "near-degeneracy" behavior for solving Program ELI^k, which in turn causes a "flat" dual objective near its optimal solution. If the required line-search accuracy is high, e.g., 10^{-7} or smaller, a slow and lengthy sequence of insignificant improvements towards the optimal solution may occur. Also note that for a finer discretization, the potential numerical problem associated with this type of near-degeneracy behavior in the neighborhood of the optimal solution may become more severe.

5. The proposed computational procedure has two major components, the cutting-plane algorithm CPSIE for generating a sequence of entropy optimization problems ELI^k with a finite number of constraints and the projected gradient method for solving DELI^k, i.e., the dual of ELI^k. The former is a general conceptual algorithm and is therefore applicable to many other mathematical programs with a convex objective function. Extending the duality theory establishing the unconstrained dual program for the linear equality constrained entropy optimization problems, Ben-Tal et al. [7], Borwein and Lewis [9, 10], and Ben-Tal and Charnes [5] showed the unconstrained nature of the dual for programs with certain entropy-like objective functions as well as programs with other entropy-like features. The presence of inequality constraints led naturally to dual programs with nonnegativity constraints. Therefore, the proposed computational procedure is not restricted only to optimization of Shannon's entropy. It is actually applicable to many other cases.

6. The CPSIE Algorithm involves two optimization problems as its subproblems. In Step 1, an optimal solution of Program ELI^k is obtained. In Step 2, a maximizer t_{k+1} of $v_{k+1}(t)$ over T is sought. Although the convergence results established in Subsection 3.3.1 are based on the ability to obtain the

exact minimum and maximum of the two optimization problems, respectively, they remain valid with inexact minimization and maximization. Such inexact optimization may reduce computational effort. In Step 1, only an ϵ-optimal solution is needed, namely, a solution whose objective value is within $\epsilon_{1k} > 0$ of the true minimum. Again in Step 2, only an ϵ-optimal solution is required, e.g., a solution t_{k+1} such that $v_{k+1}(t_{k+1})$ is within ϵ_{2k} of the true maximum. It can be shown that as long as tolerances ϵ_{1k} and ϵ_{2k} converge to 0, as k tends to infinity, the convergence results established in Subsection 3.3.1 still hold. A recent application of the inexact approach can be found in [32].

REFERENCES

[1] Agmon, N., Alhassid, Y., and Levine, R.D., "An Algorithm for Determining the Lagrange Parameters in the Maximal Entropy formalism," *The Maximum Entropy Formalism*, edited by R. D. Levine and M. Tribus, MIT Press, Cambridge, Massachusetts, 1979, pp. 207–209.

[2] Anas, A., "Discrete Choice Theory, Information Theory and the Multinomial Logit and Gravity Models," *Transportation Research Part B*, Vol. 17B, 1983, pp. 13–23.

[3] Bazaraa, M. S., and Shetty, C. M., *Nonlinear Programming: Theory and Algorithms*, John Wiley, New York, 1979.

[4] Bellman, R.E., *Mathematical Methods in Medicine*, World Scientific, Singapore, 1983.

[5] Ben-Tal A., and Charnes, A., "A Dual Optimization Framework for Some Problems of Information Theory and Statistics," *Problems of Control and Information Theory*, Vol. 8, 1979, pp. 387–401.

[6] Ben-Tal, A., Melman, A., and Zowe, J., "Curved Search Methods for Unconstrained Optimization," *Optimization*, Vol. 21, 1990, pp. 669–695.

[7] Ben-Tal, A., Teboulle, M., and Charnes, A., "The Role of Duality in Optimization Problems involving Entropy Functionals with Applications to Information Theory," *Journal of Optimization Theory and Applications*, Vol. 58, 1988, pp. 209–223.

[8] Bishop, Y.M.M., "Full Contingency Tables, Logits, and Split Contingency Tables," *Biometrika*, Vol. 25, 1969, pp. 339–383.

[9] Borwein, J.M., and Lewis, A.S., "Duality Relationships for Entropy-Like Minimization Problems," *SIAM Journal on Control and Optimization*, Vol. 29, 1991, pp. 325–338.

[10] Borwein, J.M., and Lewis, A.S., "Partially-Finite Programming in L_1 and the Existence of Maximum Entropy Estimates," *SIAM Journal on Optimization*, Vol. 3, 1993, pp. 248–267.

[11] Bregman, L.M., "The Relaxation Method of Finding the Common Point of Convex Sets and Its Application to the Solution of Problems in Convex Programming," *U.S.S.R. Computational Mathematics and Mathematical Physics*, Vol. 7, 1967, pp. 200–217.

[12] Censor, Y., "Row-Action Methods for Huge and Sparse Systems and Their Applications," *SIAM Review*, Vol. 23, 1981, pp. 444–466.

[13] Censor, Y., "On Linearly Constrained Entropy Maximization," *Linear Algebra and Its Applications*, Vol. 80, 1986, pp. 191–195.

[14] Censor, Y., Elfving, T., and Herman, G.T., "Special Purpose Algorithms for Linearly Constrained Optimization," *Maximum-Entropy and Bayesian Spectral Analysis and Estimation Problems*, edited by C.R. Smith and G.J. Erickson, D. Reidel Publishing Company, 1987, pp. 241–254.

[15] Censor, Y., and Lent, A., "An Iterative Row-Action Method for Interval Convex Programming," *Journal of Optimization Theory and Applications*, Vol. 34, 1981, pp. 321–353.

[16] Censor, Y., De Pierro, A.R., Elfving, T., Herman, G.T., and Iusem, A.N., "On Iterative Methods for Linearly Constrained Entropy Maximization," *Banach Center Publications*, Vol. 24, 1990, pp. 145–163.

[17] Darroch, J.N., and Ratcliff, D., "Generalized Iterative Scaling for Log-linear Models," *Annals of Mathematical Statistics*, Vol. 43, 1972, pp. 1470–1480.

[18] Dennis, J.E., and Schnabel, R.B., *Numerical Methods for Unconstrained Optimization and Nonlinear Equations*, Prentice Hall, Englewood Cliffs, New Jersey, 1983.

[19] Dinkel, J.J., and Kochenberger, G.A., "Constrained Entropy Models: Solvability and Sensitivity," *Management Science*, Vol. 25, 1979, pp. 555–564.

[20] Dinkel, J.J., Kochenberger, G.A., and Wong, S.N., "Entropy Optimization and Geometric Programming," *Environment and Planning*, Vol. 9, 1977, pp. 419–427.

[21] Dinkel, J.J., and Wong, D.S., "External Zones in Trip Distribution Models: Characterization and Solvability," *Transportation Science*, Vol. 18, 1984, pp. 253–266.

[22] Elfving, T., "On Some Methods for Entropy Maximization and Matrix Scaling," *Linear Algebra and its Applications*, Vol. 34, 1980, pp. 321–339.

[23] Eriksson, J.R., "A Note on Solution of Large Sparse Maximum Entropy Problems with Linear Equality Constraints," *Mathematical Programming*, Vol. 18, 1980, pp. 146–154.

[24] Erlander, S., "Accessibility, Entropy and the Distribution and Assignment of Traffic," *Transportation Research*, Vol. 11, 1977, pp. 149–153.

[25] Erlander, S., "Entropy in Linear Programming," *Mathematical Programming*, Vol. 21, 1981, pp. 137–151.

[26] Fang, S.-C., "A New Unconstrained Convex Programming Approach to Linear Programming," *Zeitschrift fur Operations Research*, Vol. 36, 1992, pp. 149–161.

[27] Fang, S.-C., Peterson, E.L., and Rajasekera, J.R., "Minimum Cross-Entropy Analysis with Entropy-Type Constraints," *Journal of Computational and Applied Mathematics*, Vol. 39, 1992, pp. 165–178.

[28] Fang, S.-C., and Puthenpura, S., *Linear Optimization and Extensions: Theory and Algorithms*, Prentice Hall, Englewood Cliffs, New Jersey, 1993.

[29] Fang, S.-C., and Rajasekera, J.R., "Quadratically Constrained Minimum Cross-Entropy Analysis," *Mathematical Programming*, Vol. 44, 1989, pp. 85–96.

[30] Fang, S.-C., and Tsao, H.-S.J., "Linear Programming with Entropic Perturbation," *Zeitschrift fur Operations Research*, Vol. 37, 1993, pp. 171–186.

[31] Fang, S.-C., and Tsao, H.-S.J., "A Quadratically Convergent Global Algorithm for the Linearly-Constrained Minimum Cross-Entropy Problem," *European Journal of Operational Research*, Vol. 79, 1994, pp. 369–378.

[32] Fang, S.-C., and Wu, S.-Y., "An Inexact Approach to Solving Linear Semi-Infinite Programming Problems," *Optimization*, Vol. 28, 1994, pp. 291–299.

[33] Fiacco, A.V., and Kortanek, K.O., *Semi-Infinite Programming and Applications*, Lecture Notes in Economics and Mathematical Systems No. 215, Springer-Verlag, New York, 1983.

[34] Gordon, R., Bender, R., and Herman, G.T., "Algebraic Reconstruction Techniques (ART) for Three Dimensional Electron Microscopy and X-ray Photography," *Journal of Theoretical Biology*, Vol. 29, 1970, pp. 471–481.

[35] Grandy, W.T. Jr., and Schick, L.H., editors, *Proceedings of the 10th International Workshop on Maximum Entropy and Bayesian Methods*, Kluwer Academic Publishers, Dordrecht, The Netherlands, 1991.

[36] Guiasu, S., *Information Theory with Applications*, McGraw-Hill, New York, 1977.

[37] Guiasu, S., "Maximum Entropy Condition in Queueing Theory," *Journal of Operational Research Society*, Vol. 37, 1986, pp. 293–301.

[38] Hoshino, K., Kumar, V., Kumar, U., "On Optimizing of Generalized Iterative Scaling Method," Working Paper No. 89-22, School of Business, Carleton University, Ontario, Canada, 1989.

[39] Jan, G.W., and Fang, S.-C., "A New Variant of the Primal Affine Scaling Algorithm for Linear Programs," *Optimization*, Vol. 22, 1991, pp. 681–715.

[40] Jaynes, E.T., "Information Theory and Statistical Mechanics II," *Physics Review*, Vol. 108, 1957, pp. 171–190.

[41] Jefferson, T.R., and Scott, C.H., "The Analysis of Entropy Models with Equality and Inequality Constraints," *Transportation Research*, Vol. 138, 1979, pp. 123–132.

[42] Jornsten, K.O., and Lundgren, J.T., "An Entropy-Based Model Split Model," *Transportation Research Part B*, Vol. 23B, 1989, pp. 345–359.

[43] Kapur, J.N., *Maximum-Entropy Models in Science and Engineering*, Wiley Eastern Ltd., New Delhi, 1989.

[44] Kortanek, K.O., Potra, F., and Ye, Y., "On Some Efficient Interior Point Methods for Nonlinear Convex Programming," *Linear Algebra and Its Applications*, Vol. 152, 1991, pp. 169–189.

[45] Kumar, V., Hoshino, K., and Kumar, U., "An Application of the Entropy Maximization Approach in Shopping Area Planning," *International Journal of General Systems*, Vol. 16, 1989, pp. 25–42.

[46] Kullback, S., *Information and Statistics*, John Wiley, New York, 1959.

[47] Kullback, S., and Libler, R.A., "On Information and Sufficiency," *Annals of Mathematical Statistics*, Vol. 22, 1951, pp. 79–86.

[48] Lamond, B., and Stewart, N.F., "Bregman's Balancing Method," *Transportation Research Part B*, Vol. 15B, 1981, pp. 239–248.

[49] Lent, A., "Maximum Entropy and MART," in *Image Analysis and Evaluation*, SPSE Conference Proceedings, edited by R. Shaw, Toronto, Canada, 1976, pp. 249–257.

[50] Lent, A., "A Convergent Algorithm for Maximum Entropy Image Restoration with a Medical X-Ray Application," in *Image Analysis and Evaluation*, edited by R. Shaw, Society of Photographic Scientists and Engineers (SPSE), Washington, D.C., 1977, pp. 249-257.

[51] Luenberger, D.G., *Linear and Nonlinear Programming, 2nd Edition*, Addison-Wesley, Reading, Massachusetts, 1984.

[52] Minoux, M., *Mathematical Programming*, Wiley-Interscience, New York, 1986.

[53] Peterson, E.L., "Geometric Programming," *SIAM Review*, Vol. 19, 1976, pp. 1–45.

[54] De Pierro, A.R. and Iusem, A.N., "A Relaxed version of Bregman's Method for Convex Programming," *Journal of Optimization Theory and Applications*, Vol. 51, 1986, pp. 421–440.

[55] Potra, F., and Ye, Y., "A Quadratically Convergent Polynomial Algorithm for Solving Entropy Optimization Problems," *SIAM Journal on Optimization*, Vol. 3, 1993, pp. 843–860.

[56] Rockafellar, R.T., *Convex Analysis,* Princeton University Press, Princeton, New Jersey, 1970.

[57] Safwat, K.N.A., and Magnanti, T.L., "A Combined Trip Generation, Trip Distribution, Modal Split, and Trip Assignment Model," *Transportation Science*, Vol. 22, 1988, pp. 14–30.

[58] Shore, J.E., "Minimum Cross-entropy Spectral Analysis," *IEEE Transactions on Acoustics, Speech, and Signal Processing*, Vol. ASSP-29, 1981, pp. 230–237.

[59] Sheu, R.L., and Fang, S.-C., "On the Generalized Path-Following Methods for Linear Programming," *Optimization*, Vol. 30, 1994, pp. 235–249.

[60] Smith, C.R., and Grandy, W.T., editors, *Maximum-Entropy and Bayesian Methods in Inverse Problems*, D. Reidel Publishing Company, Dordrecht, Holland, 1985.

[61] Todd, M.J., and Ye, Y., "A Centered Projective Algorithm for Linear Programming," *Mathematics of Operations Research*, Vol. 15, 1990, pp. 508–529.

[62] Tomlin, J.A., "A Mathematical Programming Model for the Combined Distribution-Assignment of Traffic," *Transportation Science*, Vol. 5, 1971, pp. 122–140.

[63] Tomlin, J.A., and Tomlin, S.G., "Traffic Distribution and Entropy," *Nature*, Vol. 220, 1968, pp. 974–976.

[64] Tsao, H.-S.J., and Fang, S.-C., "Linear Programming with Inequality Constraints via Entropic Perturbation," *International Journal of Mathematics and Mathematical Sciences*, Vol. 19, 1996, pp. 177–184.

[65] Tsao, H.-S.J., Fang, S.-C., and Lee, D.N., "On the Optimal Entropy Analysis," *European Journal of Operational Research*, Vol. 59, 1992, pp. 324–329.

[66] Tsao, H.-S.J., Fang, S.-C., and Lee, D.N., "A Bayesian Interpretation of the Linearly-Constrained Cross-Entropy Minimization Problem," *Engineering Optimization*, Vol. 22, 1993, pp. 65–75.

[67] Wilson, A.G., "A Statistical Theory of Spatial Distribution Models," *Transportation Research*, Vol. 1, 1967, pp. 253–269.

[68] Wu, J.-S., and Chan, W.C., "Maximum Entropy Analysis of Multiple-Server Queueing Systems," *Journal of Operational Research Society*, Vol. 40, 1989, pp. 815–826.

[69] Zhang, J., and Brockett, P.L., "Quadratically Constrained Information Theoretic Analysis," *SIAM Journal on Applied Mathematics*, Vol. 47, 1987, pp. 871–885.

[70] Zhu, J., and Ye, Y., "A Path-Following Algorithm for a Class of Convex Programming Problems," Working Paper No. 90-14, College of Business Administration, The University of Iowa, Iowa City, Iowa, 1990.

4

ENTROPY OPTIMIZATION METHODS: GENERAL CONVEX CASE

Let $\mathbf{q} \equiv (q_1, \ldots, q_n)^T \geq \mathbf{0}$ and $\mathbf{p} \equiv (p_1, \ldots, p_n)^T > \mathbf{0}$ be two probability distribution functions. With the convention of $0 \ln 0 = 0$, in Chapter 1, we defined the quantity $\sum_{j=1}^{n} q_j \ln(q_j/p_j)$ to be the cross-entropy of $\mathbf{q}$ with respect to $\mathbf{p}$. In this chapter, we study three classes of minimum cross-entropy problems, namely,

1) minimizing cross-entropy subject to quadratic constraints,

2) minimizing cross-entropy subject to entropic constraints, and

3) minimizing cross-entropy subject to convex constraints.

The problem of minimizing cross-entropy subject to quadratic constraints, denoted as Program EQ, has the following form:

Program EQ:

$$\min \quad g(\mathbf{q}) = \sum_{j=1}^{n} q_j \ln(q_j/p_j) \tag{4.1}$$

$$\text{s.t.} \quad g_i(\mathbf{q}) = \frac{1}{2}\mathbf{q}^T \mathbf{H}_i \mathbf{q} + \mathbf{b}_i^T \mathbf{q} + c_i \leq 0, \quad i = 1, 2, ..., m, \tag{4.2}$$

$$\sum_{j=1}^{n} q_j = 1, \tag{4.3}$$

$$\mathbf{q} \geq \mathbf{0}, \tag{4.4}$$

where $\mathbf{H}_i$ is an $n \times n$, symmetric, positive semidefinite matrix, $\mathbf{q}$ and $\mathbf{b}_i$ are n-vectors, and c_i is a constant, for $i = 1, 2, ..., m$. We assume that $\mathbf{p} > \mathbf{0}$ is a known probability distribution.

Note that the decision variables are represented by $q_j's$ instead of $x_j's$. This is different from the setting used in Chapter 3. However, when we treat the problem, it will be transformed to a new problem using $x_j's$ as decision variables. Then it becomes consistent with the notation used in Chapter 3.

A form of the quadratically constrained minimum cross-entropy problem was originally formulated by Zhang and Brockett [26], with the omission of Constraint (4.3), for an application in actuarial science. It was treated as a direct extension of the linearly constrained cross-entropy optimization problem. Since then, the problem has been studied by a number of researchers, including Fang and Rajasekera [12], Teboulle [24], and Ji and Potra [15].

Basically, Fang and Rajasekera proposed a geometric dual perturbation approach to solve the problem, while Teboulle proposed a Lagrangian based dual method, and Ji and Potra proposed an interior point method. These approaches and methods will be discussed in Sections 4.1 and 4.2.

The problem of minimizing cross-entropy subject to entropic constraints, denoted as Program EE, has the following form:

Program EE:

$$\min \quad g(\mathbf{q}) = \sum_{j=1}^{n} q_j \ln(q_j/p_j^0) \tag{4.5}$$

$$\text{s.t.} \quad g_i(\mathbf{q}) = \sum_{j=1}^{n} q_j \ln(q_j/p_j^i) \leq e_i, \quad i = 1, 2, ...m, \tag{4.6}$$

$$\sum_{j=1}^{n} q_j = 1, \tag{4.7}$$

$$\mathbf{q} \geq \mathbf{0}, \tag{4.8}$$

where e_i is a given tolerance, for $i = 1, 2, ..., m$, and $\mathbf{p}^i = (p_1^i, \ldots, p_n^i) > \mathbf{0}$ is a known probability distribution, for $i = 0, 1, 2, ..., m$.

As pointed out by Ben-Tal et al. [2], since the entropy function is closely related to the content of information, many problems in information theory involve entropy-type constraints. They studied the minimum cross-entropy problem subject to entropic constraints. However, their work emphasized on the continuous functional spaces. The finite dimensional case, i.e., Program EE, was explicitly treated by Fang et al. in [7]. Their method, similar to its counterpart for solving Program EQ, is based on a geometric dual approach. Details of this approach with a computational algorithm and numerical examples are given in Section 4.3.

The problem of minimizing cross-entropy subject to convex constraints, denoted as Program EC, has the following form:

Program EC:

$$\min \quad g(\mathbf{q}) = \sum_{j=1}^{n} q_j \ln(q_j/p_j) \tag{4.9}$$

$$\text{s.t.} \quad g_i(\mathbf{q}) = h_i(\mathbf{A}_i\mathbf{q}) + \mathbf{b}_i^T\mathbf{q} + c_i \leq 0, \quad i = 1, 2, ..., m, \tag{4.10}$$

$$\sum_{j=1}^{n} q_j = 1, \tag{4.11}$$

$$\mathbf{q} \geq \mathbf{0}, \tag{4.12}$$

where $\mathbf{A}_i$ is a $k_i \times n$ matrix with full row-rank, $\mathbf{b}_i$ an n-dimensional column vector, c_i a constant, and h_i a convex function on R^{k_i}, for $i = 1, 2, ..., m$. We assume that $\mathbf{p} > \mathbf{0}$ is a known probability distribution.

Note that Program EC is a very general setting. For example, if $h_i(\mathbf{A}_i\mathbf{q}) \equiv 0$, then $g_i(\mathbf{q}) \leq 0$ becomes a linear inequality constraint. If $h_i(\mathbf{A}_i\mathbf{q}) \equiv 1/2(\mathbf{q}^T\mathbf{A}_i^T\mathbf{A}_i\mathbf{q})$, then $g_i(\mathbf{q}) \leq 0$ becomes a quadratic constraint. Moreover, if $\mathbf{A}_i \equiv \mathbf{I}_n$, $h_i(\mathbf{q}) \equiv \sum_{j=1}^{n} q_j \ln(q_j/p_j^i)$ (for a given probability distribution $\{p_j^i > 0\}$, $j = 1, 2, ..., n$) and $\mathbf{b}_i \equiv \mathbf{0}$, then $g_i(\mathbf{q}) \leq 0$ becomes an entropic constraint.

By extending the minimum cross-entropy problem subject to quadratic constraints to the case with convex constraints, Teboulle considered a problem similar to Program EC [24]. Although a Lagrangian based dual problem was constructed, neither the specific duality properties nor computational issues were discussed in the paper. Program EC was formally addressed by Fang and Rajasekera in [9]. Their method, similar to its counterpart for solving Programs EQ and EE, is based on a geometric dual approach. This is the subject to be studied in Section 4.4.

4.1 EXISTING METHODS

As we mentioned before, among the three classes of minimum cross-entropy problems, most known work centered around the one with quadratic constraints. To solve such problems, there are two basic approaches, namely, the primal approach and the dual approach. The former works on Program EQ directly, while the latter converts Program EQ to a dual problem and then works on the dual.

For the primal approach, we shall introduce a newly developed interior point method, based on Ji and Potra's work [15], in Subsection 4.1.2. As to the dual approach, we can either construct a Lagrangian dual or a geometric dual problem. Subsection 4.1.1 is devoted to describe the Lagrangian dual approach while the geometric dual approach will be studied in Section 4.2.

4.1.1 Lagrangian Dual Approach for Quadratically Constrained Problems

The idea of using the Lagrangian dual to solve minimum cross-entropy problems with quadratic constraints can be found in Zhang and Brockett [26]. The problem they considered has the following form:

Program $\overline{\mathbf{EQ}}$:

$$\min \quad g(\mathbf{q}) = \sum_{j=1}^{n} q_j \ln(q_j/ep_j) \tag{4.13}$$

$$\text{s.t.} \quad g_i(\mathbf{q}) = \frac{1}{2}\mathbf{q}^T \mathbf{H}_i \mathbf{q} + \mathbf{b}_i^T \mathbf{q} + c_i \leq 0, \quad i = 1, 2, ..., m, \tag{4.14}$$

$$\mathbf{q} \geq \mathbf{0}, \tag{4.15}$$

where e is the base of the natural logarithm, $p_j > 0$, for $j = 1, 2, ..., n$, are given, $\mathbf{H}_i$ is an $n \times n$, symmetric, positive semidefinite matrix, $\mathbf{b}_i$ is an n-vector, and c_i is a constant, for $i = 1, 2, \ldots, m$. Note here the absence of the probability constraints $\sum_{j=1}^{n} q_j = 1$ and $\sum_{j=1}^{n} p_j = 1$ from the original formulation of Program EQ. Also note the presence of e in the objective function.

By noticing that *"it is quite difficult to derive a dual problem by using the Lagrange duality method"* ([26], page 876), Zhang and Brockett used a set of supporting hyperplanes to approximate the quadratically constrained feasible region of Program $\overline{\text{EQ}}$. In this way, Program $\overline{\text{EQ}}$ becomes a limit case of a sequence of linearly constrained entropy optimization problems. This allowed them to use the already established duality theory for linearly constrained entropy problems [5] to derive the corresponding sequence of dual programs. By taking the limit of the dual program sequence, they constructed a "dual program" of Program $\overline{\text{EQ}}$. Later, Teboulle [24], showed that Zhang and Brockett's result was not quite accurate and constructed a Lagrange dual program by directly using the well-established duality theory of convex programming.

To achieve this objective, we focus on Program $\overline{\text{EQ}}$. Since $\mathbf{H}_i$ is a symmetric, positive semi-definite matrix, we can write $\mathbf{H}_i = \mathbf{A}_i^T \mathbf{A}_i$, where $\mathbf{A}_i$ is a $k_i \times n$ matrix and k_i is the row-rank of $\mathbf{H}_i$, for $i = 1, 2, ..., m$. Moreover, for $i = 1, ..., m$, we let

$$\mathbf{u}_i = \mathbf{A}_i \mathbf{q}. \tag{4.16}$$

Then, Constraint (4.14) of Program $\overline{\text{EQ}}$ becomes

$$g_i(\mathbf{q}) = \frac{1}{2}\mathbf{u}_i^T \mathbf{u}_i + \mathbf{b}_i^T \mathbf{q} + c_i \leq 0, \quad i = 1, 2, ..., m, \tag{4.17}$$

and Program $\overline{\text{EQ}}$ can be restated in the following form:

Program $\overline{\text{EQ}}'$:

$$\min \quad g(\mathbf{q}) = \sum_{j=1}^{n} q_j \ln(q_j / e p_j) \tag{4.18}$$

$$\text{s.t.} \quad \frac{1}{2}\mathbf{u}_i^T \mathbf{u}_i + \mathbf{b}_i^T \mathbf{q} + c_i \leq 0, \quad i = 1, 2, ..., m, \tag{4.19}$$

$$\mathbf{A}_i \mathbf{q} - \mathbf{u}_i = \mathbf{0}, \quad i = 1, 2, ..., m, \tag{4.20}$$

$$\mathbf{q} \geq \mathbf{0}, \tag{4.21}$$

$$\mathbf{u}_i \in R^{k_i}, \quad i = 1, 2, ..., m. \tag{4.22}$$

Let $\lambda_i \geq 0$ be the Lagrange multiplier associated with Constraint (4.19), for $i = 1, 2, ..., m$. Similarly, let $\mathbf{w}_i \in R^{k_i}$ be the Lagrange vector associated with Constraint (4.20), for $i = 1, 2, ..., m$. Then, we have the Lagrangian function

$$L(\mathbf{q}, \mathbf{u}; \boldsymbol{\lambda}, \mathbf{w}) \equiv \sum_{j=1}^{n} q_j \ln(q_j/ep_j) \quad + \quad \sum_{i=1}^{m} \lambda_i \{\frac{1}{2}\mathbf{u}_i^T \mathbf{u}_i + \mathbf{b}_i^T \mathbf{q} + c_i\}$$
$$+ \quad \sum_{i=1}^{m} \mathbf{w}_i^T (\mathbf{A}_i \mathbf{q} - \mathbf{u}_i), \tag{4.23}$$

where $\mathbf{u} = \times_{i=1}^{m} \mathbf{u}_i \in R^{k_1+k_2+...+k_m}$, $\boldsymbol{\lambda} = (\lambda_1, ..., \lambda_m)^T \in R_+^m$, $\mathbf{w} = \times_{i=1}^{m} \mathbf{w}_i \in R^{k_1+k_2+...+k_m}$.

Rearranging the terms results in

$$L(\mathbf{q}, \mathbf{u}; \boldsymbol{\lambda}, \mathbf{w}) \equiv \mathbf{c}^T \boldsymbol{\lambda} \quad + \quad \sum_{j=1}^{n} q_j \ln(q_j/ep_j) + \sum_{i=1}^{m} (\mathbf{w}_i^T \mathbf{A}_i + \lambda_i \mathbf{b}_i^T)\mathbf{q}$$
$$+ \quad \sum_{i=1}^{m} (\frac{1}{2}\lambda_i \mathbf{u}_i^T \mathbf{u}_i - \mathbf{w}_i^T \mathbf{u}_i). \tag{4.24}$$

Then we define

$$h(\boldsymbol{\lambda}, \mathbf{w}) \equiv \inf\{L(\mathbf{q}, \mathbf{u}; \boldsymbol{\lambda}, \mathbf{w}) | \mathbf{q} \geq \mathbf{0}, \mathbf{u} \in R^{k_1+k_2+...+k_m}\}, \tag{4.25}$$

as the Lagrangian dual objective function [23]. Correspondingly, we have a Lagrangian dual program:

Program $\overline{\mathbf{DEQ}'}$:

$$\sup \quad h(\boldsymbol{\lambda}, \mathbf{w}) \tag{4.26}$$

$$\text{s.t.} \quad \boldsymbol{\lambda} \in R_+^m, \tag{4.27}$$

$$\mathbf{w}_i \in R^{k_i}, \quad i = 1, 2, ..., m. \tag{4.28}$$

Due to the separability of variables $\mathbf{q}$ and $\mathbf{u}$, we know

$$h(\boldsymbol{\lambda}, \mathbf{w}) = \mathbf{c}^T \boldsymbol{\lambda} \quad + \quad \sum_{j=1}^{n} \inf_{q_j \geq 0} \{q_j \ln(q_j/ep_j) + (\mathbf{A}_i^T \mathbf{w}_i + \lambda_i \mathbf{b}_i)_j q_j\}$$

$$+ \sum_{i=1}^{m} \inf_{\mathbf{u}_i \in R^{k_i}} (\frac{1}{2}\lambda_i \mathbf{u}_i^T \mathbf{u}_i - \mathbf{w}_i^T \mathbf{u}_i), \tag{4.29}$$

where $\mathbf{c} = (c_1, c_2, ..., c_m)^T$ and $(\mathbf{A}_i^T \mathbf{w}_i + \lambda_i \mathbf{b}_i)_j$ is the jth component of the vector $(\mathbf{A}_i^T \mathbf{w}_i + \lambda_i \mathbf{b}_i)$.

To obtain the infimum with respect to q_j, we set the derivative of the corresponding portion of Equation (4.29) to be zero for finding an optimum

$$q_j^* = p_j \exp(-\sum_{i=1}^{m} (\mathbf{A}_i^T \mathbf{w}_i + \lambda_i \mathbf{b}_i)_j), \quad j = 1, 2, ..., n. \tag{4.30}$$

Then, the corresponding infimum (achieved at $\mathbf{q}^*$) becomes

$$-\sum_{j=1}^{n} p_j \exp(-\sum_{i=1}^{m} (\mathbf{A}_i^T \mathbf{w}_i + \lambda_i \mathbf{b}_i)_j). \tag{4.31}$$

Likewise, to obtain the infimum with respect to $\mathbf{u}_i$, we have

$$\lambda_i \mathbf{u}_i^* = \mathbf{w}_i, \quad i = 1, 2, ..., m. \tag{4.32}$$

Therefore, the corresponding infimum (achieved at $\mathbf{u}^*$) becomes

$$-\frac{1}{2} \sum_{i=1}^{m} \frac{||\mathbf{w}_i||^2}{\lambda_i}. \tag{4.33}$$

Putting things together, we can write the Lagrangian dual of Program $\overline{\text{EQ}'}$ in the following form:

Program $\overline{\textbf{DEQ}'}$:

$$\sup \quad -\sum_{j=1}^{n} p_j \exp(-\sum_{i=1}^{m} (\mathbf{A}_i^T \mathbf{w}_i + \lambda_i \mathbf{b}_i)_j) + \mathbf{c}^T \boldsymbol{\lambda} - \frac{1}{2} \sum_{i=1}^{m} \frac{||\mathbf{w}_i||^2}{\lambda_i} \tag{4.34}$$

$$\text{s.t.} \quad \boldsymbol{\lambda} \in R_+^m, \tag{4.35}$$

$$\mathbf{w}_i \in R^{k_i}, \quad i = 1, 2, ..., m. \tag{4.36}$$

Note that the dual program $\overline{\text{DEQ}'}$ has a concave objective function and nonnegativity constraints. Hence, it is computationally attractive. Moreover, the known duality theory can be applied to establish the relationship between programs $\overline{\text{DEQ}'}$ and $\overline{\text{EQ}'}$. The conditions under which Program $\overline{\text{DEQ}'}$ has an optimum solution can be stated in the following theorem:

Theorem 4.1. If Program $\overline{\text{EQ}'}$ has a feasible solution $\mathbf{q}$ with $\mathbf{q} > \mathbf{0}$ and $g_i(\mathbf{q}) < 0$, for $i = 1, 2, ..., m$, then the dual Program $\overline{\text{DEQ}'}$ has an optimum solution $(\boldsymbol{\lambda}^*, \mathbf{w}^*)$. Moreover, the dual optimal solution $(\boldsymbol{\lambda}^*, \mathbf{w}^*)$ defines a primal optimal solution $\mathbf{q}^*$ with

$$q_j^* = p_j \exp(-\sum_{i=1}^{m}(\mathbf{A}_i^T \mathbf{w}_i^* + \lambda_i^* \mathbf{b}_i)_j), \quad j = 1, 2, ..., n, \tag{4.37}$$

to Program $\overline{\text{EQ}'}$.

A simple proof of this theorem can be found in [24]. Since Programs $\overline{\text{EQ}}$ and $\overline{\text{EQ}'}$ are equivalent, we can easily see that the solution $\mathbf{q}^*$ defined by Equation (4.37) is an optimum solution of Program $\overline{\text{EQ}}$.

A Lagrangian dual based method which contains two major steps in computation is listed below.

Algorithm 4.1: (Lagrangian Dual Based Method - Quadratic Constraints)

Step 1: Find the optimal solution $(\boldsymbol{\lambda}^*, \mathbf{w}^*)$ of Program $\overline{\text{DEQ}'}$.
Step 2: Use the dual-to-primal conversion formula (4.37) to compute a primal optimal solution $\mathbf{q}^*$.

Although the scheme of Algorithm 4.1 looks simple, there is an important issue to be noted here. Notice that the objective function of the dual program $\overline{\text{DEQ}'}$, i.e., Equation (4.34), is neither differentiable nor continuous at the boundary of the nonnegativity constraints $\lambda_i \geq 0$, $i = 1, 2, ..., m$. Since most optimization routines and algorithms require at least the first-order information of the objective function, this can cause major problems in Step 1. Therefore, special computational efforts are needed for this approach.

4.1.2 Interior Point Method for Quadratically Constrained Problem

Unlike the Lagrangian dual approach, the interior point method deals with the quadratically constrained minimum cross-entropy problems from the primal side. Encouraged by the earlier work on solving linearly constrained quadratic optimization problems [25], Potra and Ye [21] extended the interior point method to solve linearly constrained entropy problems, as discussed in Subsection 3.1.5. At about the same time, an interior-point method had also been developed for solving quadratically constrained quadratic programs by Mehrotra and Sun [19]. Motivated by these developments, Ji and Potra [15] proposed an interior point method for solving quadratically constrained minimum cross-entropy problems in the following form:

Program $\widehat{\mathbf{EQ}}$:

$$\min \quad g(\mathbf{q}) = \sum_{j=1}^{n} q_j \ln(q_j/p_j) \tag{4.38}$$

$$\text{s.t.} \quad g_i(\mathbf{q}) = \frac{1}{2}\mathbf{q}^T\mathbf{H}_i\mathbf{q} + \mathbf{b}_i^T\mathbf{q} + c_i \le 0, \quad i = 1, 2, ..., m, \tag{4.39}$$

$$\mathbf{q} \ge \mathbf{0}, \tag{4.40}$$

where $\mathbf{H}_i$ is an $n \times n$, symmetric, positive semidefinite matrix, $\mathbf{q}$ and $\mathbf{b}_i$ are n-vectors, and c_i is a constant, for $i = 1, 2, ..., m$. Moreover, $p_j > 0$ is given, for $j = 1, 2, ..., n$.

A solution $\mathbf{q} \in R^n$ of Program $\widehat{\text{EQ}}$ is called an interior solution if $\mathbf{q} > \mathbf{0}$ and $g_i(\mathbf{q}) < 0$, for $i = 1, ..., m$. The interior point method assumes that Program $\widehat{\text{EQ}}$ has at least one interior solution. With this assumption, we know that Program $\widehat{\text{EQ}}$ has a unique optimal solution $\mathbf{q}^*$.

Let z^* denote the optimal value of Program $\widehat{\text{EQ}}$. For any $z > z^*$, define

$$F_z \equiv \{\mathbf{q} \in R^n |\ g(\mathbf{q}) \le z,\ g_i(\mathbf{q}) \le 0, i = 1, 2, ..., m,\ \mathbf{q} \ge \mathbf{0}\}. \tag{4.41}$$

Since the objective function $g(\mathbf{q})$ is convex, for $z > z^*$, F_z is a nonempty bounded and convex set contained in the the feasible region of the program. It is also easy to check that, if $z^* < z^{k+1} < z^k$ then $F_{z^{k+1}} \subseteq F_{z^k}$. The interior point method proposed by Ji and Potra searches for the optimal solution by shrinking the convex sets F_z inside the feasible region of Program $\widehat{\text{EQ}}$.

Intuitively, we see that by bringing z^k closer to z^*, we can approach q^*, which is the optimal solution to Program $\widehat{\text{EQ}}$ and belongs to F_{z^*}. However, without the knowledge of z^*, it is not clear how to find z^{k+1} that is closer to z^* than z^k while remaining larger than z^*. To help find such z^{k+1}, Ji and Potra introduced a "potential function"

$$\Phi(\mathbf{q}, z) \equiv (m+n)\ln(z - g(\mathbf{q})) + \sum_{i=1}^{m} \ln(-g_i(\mathbf{q})) + \sum_{j=1}^{n} \ln q_j. \qquad (4.42)$$

Moreover, we define

$$Int(F_z) \equiv \{\mathbf{q} \in R^n | \; g(\mathbf{q}) < z, \; g_i(\mathbf{q}) < 0, \quad i = 1, 2, ..., m, \; \mathbf{q} > \mathbf{0}\}. \qquad (4.43)$$

Then a point $\mathbf{w} \in R^n$ is called an "analytic center" of F_z if it maximizes $\Phi(\mathbf{q}, z)$ over $Int(F_z)$. Because $\Phi(\mathbf{q}, z)$ involves the term $(z - g(\mathbf{q}))$, we see intuitively that maximizing $\Phi(\mathbf{q}, z)$, with respect to $\mathbf{q}$, is related to minimizing $g(\mathbf{q})$.

A "normalized potential function" of $\Phi(\mathbf{q}, z)$ is defined as

$$\phi(\mathbf{q}, z) \equiv \Phi(\mathbf{w}, z) - \Phi(\mathbf{q}, z). \qquad (4.44)$$

In this way, $\mathbf{w}$ is an analytic center if and only if it minimizes $\phi(\mathbf{q}, z)$ over $\mathbf{q} \in Int(F_z)$ with the minimum value $\phi(\mathbf{w}, z) = 0$.

With the above definition, we can easily derive the gradient $\nabla\phi(\mathbf{q}, z)$ and the Hessian $\nabla^2\phi(\mathbf{q}, z)$ of $\phi(\mathbf{q}, z)$, with respect to $\mathbf{q}$, as below.

$$\nabla\phi(\mathbf{q}, z) = (m+n)\frac{\nabla g(\mathbf{q})}{(z - g(\mathbf{q}))} - \sum_{i=1}^{m} \frac{\nabla g_i(\mathbf{q})}{g_i(\mathbf{q})} - \mathbf{Q}^{-1}\mathbf{e}, \qquad (4.45)$$

$$\begin{aligned}\nabla^2\phi(\mathbf{q}, z) = \quad & (m+n)\frac{\nabla^2 g(\mathbf{q})}{(z - g(\mathbf{q}))} + (m+n)\frac{\nabla g(\mathbf{q})(\nabla g(\mathbf{q}))^T}{(z - g(\mathbf{q}))^2} \\ & - \sum_{i=1}^{m} \frac{\nabla^2 g_i(\mathbf{q})}{g_i(\mathbf{q})} + \sum_{i=1}^{m} \frac{\nabla g_i(\mathbf{q})(\nabla g_i(\mathbf{q}))^T}{(g_i(\mathbf{q}))^2} + \mathbf{Q}^{-2},\end{aligned} \qquad (4.46)$$

where

$$\nabla g(\mathbf{q}) = (\ln(q_1/p_1),\ \ln(q_2/p_2)), ...,\ \ln(q_n/p_n))^T - \mathbf{e}^T, \tag{4.47}$$

$$\nabla^2 g(\mathbf{q}) = \mathbf{Q}^{-1}, \tag{4.48}$$

$$\nabla g_i(\mathbf{q}) = H_i\mathbf{q} + b_i, \quad i = 1, 2, ..., m, \quad \text{and} \tag{4.49}$$

$$\nabla^2 g_i(\mathbf{q}) = H_i, \quad i = 1, 2, ..., m, \tag{4.50}$$

with $\mathbf{Q}$ being the diagonal matrix formed by the components of the vector $\mathbf{q}$ and $\mathbf{e}$ being an n-vector of all 1's.

The motivation of Ji and Potra's algorithm is the following. Suppose that, at the beginning of the kth iteration, we have a value z^k such that $z^* < z^k$ and an approximation $\mathbf{q}^k \in Int(F_{z^k})$ of the analytic center $\mathbf{w}^k$. We may take a partial Newton step to reduce the normalized potential function and move to $\mathbf{q}^{k+1}$ which is closer to $\mathbf{w}^k$ than $\mathbf{q}^k$. Then z^k is updated to be a smaller value z^{k+1} such that $\mathbf{q}^{k+1}$ also approximates $\mathbf{w}^{k+1}$. In this way, the sequence $\{\mathbf{q}^k\}$ converges to the optimal solution as the sequence of nested convex sets $\{F_{z^k}\}$ shrinks.

To be more precise, by minimizing $\phi(\mathbf{q}, z)$ at $(\mathbf{q}^k, z^k)$, the Newton direction $\gamma \in R^n$ is obtained by solving

$$\nabla^2\phi(\mathbf{q}^k, z^k)\gamma = -\nabla\phi(\mathbf{q}^k, z^k). \tag{4.51}$$

Then the updates are done by assigning

$$\mathbf{q}^{k+1} = \mathbf{q}^k + \lambda_k\gamma, \tag{4.52}$$

$$z^{k+1} = z^k + \mu_k(g(\mathbf{q}^{k+1}) - z^k), \tag{4.53}$$

with λ_k and μ_k being properly determined positive step sizes.

With careful analysis of the desired step sizes, the interior-point method can be stated as below.

Algorithm 4.2: (Interior-Point Method - Quadratic Constraints)

Initialization:

Given a small number $\epsilon > 0$, an initial value z^0 $(> z^*)$, and an initial solution $\mathbf{q}^0 \in Int(F_{z^0})$ such that $\phi(\mathbf{q}^0, z^0) \leq 0.0026$, calculate $\delta^0 = z^0 - g(\mathbf{q}^0)$, where $\phi(\mathbf{q}, z)$ is defined according to Equation (4.44), and set $k = 0$.

Iterations:

If $\delta^k < \epsilon$, then stop. Otherwise,

compute $\nabla(\mathbf{q}^k, z^k)$ and $\nabla^2(\mathbf{q}^k, z^k)$ using Equations (4.45) and (4.46);

compute the Newton direction γ by solving

$$\nabla^2\phi(\mathbf{q}^k, z^k)\gamma = -\nabla\phi(\mathbf{q}^k, z^k); \tag{4.54}$$

compute

$$\lambda_k = \frac{0.03}{\sqrt{\gamma^T \nabla\phi(\mathbf{q}^k, z^k)\gamma}}; \tag{4.55}$$

$$\mu_k = \mu = \frac{0.002}{\sqrt{(m+n)}}; \tag{4.56}$$

compute

$$\mathbf{q}^{k+1} = \mathbf{q}^k + \lambda_k \gamma; \tag{4.57}$$

$$z^{k+1} = z^k + \mu_k(g(\mathbf{q}^{k+1}) - z^k); \tag{4.58}$$

compute

$$\delta^{k+1} = z^{k+1} - g(\mathbf{q}^{k+1}); \tag{4.59}$$

reset $k \leftarrow k+1$ for the next iteration.

Ji and Potra showed that Algorithm generates sequence $\{\mathbf{q}^k\} \subset Int(F_{z^0})$ and $\{z^k\}$ such that

$$z^* \leq g(\mathbf{q}^{k+1}) \leq z^k, \tag{4.60}$$

$$(z^{k+1} - z^*)/(z^k - z^*) \leq 1 - 0.0009/\sqrt{m+n}. \tag{4.61}$$

The second inequality shows that $\{z^k\}$ is Q-linearly convergent to z^*. The first inequality implies that $\{g(\mathbf{q}^k)\}$ is R-linearly convergent to z^*. A standard argument show that we can have $|z^k - z^*| \leq \epsilon$ after taking at most $O(\sqrt{(m+n)}|\ln \epsilon|)$ iterations. This fact can be stated in the following theorem:

Theorem 4.2. Given an initial $z^0 > z^*$ and $\mathbf{q}^0 \in Int(F_{z^0})$ such that $\phi(\mathbf{q}^0, z^0) \leq 0.0026$, Algorithm 4.2 converges in at most $O(\sqrt{(m+n)}|\ln \epsilon|)$ iterations.

Note that, in order to make the algorithm work, an initial pair $(\mathbf{q}^0, z^0)$ such that $\mathbf{q}^0 \in Int(F_{z^0})$ and $\phi(\mathbf{q}^0, z^0) \leq 0.0026$ is needed. Ji and Potra [15] suggested using the Big-M Method to construct an artificial problem having a trivial initial pair satisfying the requirement. Details can be found in [15].

4.2 ENTROPY OPTIMIZATION WITH QUADRATIC CONSTRAINTS

As we discussed in Section 4.1, one form of the quadratically constrained minimum cross-entropy problems was originally formulated by Zhang and Brockett [26] by extending a linearly constrained minimum cross-entropy problem to include the quadratic constraints. The Lagrangian dual program was later explicitly derived by Teboulle [24]. However, neither of them addressed the computational issues associated with the Lagrangian dual approach. In this section, we present a dual approach based on geometric programming [6] and devise a dual perturbation method to solve problems in the following form:

Program EQ:

$$\min \quad g(\mathbf{q}) = \sum_{j=1}^{n} q_j \ln(q_j/p_j) \tag{4.62}$$

$$\text{s.t.} \quad g_i(\mathbf{q}) = \frac{1}{2}\mathbf{q}^T \mathbf{H}_i \mathbf{q} + \mathbf{b}_i^T \mathbf{q} + c_i \leq 0, \quad i = 1, 2, ..., m, \tag{4.63}$$

$$\sum_{j=1}^{n} q_j = 1, \tag{4.64}$$

$$\mathbf{q} \geq \mathbf{0}, \tag{4.65}$$

where $\mathbf{H}_i$ is an $n \times n$, symmetric, positive semidefinite matrix, $\mathbf{q}$ and $\mathbf{b}_i$ are n-vectors, c_i is a constant, for $i = 1, 2, \ldots, m$, and $\mathbf{p} > \mathbf{0}$ is a known probability distribution.

Notice that, unlike Zhang and Brockett's original formulation, Program $\overline{\text{EQ}}$ defined in Subsection 4.1.1, the probability constraint is explicitly stated. This would help us construct a geometric dual problem rather easily.

In Subsection 4.2.1, two simple inequalities will be used to derive a geometric dual problem which has a concave objective function subject to linear constraints. A "weak duality theorem" can also be derived. In Subsection 4.2.2, we prove the "strong duality theorem" (or "existence theorem") for the quadratically constrained minimum cross-entropy problems. Then a dual-to-primal conversion formula is obtained in Subsection 4.2.3. The conversion formula provides an ϵ-optimal solution to the original problem once a corresponding dual problem is solved. Finally, the computational issues are discussed in Subsection 4.2.4 and concluding remarks are given in Subsection 4.2.5.

4.2.1 Geometric Dual

For easy cross-reference purpose, we adhere to the following notation that has been used extensively in the geometric programming literature [6, 10]. Since the matrix $\mathbf{H}_i$ is symmetric and positive semi-definite, we can write it as $\mathbf{H}_i = \mathbf{A}_i^T \mathbf{A}_i$, where $\mathbf{A}_i$ is a $k_i \times n$ matrix and k_i is the row-rank of $\mathbf{H}_i$, for $i = 1, 2, \ldots, m$. Then define $\mathbf{I}_n$ to be the $n \times n$ identity matrix and the index notation

$$[0] = \{1, 2, \ldots, n\}, \quad]0[= n, \tag{4.66}$$

$$[i] = \{\ ]i-1[+1, \ldots,]i-1[+k_i\}, \ ]i[\ = \ ]i-1[+k_i+1, \quad \text{for} \quad i = 1, \ldots, m, \tag{4.67}$$

$$]m[= k. \tag{4.68}$$

Moreover, we let X denote the column space of the $k \times n$ matrix $\mathbf{M}$ defined by

$$\mathbf{M}^T \equiv (\mathbf{I}_n, \mathbf{A}_1^T, \mathbf{b}_1, \ldots, \mathbf{A}_m^T, \mathbf{b}_m). \tag{4.69}$$

Hence, each $\mathbf{x} \in X$ is a k-dimensional column vector that can be expressed in the form of the Cartesian product $\mathbf{x} = \times_{i=0}^{m} \mathbf{x}^i$, where $\mathbf{x}^0 = (x_1, ..., x_{]0[})^T$, and $\mathbf{x}^i = (x_{]i-1[+1}, ..., x_{]i[})^T$, for $i = 1, ..., m$. Then it is a simple routine to check that Program EQ is equivalent to the following "primal" geometric program:

Program EQ$'$:

$$\min \quad G(\mathbf{x}) = \sum_{j\epsilon[0]} x_j \ln(x_j/p_j) \tag{4.70}$$

$$\text{s.t.} \quad G_i(\mathbf{x}) = \frac{1}{2}\sum_{j\epsilon[i]} x_j^2 + x_{]i[} + c_i \leq 0, \quad i = 1, 2, ..., m, \tag{4.71}$$

$$\sum_{j\epsilon[0]} x_j = 1, \tag{4.72}$$

$$x_j \geq 0, \quad j \in [0], \tag{4.73}$$

$$\mathbf{x} \epsilon X. \tag{4.74}$$

In order to construct a geometric dual problem, corresponding to the primal vector $\mathbf{x}$, we define a dual vector $\mathbf{w} = \{\mathbf{w}_0\} \times_{i=1}^{m} \{\mathbf{w}_i, w_{]i[}\}$, where $\mathbf{w}_0 = (w_1, ..., w_{]0[})^T$ and $\mathbf{w}_i = (w_{]i-1[+1}, ..., w_{]i-1[+k_i})^T$, for $i = 1, ...m$. In this way, $\mathbf{w}$ is also a k-dimensional column vector. Then consider the arithmetic-geometric inequality used in [6]:

$$\sum_{j\epsilon[0]} p_j \exp(w_j) \geq \prod_{j\epsilon[0]} \{p_j \exp(w_j)/x_j\}^{x_j}. \tag{4.75}$$

This inequality is true for any real numbers $w_j, x_j > 0, j \in [0]$, and $\sum_{j\epsilon[0]} x_j = 1$. By taking the logarithm on both sides and simplifying it, we have

$$\sum_{j\in[0]} x_j w_j \leq \sum_{j\epsilon[0]} x_j \ \ln(x_j/p_j) + \ln(\sum_{j\epsilon[0]} p_j \exp\,(w_j)). \tag{4.76}$$

With the convention of $0 \times \ln 0 = 0$, the above inequality is true for all $w_j, x_j \geq 0$, $j\epsilon[0]$, and $\sum_{j\epsilon[0]} x_j = 1$. The equality holds if and only if there is a constant K such that $p_j \exp\,(w_j) = K \times x_j$, for $j \in [0]$.

Next consider the quadratic inequality used in [10]:

$$\sum_{j\epsilon[i]} x_j z_j \;\le\; \frac{1}{2}\sum_{j\epsilon[i]}(x_j^2+z_j^2), \text{ for } x_j, z_j \in R. \tag{4.77}$$

After replacing z_j by $w_j/w_{]i[}$, multiplying both sides by $w_{]i[}$, adding $x_{]i[}w_{]i[}$ to both sides, and then adding and subtracting the term $c_i w_{]i[}$ on the right hand side, we have

$$\begin{aligned}\sum_{j\epsilon[i]} x_j w_j \;+\; x_{]i[}w_{]i[} \;\le\;\; & w_{]i[}\{\frac{1}{2}\sum_{j\epsilon[i]} x_j^2 + \; x_{]i[} \;+\; c_i\} \\ + \;\; & \{\frac{1}{2}\sum_{j\epsilon[i]} w_j^2 \;/w_{]i[} - \; c_i w_{]i[}\}.\end{aligned} \tag{4.78}$$

With the understanding that $w_{]i[} = 0$ only if $w_j = 0$, for each $j \in [i]$, this expression holds for all $w_{]i[} \ge 0$ and $i = 1, 2, ..., m$.

Combining Inequalities (4.76) and (4.78), for $i \;=\; 1, 2, ..., m$, we obtain

$$\begin{aligned}\sum_{j=1}^{k} x_j w_j \;\le\; & \sum_{j\epsilon[0]} x_j \ln(x_j/p_j) + \sum_{i=1}^{m} w_{]i[}\{\frac{1}{2}\sum_{j\epsilon[i]} x_j^2 + x_{]i[} + c_i\} \\ & + \ln(\sum_{j\epsilon[0]} p_j \exp\,(w_j)) \;+ \sum_{i=1}^{m}\{\frac{1}{2}\sum_{j\epsilon[i]} w_j^2/w_{]i[} - c_i w_{]i[}\}.\end{aligned} \tag{4.79}$$

If $\mathbf{x}$ is primal feasible, $w_{]i[} \ge 0$ (and $w_{]i[} = 0$ only if $w_j = 0$ for each $j \in [i]$), for $i \;=\; 1, 2, ..., m$, and $\mathbf{M}^T\mathbf{w} \;=\; \mathbf{0}$, then we know that $\sum_{j=1}^{k} x_j w_j \;=\; 0$ and $w_{]i[}\{\frac{1}{2}\sum_{j\epsilon[i]} x_j^2 + \; x_{]i[} \;+ c_i\} \;\le\; 0$, for $i = 1, ..., m$. Consequently,

$$-\ln(\sum_{j\epsilon[0]} p_j \exp\,(w_j)) \;+ \sum_{i=1}^{m} c_i w_{]i[} - \frac{1}{2}\sum_{i=1}^{m}\sum_{j\epsilon[i]} w_j^2/w_{]i[} \le \sum_{j\epsilon[0]} x_j \ln(x_j/p_j). \tag{4.80}$$

Based on this inequality, we can define the "geometric dual program" of Program EQ$'$ as follows:

Program DEQ$'$:

$$\begin{aligned}\max \qquad V(\mathbf{w}) \;=\; & -\ln(\sum_{j\epsilon[0]} p_j \exp\,(w_j)) \;+\; \sum_{i=1}^{m} c_i w_{]i[} \\ & - \;(1/2)\sum_{i=1}^{m}\sum_{j\epsilon[i]} w_j^2/w_{]i[}\end{aligned} \tag{4.81}$$

$$\text{s.t.} \quad \mathbf{M}^T\mathbf{w} = \mathbf{0}, \tag{4.82}$$

$$w_{]i[} \geq 0, \quad i = 1, 2, ..., m, \tag{4.83}$$

where $\mathbf{w} \in R^k$ and $\mathbf{M}$, $[i]$, $]i[$ and c_i are defined as before.

Notice that we were able to replace the constraint "$w_{]i[} \geq 0$ and $w_{]i[} = 0$ only if $w_j = 0$, for each $j \in [i]$" by a simple constraint "$w_{]i[} \geq 0$," since if $w_{]i[} = 0$ and $w_j \neq 0$, for some $j \in [i]$, then $V(\mathbf{w})$ would become $-\infty$.

Several observations can be made here.

Observation 4.1: As an immediate consequence of Inequality (4.80), we can easily show the "weak duality theorem" as follows:

Theorem 4.3. If $\mathbf{x}$ is a primal feasible solution of Program EQ$'$ and $\mathbf{w}$ is a dual feasible solution of Program DEQ$'$, then $V(\mathbf{w}) \leq G(\mathbf{x})$.

Observation 4.2: We follow the well-established geometric programming theory to define a "canonical program" as a program which has a feasible dual solution $\mathbf{w}^+$ such that $w^+{}_{]i[} > 0$, for $i = 1, ..., m$. Then, since matrix M is defined by Equation (4.69), if we take vector $\mathbf{w}^+$ with components $(\mathbf{w}^+)_0 = -\sum_{i=1}^m \mathbf{b}_i$, $w^+{}_{]i-1[+1} = ... = w^+{}_{]i-1[+k_i} = 0$, and $w^+_{]i[} = 1$, for $i = 1, ..., m$, Equation (4.82) is automatically satisfied by $\mathbf{w}^+$. Therefore, we have the following result.

Theorem 4.4. Program DEQ$'$ is a canonical program that is always consistent.

Observation 4.3: As a direct consequence of Observations 4.1 and 4.2, we have the next theorem.

Theorem 4.5. If Program EQ$'$ is consistent, then Program DEQ$'$ has a finite optimum value.

Moreover, just like most geometric duals [10, 11], it is easy to see that $V(\mathbf{w})$ is concave over the dual feasible region. However, it is not differentiable at the boundary of the constraints $w_{]i[} \geq 0$, for $i = 1, 2, ..., m$. This issue will be explicitly addressed in the next section.

4.2.2 Existence Theorem

The dual program DEQ$'$ has a concave objective function with linear constraints. Compared to the primal program EQ$'$, this makes the dual approach computationally attractive. However, as we observed in the last subsection, the dual problem is nondifferentiable at some of its boundary points. This would cause at least two difficulties: (1) applying Lagrangian theory is difficult without resorting to subgradient techniques as developed in [23]; (2) developing a dual based solution procedure becomes more troublesome.

Fortunately, the recently developed "controlled dual perturbation" approach [10, 11] can be applied here. We can perturb Constraint (4.83) by a well-controlled positive amount l_i, for $i = 1, 2, ..., m$, and construct the following "perturbed dual" program:

Program DEQ$'_l$:

$$\max \quad V(\mathbf{w}) = -\ln\Big(\sum_{j\epsilon[0]} p_j \exp(w_j)\Big) + \sum_{i=1}^{m} c_i w_{]i[} - (1/2)\sum_{i=1}^{m}\sum_{j\epsilon[i]} w_j^2/w_{]i[} \qquad (4.84)$$

$$\text{s.t.} \quad \mathbf{M}^T\mathbf{w} = \mathbf{0}, \qquad (4.85)$$

$$w_{]i[} \geq l_i, \quad i = 1, 2, ..., m, \qquad (4.86)$$

where the notation is defined as before and the positive l_i's are called "perturbations."

One interesting observation about this perturbed dual program is as follows. Since Theorem 4.4 indicates that Program DEQ$'$ is canonical with a feasible solution $\mathbf{w}^+$ such that $w^+{}_{]i[} > 0$, for $i = 1, ..., m$, we can always choose a perturbation vector $\boldsymbol{l} = (l_1, ..., l_m)$ with $w^+{}_{]i[} > l_i > 0$, for $i = 1, ..., m$. In this way, the corresponding Program DEQ$'_l$ is always feasible. Moreover, Program DEQ$'_l$ has a concave differentiable objective function over its linearly constrained nonempty feasible region. This will make the dual approach very favorable since any general purpose optimizer like MINOS [20] can be used to solve it without much difficulty. However, we have to first show that there is no duality gap.

The following theorem will lead us to this goal:

Theorem 4.6. If Program DEQ$'$ has a finite supremum, then Program EQ$'$ is consistent. Moreover, for any given $\epsilon > 0$, we can choose a proper perturbation vector $\boldsymbol{l}(\epsilon) = (l(\epsilon)_1, ..., l(\epsilon)_m)$ with $l(\epsilon)_i > 0$, for $i = 1, ..., m$, such that the perturbed dual program DEQ$'_{l(\epsilon)}$ has an optimal solution $\mathbf{w}$ and the Program EQ$'$ has a feasible solution $\mathbf{x}$ satisfying the condition that $0 \leq G(\mathbf{x}) - V(\mathbf{w}) \leq \epsilon$.

Proof. Let $\mathbf{w}^+$ be a feasible solution of Program DEQ$'$ with $w_{]i[}{}^+ > 0$, for $i = 1, ..., m$. Consider Program DEQ$'_l$ with $l_i < w_{]i[}{}^+$, for $i = 1, ...m$. Since Program DEQ$'$ has a finite supremum, so does Program DEQ$'_l$. Now, consider the Lagrangian defined as

$$L = V(\mathbf{w}) + \sum_{i=1}^{m} \lambda_i (w_{]i[} - l_i) + \gamma^T \mathbf{M}^T \mathbf{w}, \tag{4.87}$$

where γ (a column vector of size n) and λ_i, $i = 1, 2, ..., m$, are Lagrange multipliers of the Constraints (4.85) and (4.86), respectively.

Because Program DEQ$'_l$ is differentiable and concave, there exist solutions to the following Karush-Kuhn-Tucker (K.K.T.) conditions:

$$\partial L / \partial w_j = -(p_j \exp{(w_j)}) / \sum_{k \in [0]} (p_k \exp{(w_k)}) + \gamma^T \mathbf{M}_j = 0, \text{ for } j \in [0], \tag{4.88}$$

$$\partial L / \partial w_j = -w_j / w_{]i[} + \gamma^T \mathbf{M}_j = 0, \quad \text{for } j \in [i], \quad i = 1, 2, ..., m, \tag{4.89}$$

$$\partial L / \partial w_{]i[} = c_i + \frac{1}{2} \sum_{j \in [i]} w_j^2 / w_{]i[}^2 + \lambda_i + \gamma^T \mathbf{M}_{]i[} = 0, \quad i = 1, 2, ..., m, \tag{4.90}$$

$$\lambda_i (w_{]i[} - l_i) = 0, \quad i = 1, 2, ..., m, \tag{4.91}$$

$$\lambda_i \geq 0, \quad i = 1, 2, ..., m, \tag{4.92}$$

plus conditions (4.85) and (4.86), where $\mathbf{M}_j$ (a column vector of size n) is the jth column of matrix $\mathbf{M}^T$.

If we let

$$x_j = \gamma^T \mathbf{M}_j, \text{ for } j = 1, ..., k, \tag{4.93}$$

then, by definition, we know $\mathbf{x} \in X$ (the column space of $\mathbf{M}$). By condition (4.88), we know $x_j \geq 0$ and $\sum_{j\epsilon[0]} x_j = 1$. Moreover, by condition (4.89), we have $w_j/w_{]i[} = x_j$, for $j \in [i]$, $i = 1, ..., m$. Plugging this equation into condition (4.90), we have

$$\frac{1}{2}\sum_{j\epsilon[i]} x_j^2 + x_{]i[} + c_i + \lambda_i = 0, \quad i = 1, 2, ..., m. \tag{4.94}$$

Since $\lambda_i \geq 0$ is given by condition (4.92), we know condition (4.71) is satisfied by this vector $\mathbf{x}$. Hence Program EQ' has a feasible solution $\mathbf{x}$.

To prove the second part of this theorem, let us consider the right-hand side of the Lagrangian defined by Equation (4.87). Since condition (4.91) holds and $\mathbf{x}$ and $\mathbf{w}$ are complementary, we have

$$L = V(\mathbf{w}).$$

On the other hand, by substituting $V(\mathbf{w})$ (defined by Equation (4.84)) on the right-hand side of Equation (4.87) and rearranging the terms, we obtain:

$$\begin{aligned} L = & - \ln(\sum_{j\epsilon[0]} p_j \exp{(w_j)}) + \sum_{j\epsilon[0]} \gamma^T \mathbf{M}_j w_j \\ & + \sum_{i=1}^{m} w_{]i[}\{c_i - \frac{1}{2}\sum_{j\epsilon[i]} w_j^2/w_{]i[}^2 + \lambda_i + \gamma^T M_{]i[}\} \\ & + \sum_{i=1}^{m}\sum_{j\epsilon[i]} \gamma^T \mathbf{M}_j w_j - \sum_{i=1}^{m} \lambda_i l_i. \end{aligned}$$

In other words,

$$\begin{aligned} V(\mathbf{w}) = & - \ln(\sum_{j\epsilon[0]} p_j \exp{(w_j)}) + \sum_{j\epsilon[0]} \gamma^T \mathbf{M}_j w_j \\ & + \sum_{i=1}^{m} w_{]i[}\{c_i - \frac{1}{2}\sum_{j\epsilon[i]} w_j^2/w_{]i[}^2 + \lambda_i + \gamma^T \mathbf{M}_{]i[}\} \\ & + \sum_{i=1}^{m}\sum_{j\epsilon[i]} \gamma^T \mathbf{M}_j w_j - \sum_{i=1}^{m} \lambda_i l_i. \end{aligned} \tag{4.95}$$

Now, by condition (4.88), we know that $p_j \exp(w_j) = x_j \times \sum_{j \epsilon [0]}(p_j \exp(w_j))$, for each $j \epsilon [0]$. But $\sum_{j \epsilon [0]}(p_j \exp(w_j))$ is a constant. Hence, Inequality (4.76) becomes an equality and

$$\sum_{j \epsilon [0]} \gamma^T \mathbf{M}_j w_j = \sum_{j \epsilon [0]} x_j w_j = \ln(\sum_{j \epsilon [0]} p_j \exp\ w_j) + \sum_{j \epsilon [0]} x_j \ln(x_j/p_j). \quad (4.96)$$

In addition, by condition (4.89), we have

$$\sum_{i=1}^{m} \sum_{j \epsilon [i]} \gamma^T \mathbf{M}_j w_j = \sum_{i=1}^{m} \sum_{j \epsilon [i]} w_j^2 / w_{]i[}. \quad (4.97)$$

Substituting these values into Equation (4.95) and noticing condition (4.90), we see that

$$V(\mathbf{w}) = G(\mathbf{x}) - \sum_{i=1}^{m} \lambda_i l_i$$

By Theorem 4.3, we further have

$$0 \leq G(\mathbf{x}) - V(\mathbf{w}) = \sum_{i=1}^{m} \lambda_i l_i. \quad (4.98)$$

Now substitute the components of the vector $\mathbf{w}^+$ in Inequalities (4.76) and (4.78). By adding these inequalities together and using condition (4.94), for $i = 1, ..., m$, we have

$$0 \leq -\sum_{i=1}^{m} \lambda_i w_{]i[}^+ - V(\mathbf{w}^+) + G(\mathbf{x}). \quad (4.99)$$

Combining with Inequality (4.98), we get

$$0 \leq \sum_{i=1}^{m} \lambda_i (w_{]i[}^+ - l_i) \leq V(\mathbf{w}) - V(\mathbf{w}^+) \leq V - V(\mathbf{w}^+), \quad (4.100)$$

where V is an upper bound for the dual program as assumed for this theorem.

Now, for any given $\epsilon > 0$, if we choose l_i such that

$$l_i = l_i(\epsilon) = \epsilon \delta w_{]i[}^+ / (V - V(\mathbf{w}^+) + \epsilon), \quad (4.101)$$

for any $0 < \delta < 1$ and $i = 1, 2, ..., m$, then there are two possible cases for consideration:

Case 1: If $V - V(\mathbf{w}^+) = 0$, then since $l_i = \delta y^+_{]i[}$ and $w^+_{]i[} - l_i = (1-\delta)y^+_{]i[} > 0$, for every i, λ_i in Inequality (4.100) has to be 0. Consequently, by Inequality (4.98), we have $G(\mathbf{x}) - V(\mathbf{w}) = 0$.

Case 2: If $V - V(\mathbf{w}^+) > 0$, by Equation (4.101), $w^+_{]i[} = [(V - V(\mathbf{w}^+) + \epsilon)/\epsilon\delta]l_i$. Plugging this into Inequality (4.100), we obtain $\sum_{i=1}^m \lambda_i[((V - V(\mathbf{w}^+))/\delta\epsilon) + (1/\delta) - 1]l_i \leq V - V(\mathbf{w}^+)$. This implies that $\sum_{i=1}^m \lambda_i l_i \leq \delta\epsilon < \epsilon$. Therefore, by Inequality (4.98), we have $0 \leq G(\mathbf{x}) - V(\mathbf{w}) < \epsilon$.

This completes the proof.

□

We now are ready to prove the "strong duality theorem" (or "existence theorem") for the quadratically constrained minimum cross-entropy problems.

Theorem 4.7. Program EQ$'$ is consistent if and only if its dual Program DEQ$'$ has a finite optimum value. In this case, Programs EQ$'$ and DEQ$'$ attain a common optimum value.

Proof. By combining Theorems 4.5 and 4.6, we know that Program EQ$'$ has a feasible solution if and only if Program DEQ$'$ has a finite optimum value. Moreover, for any given $\epsilon > 0$, we can find feasible $\mathbf{x}$ and $\mathbf{w}$ such that $0 \leq G(\mathbf{x}) - V(\mathbf{w}) \leq \epsilon$. Hence we have $V(\mathbf{w}) \leq V^* \leq G^* \leq G(\mathbf{x})$, where V^* and G^* are optimum values of Programs DEQ$'$ and EQ$'$, respectively. Consequently, $0 \leq G^* - V^* \leq G(\mathbf{x}) - V(\mathbf{w}) \leq \epsilon$, for any $\epsilon > 0$. Therefore. $G^* = V^*$ and the proof is complete.

□

4.2.3 Dual-to-Primal Conversion

In this subsection, we show that an ϵ-optimal solution to the original quadratically constrained cross-entropy problem (Program EQ) can be derived by solving Program $\mathrm{DEQ}'_{l(\epsilon)}$ with $l(\epsilon)$ being properly selected according to Equation (4.101). The first result is as follows:

Theorem 4.8. If γ is a vector of Lagrange multipliers associated with the constraint set $\mathbf{M}^T\mathbf{w} = \mathbf{0}$ of Program $\mathrm{DEQ}'_{l(\epsilon)}$, then γ is a feasible solution for Program EQ and $0 \le g(\gamma) - V(\mathbf{w}) \le \epsilon$.

Proof. Since the first n columns of matrix $\mathbf{M}^T$ form an identity matrix $\mathbf{I}_n$, by Equation (4.93), we conclude that $x_j = \gamma_j$, for $j \in [0]$, and $G(\mathbf{x}) = g(\gamma)$. By condition (4.88), we know $\gamma_j \ge 0$ and $\sum_{j\in[0]} \gamma_j = 1$. Moreover, by putting the structure of matrix $\mathbf{M}$ and Equation (4.93) together, then substituting for x_j $(j \in [i])$ and $x_{]i[}$ $(i = 1, ..., m)$ in Equation (4.94), and finally noticing the fact that $\lambda_i \ge 0$, we can conclude that $g_i(\gamma) = (1/2)\gamma^T\mathbf{H}_i\gamma + \mathbf{b}_i^T\gamma + c_i \le 0$, for $i = 1, 2, ..., m$. Hence γ is a solution to Program EQ. Then, by Theorem 4.6, it is clear that $0 \le g(\gamma) - V(\mathbf{w}) \le \epsilon$. This completes the proof.

□

The second result is to express the primal solution γ in terms of dual solutions. To do so, we define

$$\mu = \ln(\sum_{j\in[0]} p_j \exp(w_j)). \tag{4.102}$$

The constraint $\mathbf{M}^T\mathbf{w} = \mathbf{0}$ implies that

$$\mathbf{w}_0 = -\sum_{i=1}^{m} \{\mathbf{A}_i^T\bar{\mathbf{w}}_i + \mathbf{b}_i w_{]i[}\}, \tag{4.103}$$

where $\bar{\mathbf{w}}_i$ is a column vector with elements w_j such that $j \in [i]$, for $i = 1, ..., m$. In other words, $\bar{\mathbf{w}}_i$ is the same as $\mathbf{w}_i$ but with one less element $w_{]i[}$. Hence,

$$w_j = -\sum_{i=1}^{m} \{\mathbf{A}_{i,j}^T\bar{\mathbf{w}}_i + b_{i,j} w_{]i[}\}, \quad \text{for } j \in [0], \tag{4.104}$$

where $\mathbf{A}_{i,j}^T$ is the jth row of $\mathbf{A}_i^T$ and $b_{i,j}$ is the jth component of $\mathbf{b}_i$.

By Equations (4.93) and (4.88), we have $p_j \exp(w_j) = \gamma_j \exp\mu$, for $j \in [0]$. Now, substituting for w_j in this expression, we have

$$\gamma_j = p_j \exp(-\mu - \sum_{i=1}^{m}\{\mathbf{A}_{i,j}^T\bar{\mathbf{w}}_i + b_{i,j} w_{]i[}\}), \quad \text{for } j \in [0]. \tag{4.105}$$

Hence we have the following theorem:

Theorem 4.9. For any given $\epsilon > 0$, if $\boldsymbol{l}(\epsilon)$ is defined by Equation (4.101) and $\mathbf{w}$ solves Program $\text{DEQ}'_{l(\epsilon)}$, then Equations (4.102) and (4.105) generate an ϵ-optimal solution to the original quadratically constrained minimum cross-entropy problem (Program EQ).

4.2.4 Computational Issues

As a by-product of the theoretical insights we obtained in previous subsections, we can outline a dual based computation procedure as follows:

Algorithm 4.3: (Dual Perturbation Method – Quadratic Constraints)

Step 1: Given any $\epsilon > 0$, a dual feasible vector $\mathbf{w}^+$ with $w^+_{]i[} > 0$, for $i = 1, ..., m$, and a dual upper bound V of $V(\mathbf{w})$.

Step 2: Choose a perturbation vector $\boldsymbol{l}(\epsilon) = (l(\epsilon)_1, ..., l(\epsilon)_n)$ according to Equation (4.101).

Step 3: Find an optimal solution $\mathbf{w}$ for the perturbed dual program $\text{DEQ}'_{l(\epsilon)}$.

Step 4: Generate an $\epsilon-$optimal solution γ according to Equations (4.102) and (4.105). Then,

$$0 \leq g(\gamma) - V(\mathbf{w}) \leq \epsilon.$$

Step 5: Stop.

Several observations can be made here.

Observation 4.4: Compared to the primal program EQ, the perturbed dual program $\text{DEQ}'_{l(\epsilon)}$ is relatively simple since it possesses a concave differentiable objective function and linear constraints. We may use a commercial nonlinear optimizer, eg., MINOS, to solve this dual problem. The additional customization of computer codes is minimal.

Observation 4.5: In Step 1, $\mathbf{w}^+$ can be easily obtained, as derived in the proof of Theorem 4.4, i.e.,

$$\mathbf{w}_0{}^+ = -\sum_{i=1}^{m} \mathbf{b}_i,\ w^+{}_{]i-1[+1} = \ldots = w^+{}_{]i-1[+k_i} = 0, \text{ and } w^+_{]i[} = 1, \text{ for } i = 1, \ldots, m.$$

Observation 4.6: Usually a dual bound V can be estimated easily. If it is not provided, we can always solve a Program DEQ'_l with $l_i = \delta w^+_{]i[}$ first, then use Equations (4.102) and (4.105) to find a corresponding primal solution γ^*. Then, we know $V \leq g(\gamma^*)$.

Observation 4.7: To compare with the Lagrangian dual discussed in Subsection 4.1.1, let us consider Program DEQ'. If we define $\lambda_i = w_{]i[}$, for $i = 1, 2, \ldots, m$, then Constraint (4.82) (i.e., $\mathbf{M}^T\mathbf{w} = \mathbf{0}$) implies that

$$\mathbf{w}_0 = -\sum_{i=1}^{m}(\mathbf{A}_i^T\mathbf{w}_i + \lambda_i\mathbf{b}_i). \tag{4.106}$$

In other words,

$$w_j = -\sum_{i=1}^{m}(\mathbf{A}_i^T\mathbf{w}_i + \lambda_i\mathbf{b}_i)_j, \quad \text{for } j\epsilon[0], \tag{4.107}$$

where $(\mathbf{A}_i^T\mathbf{w}_i + \mathbf{b}_i)_j$ is the jth component of the $n \times 1$ vector $(\mathbf{A}_i^T\mathbf{w}_i + \mathbf{b}_i)$.

Consequently, since $[0] = \{1, 2, \ldots, n\}$ and $\sum_{j\epsilon[i]} w_j^2 = ||\mathbf{w}_i||^2$, Program DEQ' is equivalent to

$$\sup \quad -\ln\left[\sum_{j=1}^{n} p_j \exp\left(-\sum_{i=1}^{m}(\mathbf{A}_i^T\mathbf{w}_i + \lambda_i\mathbf{b}_i)_j\right)\right] + \mathbf{c}^T\boldsymbol{\lambda} - (1/2)\sum_{i=1}^{m}\frac{||\mathbf{w}_i||^2}{\lambda_i} \tag{4.108}$$

$$\text{s.t.} \quad \lambda_i \geq 0, \quad i = 1, 2, \ldots, m, \tag{4.109}$$

$$\mathbf{w}_i \in R^{k_i}, \quad i = 1, 2, \ldots, m. \tag{4.110}$$

This formulation is very close to the Lagrangian dual program $\overline{\mathrm{DEQ}'}$ defined in Subsection 4.1.1 except the first part of the objective function is in a "log" form. This can be attributed to the fact that our formulation of Program EQ, unlike [26] or [24], explicitly considers the probability constraint. But the basic characteristics of these two duals are essentially the same, and there is no difference in computational complexity.

4.2.5 Concluding Remarks

In this section, we have provided a geometric programming based dual approach to solving quadratically constrained minimum cross-entropy problems. A well-controlled dual perturbation method not only showed its theoretical value in proving the weak and strong duality theorems, but also leads to a computational procedure that would allow us to use a general purpose optimizer for solving the dual problem. Once the optimal solution to an appropriately perturbed dual problem has been obtained, a simple conversion formula effortlessly produces an ϵ-optimal solution to the original Program EQ.

4.3 ENTROPY OPTIMIZATION WITH ENTROPIC CONSTRAINTS

In this section, we extend the scope of minimum cross-entropy analysis to include entropy-type constraints [2]. Assume that $\{ p_j^i > 0 : j = 1, 2, ..., n \}$ is a given probability distribution and e_i is a real number, for $i = 0, 1, ..., m$. The minimum cross-entropy problem to be studied in this section has the following form:

Program EE:

$$\min \quad g(\mathbf{q}) = \sum_{j=1}^{n} q_j \ln(q_j / p_j^0) \tag{4.111}$$

$$\text{s.t.} \quad g_i(\mathbf{q}) = \sum_{j=1}^{n} q_j \ln(q_j / p_j^i) \leq e_i, \; i = 1, 2, ...m, \tag{4.112}$$

$$\sum_{j=1}^{n} q_j = 1, \tag{4.113}$$

$$\mathbf{q} \geq \mathbf{0}. \tag{4.114}$$

In this setting, we are looking for the probability distribution that is "closest" to the *a priori* distribution $\{p_j^0 > 0 : j = 1, ..., n\}$ and within "e_i-distance" of the known distributions $\{p_j^i > 0\}$ in terms of cross-entropy. This type of problem in continuous form of abstract spaces can be found in Charnes et al. [5] and Ben-Tal and Charnes [1]. Although corresponding Lagrangian dual problems were derived in literature, the computational issues were not particularly addressed before.

In Subsection 4.3.1, one simple geometric inequality is used to derive the geometric dual problem of Program EE. The geometric dual is shown to be a maximization problem with a concave objective function and linear constraints. The so-called "weak duality theorem" is easy to follow. In Subsection 4.3.2, we prove the "strong duality theorem" (or "existence theorem") for the minimum cross-entropy problem with entropy-type constraints, while a dual-to-primal conversion formula is provided in Subsection 4.3.3. The conversion formula can be used to construct an "ϵ-optimal" solution for the original problem once a corresponding dual problem is solved. Then a computational procedure is proposed in Subsection 4.3.4. This procedure allows us to use a general purpose optimizer like MINOS [20] to solve the problem. We demonstrate the applicability of the computational procedure via two examples. Concluding remarks are given in Subsection 4.3.5.

4.3.1 Geometric Dual

For the sake of easy cross-reference, we shall use some common notation that have been used extensively in the geometric programming literature [6]. First we let $\mathbf{I}_n$ be the $n \times n$ dimensional identity matrix and X be the column space of the $(m+1)n \times n$ dimensional matrix $\mathbf{M}$ defined by

$$\mathbf{M}^T \equiv (\mathbf{I}_n, \mathbf{I}_n, ..., \mathbf{I}_n). \tag{4.115}$$

Hence each $\mathbf{x} \in X$ is an $(m+1)n$-vector that can be expressed in the Cartesian product form $\mathbf{x} = \times_{i=0}^{m} \mathbf{x}^i$, where $\mathbf{x}^i$ is an n-vector, for $i = 0, 1, ..., m$. Then it is a simple routine to check that Program EE is equivalent to the following "primal" geometric program EE′:

Program EE′ :

$$\min \quad G_0(\mathbf{x}) = \sum_{j=1}^{n} x_j^0 \ln\left(\frac{x_j^0}{p_j^0}\right) \tag{4.116}$$

$$\text{s.t.} \quad G_i(\mathbf{x}) = \sum_{j=1}^{n} x_j^i \ln\left(\frac{x_j^i}{p_j^i}\right) - e_i \leq 0, \quad i = 1, 2, ..., m, \tag{4.117}$$

$$\sum_{j=1}^{n} x_j^i = 1, \quad i = 1, 2, ..., m, \tag{4.118}$$

$$x_j^i \geq 0, \quad i = 0, 1, ..., m, \quad j = 1, 2, ..., n, \tag{4.119}$$

$$\mathbf{x} \in X. \tag{4.120}$$

To construct its geometric dual, we define, corresponding to each primal vectors $\mathbf{x}$, a dual vector $\mathbf{w} = \times_{i=0}^{m} \mathbf{w}^i$, where $\mathbf{w}^i$ is an n-vector, for $i = 0, 1, ...m$. In this way, $\mathbf{w}$ is a column vector of size $(m+1)n$.

Consider the arithmetic-geometric inequality used in [6]:

$$\sum_{j=1}^{n} p_j \exp(w_j) \geq \prod_{j=1}^{n} \left(\frac{p_j \exp(w_j)}{x_j}\right)^{x_j}. \tag{4.121}$$

This inequality is true for any $x_j > 0$, $\sum_{j=1}^{n} x_j = 1$ and $w_j \in R$. Taking the logarithm on both sides and simplifying the resulting inequality give

$$\sum_{j=1}^{n} x_j w_j \leq \sum_{j=1}^{n} x_j \ \ln\left(\frac{x_j}{p_j}\right) + \ln\left(\sum_{j=1}^{n} p_j \exp\,(w_j)\right). \tag{4.122}$$

With the convention of $0 \times \ln 0 = 0$, the above inequality is true for all $x_j \geq 0$, $\sum_{j=1}^{n} x_j = 1$ and $w_j \in R$. The equality holds if and only if there exists a constant K such that $p_j \exp\,(w_j) = K \times x_j$, for $j = 1, ..., n$.

Replacing x_j by x_j^0, w_j by w_j^0, and p_j by p_j^0 gives

$$\sum_{j=1}^{n} x_j^0 w_j^0 \leq \sum_{j=1}^{n} x_j^0 \ \ln\left(\frac{x_j^0}{p_j^0}\right) + \ln\left(\sum_{j=1}^{n} p_j^0 \exp\,(w_j^0)\right). \tag{4.123}$$

Moreover, for any $\lambda_i > 0$, replacing x_j by x_j^i, w_j by w_j^i/λ_i, and p_j by p_j^i produces

$$\sum_{j=1}^{n} x_j^i w_j^i \leq \lambda_i \sum_{j=1}^{n} x_j^i \ \ln\left(\frac{x_j^i}{p_j^i}\right) + \lambda_i \ln\left(\sum_{j=1}^{n} p_j^i \exp\left(\frac{w_j^i}{\lambda_i}\right)\right), \tag{4.124}$$

for $i = 1, 2, ..., m$.

As λ_i decreases to 0, we have the following result:

Lemma 4.1. $\lim\limits_{\lambda_i \to 0+} \lambda_i \ln\left(\sum\limits_{j=1}^{n} p_j^i \exp\left(\frac{w_j^i}{\lambda_i}\right)\right) = \max\limits_j w_j^i$, for every $\mathbf{w}^i \in R^n$, where $\max\limits_j w_j^i$ is defined as the maximum of $w_1^i, w_2^i, ..., w_n^i$.

Proof. Without loss of generality, we may assume that $w_1^i = \max\limits_j w_j^i$. Then,

$$\begin{aligned}
\lambda_i \ln(\sum_{j=1}^{n} p_j^i \exp(\frac{w_j^i}{\lambda_i})) &= \lambda_i \ln\left[\exp\left(\frac{w_1^i}{\lambda_i}\right)\left(\sum_{j=2}^{n} p_j^i \exp(\frac{w_j^i - w_1^i}{\lambda_i}) + p_1^i\right)\right] \\
&= \lambda_i \ln(\exp(\frac{w_1^i}{\lambda_i})) + \lambda_i \ln\left[\sum_{j=2}^{n} p_j^i \exp(\frac{w_j^i - w_1^i}{\lambda_i}) + p_1^i\right] \\
&= w_1^i + \lambda_i \ln\left[\sum_{j=2}^{n} p_j^i \exp\left(\frac{w_j^i - w_1^i}{\lambda_i}\right) + p_1^i\right].
\end{aligned}$$

Now, for any given $\mathbf{w}^i \in R^n$ and $\lambda_i \geq 0$, since $w_1^i = \max\limits_j w_j^i$, we have $0 \leq \exp(\frac{w_j^i - w_1^i}{\lambda_i}) \leq 1$. Hence

$$p_1^i \leq \sum_{j=2}^{n} p_j^i \exp\left(\frac{w_j^i - w_1^i}{\lambda_i}\right) + \ p_1^i \leq \sum_{j=1}^{n} p_j^i = 1.$$

Consequently,

$$\lim_{\lambda_i \to 0+} \lambda_i \ln\left[\sum_{j=2}^{n} p_j^i \exp\left(\frac{w_j^i - w_1^i}{\lambda_i}\right) + p_1^i\right] = 0,$$

and the lemma follows.

□

Now we can claim that Inequality (4.124) remains valid for $\lambda_i \to 0+$, since in this case the left-hand side is the average value of $\{w_j^i\}$ and the right-hand side becomes the maximum value of $\{w_j^i\}$. In other words, we have shown that, for $i = 1, 2, ..., m$,

$$\sum_{j=1}^{n} x_j^i w_j^i \leq \lambda_i \sum_{j=1}^{n} x_j^i \ln\left(\frac{x_j^i}{p_j^i}\right) + \lambda_i \ln\left(\sum_{j=1}^{n} p_j^i \exp\left(\frac{w_j^i}{\lambda_i}\right)\right), \tag{4.125}$$

for any $\mathbf{w}^i \in R^n$, $x_j^i \geq 0$, $\sum_{j=1}^{n} x_j^i = 1$, and $\lambda_i \geq 0$.

Suppose that W is the null space of the matrix $\mathbf{M}^T$, i.e.,

$$W \equiv \{\mathbf{w} \epsilon R^{(m+1)n} | \mathbf{M}^T \mathbf{w} = \mathbf{0}\} = \{\mathbf{w} \epsilon\ R^{(m+1)n} | w_j^0 = -\sum_{i=1}^{m} w_j^i,\ j = 1, ..., n\}.$$

If we let $\mathbf{x}$ be a feasible solution to Program EE′ and $\mathbf{w} \in W$, $\boldsymbol{\lambda} \geq \mathbf{0}$, then we have

$$\langle \mathbf{x}, \mathbf{w} \rangle = \sum_{i=0}^{m} \sum_{j=1}^{n} x_j^i w_j^i = 0, \tag{4.126}$$

and

$$G_i(\mathbf{x}) = \sum_{j=1}^{n} x_j^i \ln\left(\frac{x_j^i}{p_j^i}\right) - e_i \leq 0, \quad i = 1, 2, ..., m. \tag{4.127}$$

Starting with Inequality (4.123) and applying Inequalities (4.126), (4.125) and (4.127) sequentially, we can show that

$$G_0(\mathbf{x}) = \sum_{j=1}^{n} x_j^0 \ln\left(\frac{x_j^0}{p_j^0}\right)$$

$$\geq -\ln\left(\sum_{j=1}^{n} p_j^0 \exp(w_j^0)\right) - \sum_{i=1}^{m} \lambda_i \left(\ln\left(\sum_{j=1}^{n} p_j^i \exp\left(\frac{w_j^i}{\lambda_i}\right)\right) + e_i\right). \tag{4.128}$$

Based on this inequality, we can define a "geometric dual program" of Program EE′ as follows:

Program DEE′ :

$$\max V(\mathbf{w}, \boldsymbol{\lambda}) = -\ln\left(\sum_{j=1}^{n} p_j^0 \exp(w_j^0)\right) - \sum_{i=1}^{m} \lambda_i \left[\ln\left(\sum_{j=1}^{n} p_j^i \exp\left(\frac{w_j^i}{\lambda_i}\right)\right) + e_i\right] \tag{4.129}$$

$$\text{s.t.} \quad \mathbf{M}^T \mathbf{w} = \mathbf{0}, \tag{4.130}$$

$$\lambda_i \geq 0, \quad i = 1, 2, ..., m. \tag{4.131}$$

Several observations can be made here.

Observation 4.8: As an immediate consequence of Inequality (4.128), we can easily show the following "weak duality theorem."

Theorem 4.10. If $\mathbf{x}$ is a primal feasible solution of Program EE′ and $(\mathbf{w}, \boldsymbol{\lambda})$ is a dual feasible solution of Program DEE′, then $V(\mathbf{w}, \boldsymbol{\lambda}) \leq G(\mathbf{x})$.

Observation 4.9: By the definition of $\mathbf{w} \in W$, we can easily see that Program DEE′ is actually equivalent to the following program of $m(n+1)$ variables with simple lower bound constraints:

Program $\overline{\text{DEE}'}$:

$$\max \bar{V}(\bar{\mathbf{w}}, \boldsymbol{\lambda}) = -\ln\left(\sum_{j=1}^{n} p_j^0 \exp\left(-\sum_{i=1}^{m} w_j^i\right)\right) - \sum_{i=1}^{m} \lambda_i \left[\ln\left(\sum_{j=1}^{n} p_j^i \exp\left(\frac{w_j^i}{\lambda_i}\right)\right) + e_i\right] \tag{4.132}$$

$$\text{s.t.} \quad w_j^i \in R, \quad i = 1, 2, ..., m, \quad j = 1, 2, ..., n, \tag{4.133}$$

$$\lambda_i \geq 0, \quad i = 1, 2, ..., m, \tag{4.134}$$

where $\bar{\mathbf{w}} = \times_{i=1}^{m} \mathbf{w}^i$.

Therefore, there is always a feasible solution that is in the relative interior of the dual feasible domain. Such a solution is called a "canonical solution" as defined in [6]. Therefore, we know that the following result is true.

Theorem 4.11. Program DEE$'$ is a canonical program that is always consistent.

Observation 4.10: The following theorem is a direct consequence of Observations 4.8 and 4.9:

Theorem 4.12. If Program EE$'$ is consistent, then Program DEE$'$ has a finite optimum value.

Notice that Program $\overline{\text{DEE}'}$ has trivial constraints, but its objective function $\bar{V}(\bar{\mathbf{w}}, \boldsymbol{\lambda})$ is neither convex nor concave over the feasible region. In order to utilize the theory of convex analysis [6, 10] in proving the strong duality theorem, we will focus on Program DEE$'$. Just like other geometric duals [6, 10, 7], it is easy to see that $V(\mathbf{w}, \boldsymbol{\lambda})$ is concave over the feasible region of Program DEE$'$. However, it is not differentiable at the boundaries of the constraints $\lambda_i \geq 0$, for $i = 1, 2, ..., m$. This issue could cause numerical difficulties in computation, and it will be explicitly addressed in the next subsection.

4.3.2 Strong Duality Theorem

Since the dual program DEE$'$ is a concave program with simple linear constraints, the dual approach is computationally attractive. However, as we observed in the last subsection, the dual problem is nondifferentiable at some of its boundary points. In order to take full advantage of the dual approach, we apply the recently developed "controlled dual perturbation" method [10, 7, 12] to perturb constraints (4.131) away from its boundaries by a well-controlled positive amount l_i, for $i = 1, 2, ..., m$, and construct the following "perturbed dual" program:

Program DEE'_l :

$$\max V(\mathbf{w},\boldsymbol{\lambda}) = -\ln\left(\sum_{j=1}^{n} p_j^0 \exp(w_j^0)\right) - \sum_{i=1}^{m} \lambda_i \left[\ln\left(\sum_{j=1}^{n} p_j^i \exp\left(\frac{w_j^i}{\lambda_i}\right)\right) + e_i\right] \quad (4.135)$$

$$\text{s.t.} \quad \mathbf{M}^T\mathbf{w} = \mathbf{0}, \quad (4.136)$$

$$\lambda_i \geq l_i, \quad i = 1, 2, ..., m, \quad (4.137)$$

where the notation is defined as before and the positive l_i's are called "perturbations".

Notice that Theorem 4.11 indicates that Program DEE′ is canonical with a feasible solution $(\mathbf{w}^+, \boldsymbol{\lambda}^+)$ such that $\lambda_i^+ > 0$, for $i = 1, ..., m$. Therefore, we can choose a perturbation vector $\boldsymbol{l} = (l_1, ..., l_m)$, such that $\lambda_i^+ \geq l_i > 0$, for $i = 1, ..., m$. In this way, the corresponding Program DEE'_l is always feasible. Moreover, Program DEE'_l has a concave differentiable objective function with linear constraints. This makes the dual approach very favorable since any general purpose optimizer like MINOS [20] can be used to solve it without much difficulty. However, we have to show that there is no duality gap first.

We achieve this by proving the following result:

Theorem 4.13. If Program DEE′ has a finite supremum, then Program EE′ is consistent. Moreover, for any given $\epsilon > 0$, we can choose a proper perturbation vector $\boldsymbol{l}(\epsilon) = (l(\epsilon)_1, ..., l(\epsilon)_m)$ with $l(\epsilon)_i > 0$, for $i = 1, ..., m$, such that the perturbed dual program $\text{DEE}'_{l(\epsilon)}$ has an optimal solution $(\mathbf{w}, \boldsymbol{\lambda})$ and the Program EE′ has a feasible solution $\mathbf{x}$ satisfying the condition $0 \leq G(\mathbf{x}) - V(\mathbf{w}, \boldsymbol{\lambda}) \leq \epsilon$.

Proof. Let $(\mathbf{w}^+, \boldsymbol{\lambda}^+)$ be a feasible solution of Program DEE′ with $\lambda_i^+ > 0$, for $i = 1, ..., m$, and consider Program DEE'_l with $l_i < \lambda_i^+$. Since Program DEE′ has a finite supremum, so does Program DEE'_l. Now, consider the Lagrangian defined as

$$L = V(\mathbf{w},\boldsymbol{\lambda}) + \sum_{i=1}^{m} \mu_i(\lambda_i - l_i) + \gamma^T \mathbf{M}^T \mathbf{w}, \quad (4.138)$$

where γ (a column vector of size n) and $\mu_i \geq 0,\ i = 1, 2, ..., m$ are Lagrange multipliers of the constraints (4.136) and (4.137), respectively.

For the differentiable concave program DEE$'_l$, there exist solutions to the following K.K.T. conditions:

$$\frac{\partial L}{\partial w_j^0} = -\left[\frac{p_j^0 \exp\left(w_j^0\right)}{\sum_{k=1}^{n} p_k^0 \exp\left(w_k^0\right)}\right] + \gamma^T \mathbf{M}_j = 0,\ j = 1, ..., n, \tag{4.139}$$

$$\frac{\partial L}{\partial w_j^i} = -\left[\frac{p_j^i \exp(w_j^i/\lambda_i)}{\sum_{k=1}^{n} p_k^i \exp(w_k^i/\lambda_i)}\right] + \gamma^T \mathbf{M}_j = 0,\ \forall\ i, j, \tag{4.140}$$

$$\begin{aligned}\frac{\partial L}{\partial \lambda_i} &= \left[\frac{\sum_{j=1}^{n}(w_j^i/\lambda_i)p_j^i \exp(w_j^i/\lambda_i)}{\sum_{k=1}^{n} p_k^i \exp(w_k^i/\lambda_i)}\right] \\ &\quad - \ln\left(\sum_{k=1}^{n} p_k^i \exp\left(\frac{w_k^i}{\lambda_i}\right)\right) - e_i + \mu_i = 0,\ \forall\ i,\end{aligned} \tag{4.141}$$

$$\mu_i(\lambda_i - l_i) = 0, \quad i = 1, 2, ..., m, \tag{4.142}$$

$$\mu_i \geq 0, \quad i = 1, 2, ..., m, \tag{4.143}$$

plus conditions (4.136) and (4.137), where $\mathbf{M}_j$ (a column vector of size n) is the jth column of matrix $\mathbf{M}^T$. Also notice that from (4.115), we know that $\mathbf{M}^T$ is an $n \times (m+1)n$ matrix whose jth, $(n+j)$th, ..., $(mn+j)$th columns are identical.

Let $\mathbf{x} = \times_{i=0}^{m} \mathbf{x}^i$, where $\mathbf{x}^i$ is an n-vector with

$$x_j^i = \gamma^T \mathbf{M}_j, \quad \text{for } i = 0, 1, ..., m \text{ and } j = 1, 2, ..., n. \tag{4.144}$$

Then, by definition (4.115), we know $\mathbf{x} \in X$ (the column space of $\mathbf{M}$). Moreover, by conditions (4.139) and (4.140), we know $x_j^i \geq 0$ and $\sum_{j=1}^{n} x_j^i = 1$, for $i = 0, 1, ..., m$ and $j = 1, ..., n$. In addition, condition (4.140) implies that

$$x_j^i = \frac{p_j^i \exp(w_j^i/\lambda_i)}{\sum_{k=1}^{n} p_k^i \exp(w_k^i/\lambda_i)}, \quad i = 1, ..., m, \quad j = 1, ...n. \tag{4.145}$$

Therefore, we have

$$\ln\left(\frac{x_j^i}{p_j^i}\right) = \frac{w_j^i}{\lambda_i} - \ln\left[\sum_{k=1}^n p_k^i \exp\left(\frac{w_k^i}{\lambda_i}\right)\right]. \tag{4.146}$$

Plugging (4.145) and (4.146) into condition (4.141) gives

$$\sum_{j=1}^n \left[\ln(\frac{x_j^i}{p_j^i}) + \ln\sum_{k=1}^n p_k^i \exp(\frac{w_k^i}{\lambda_i})\right] x_j^i - \ln\sum_{k=1}^n p_k^i \exp(\frac{w_k^i}{\lambda_i}) - e_i \ + \ \mu_i = 0.$$

Hence,

$$\sum_{j=1}^n \left[x_j^i \ln(\frac{x_j^i}{p_j^i})\right] + \left[\ln\sum_{k=1}^n p_k^i \exp(\frac{w_k^i}{\lambda_i})\right] (\sum_{j=1}^n x_j^i)$$
$$- \ln\sum_{k=1}^n p_k^i \exp(\frac{w_k^i}{\lambda_i}) - e_i + \mu_i = 0. \tag{4.148}$$

Since $\sum_{j=1}^n x_j^i = 1$, the above equation reduces to

$$\sum_{j=1}^n \left[x_j^i \ln(\frac{x_j^i}{p_j^i})\right] \ - \ e_i \ + \ \mu_i \ = 0, \quad i = 1, 2, ...m. \tag{4.149}$$

Since $\mu_i \geq 0$, $\mathbf{x}$ satisfies condition (4.117). Therefore, Program EE$'$ has a feasible solution $\mathbf{x}$.

To prove the second part of this theorem, let us consider the right-hand side of the Lagrangian defined by Equation (4.138). Since condition (4.142) holds and $\mathbf{x}$ and $\mathbf{w}$ are complementary, we have

$$L = V(\mathbf{w}, \boldsymbol{\lambda}).$$

On the other hand, substituting the value of $V(\mathbf{w}, \boldsymbol{\lambda})$, as defined by Equation (4.135), on the right-hand side of Equation (4.138) gives

$$L = \ - \ \ln(\sum_{j=1}^n p_j^0 \exp\ (w_j^0)) \ - \sum_{i=1}^m \lambda_i \left[\ln(\sum_{j=1}^n p_j^i \exp(\frac{w_j^i}{\lambda_i})) + e_i\right]$$
$$+ \ \sum_{i=1}^m \mu_i(\lambda_i - l_i) + \gamma^T \mathbf{M}^T \mathbf{w}.$$

By Equation (4.144) and after rearranging terms, we have

$$\begin{aligned} V(\mathbf{w},\boldsymbol{\lambda}) &= L \\ &= \sum_{j=1}^{n} x_j^0 w_j^0 - \ln(\sum_{j=1}^{n} p_j^0 \exp\,(w_j^0)) \\ &\quad + \sum_{i=1}^{m} \left[\sum_{j=1}^{n} x_j^i w_j^i - \lambda_i \left(\ln(\sum_{j=1}^{n} p_j^i \exp(\frac{w_j^i}{\lambda_i})) + e_i \right) - \mu_i \right] \\ &\quad - \sum_{i=1}^{m} \mu_i l_i. \end{aligned} \tag{4.150}$$

By applying Equation (4.139) and noticing that $\sum_{j=1}^{n} p_j^0 \exp(w_j^0)$ is a constant, we know that Inequality (4.123) becomes an equality and the first bracket in Equation (4.150) becomes $G_0(\mathbf{x})$. Similarly, by applying (4.140) and noticing that $\sum_{j=1}^{n} p_j^i \exp(w_j^i)$ is a constant, Inequality (4.124) becomes an equality, for $i = 1, ..., m$. Moreover, Equation (4.149) implies that the second bracket in Equation (4.150) becomes 0. Hence,

$$V(\mathbf{w},\boldsymbol{\lambda}) = G_0(\mathbf{x}) \ - \sum_{i=1}^{m} \mu_i l_i.$$

By Theorem 4.10, we further have

$$0 \le G_0(\mathbf{x}) - V(\mathbf{w}) = \sum_{i=1}^{m} \mu_i l_i. \tag{4.151}$$

Now substitute the components of the vector $(\mathbf{w}^+, \boldsymbol{\lambda}^+)$ in Inequalities (4.123) and (4.124). By adding these inequalities together and using condition (4.149), for $i = 1, ..., m$, we have

$$0 \le -\sum_{i=1}^{m} \mu_i \lambda_i^+ - V(\mathbf{w}^+, \boldsymbol{\lambda}^+) + G_0(\mathbf{x}). \tag{4.152}$$

Substituting for $G_0(\mathbf{x})$ based on (4.151), we get

$$0 \le \sum_{i=1}^{m} \mu_i(\lambda_i^+ \ - \ l_i) \le V(\mathbf{w},\boldsymbol{\lambda}) - V(\mathbf{w}^+,\boldsymbol{\lambda}^+) \le V - V(\mathbf{w}^+,\boldsymbol{\lambda}^+), \tag{4.153}$$

where V is an upper bound for the dual program as assumed for this theorem. Now, for any given $\epsilon > 0$, if we choose l_i such that

$$l_i = l_i(\epsilon) = \frac{\epsilon\delta\lambda_i^+}{V - V(\mathbf{w}^+, \boldsymbol{\lambda}^+) + \epsilon}, \tag{4.154}$$

for any $0 < \delta < 1$ and $i = 1, 2, ..., m$, then there are two possible cases for consideration.

Case 1: If $V - V(\mathbf{w}^+, \boldsymbol{\lambda}^+) = 0$, then since $l_i = \delta\lambda_i^+$ and $\lambda_i^+ - l_i = (1-\delta)\lambda_i^+ > 0$, for every i, μ_i in (4.153) has to be 0. Consequently, by (4.151), we have $G_0(\mathbf{x}) - V(\mathbf{w}, \boldsymbol{\lambda}) = 0$.

Case 2: If $V - V(\mathbf{w}^+, \boldsymbol{\lambda}^+) > 0$, by (4.154), we obtain $\lambda_i^+ = [(V - V(\mathbf{w}^+, \boldsymbol{\lambda}^+) + \epsilon)/\epsilon\delta]l_i$. Plugging this into (4.153), we obtain

$$\sum_{i=1}^{m} \mu_i \left[\frac{V - V(\mathbf{w}^+, \boldsymbol{\lambda}^+)}{\delta\epsilon} + \frac{1}{\delta} - 1\right] l_i \leq V - V(\mathbf{w}^+, \boldsymbol{\lambda}^+).$$

This implies that $\sum_{i=1}^{m} \mu_i l_i \leq \delta\epsilon < \epsilon$. Therefore, by (4.151), we have $0 \leq G_0(\mathbf{x}) - V(\mathbf{w}, \boldsymbol{\lambda}) < \epsilon$. This completes the proof.

□

Now we can prove the "strong duality theorem" (or "existence theorem") for our minimum cross-entropy problem.

Theorem 4.14. Program EE′ is consistent if and only if its dual Program DEE′ has a finite optimum value. In this case, Programs EE′ and DEE′ attain a common optimum value.

Proof. By combining Theorems 4.12 and 4.13, we know that Program EE′ has a feasible solution if and only if Program DEE′ has a finite optimum value. Moreover, for any given $\epsilon > 0$, we can find feasible solutions $\mathbf{x}$ and $(\mathbf{w}, \boldsymbol{\lambda})$ such that $0 \leq G_0(\mathbf{x}) - V(\mathbf{w}, \boldsymbol{\lambda}) \leq \epsilon$. Hence we have $V(\mathbf{w}, \boldsymbol{\lambda}) \leq V^* \leq G^* \leq G_0(\mathbf{x})$, where V^* and G^* are optimum values of DEE′ and EE′, respectively. Consequently, $0 \leq G^* - V^* \leq G_0(\mathbf{x}) - V(\mathbf{w}, \boldsymbol{\lambda}) \leq \epsilon$, for any $\epsilon > 0$. Therefore, $G^* = V^*$.

□

4.3.3 Dual-to-Primal Conversion

A dual-to-primal conversion formula is embedded in the proof of Theorem 4.13. If $(\mathbf{w}, \boldsymbol{\lambda})$ is a solution to Program $\text{DEE}'_{l(\epsilon)}$, we know, from Equations (4.139), (4.144) and (4.145), that

$$x_j^0 = \frac{p_j^0 \exp(w_j^0))}{\sum_{k=1}^n p_k^0 \exp\left(w_k^0\right)}, \quad j = 1, ..., n, \tag{4.155}$$

and

$$x_j^i = \frac{p_j^i \exp(w_j^i/\lambda_i)}{\sum_{k=1}^n p_k^i \exp(w_k^i/\lambda_i)}, \quad i = 1, ..., m, \quad j = 1, ...n. \tag{4.156}$$

form an ϵ-optimal solution to Program EE$'$. Hence we have the following theorem:

Theorem 4.15. For any given $\epsilon > 0$, if $\boldsymbol{l}(\epsilon)$ is defined by Equation (4.154) and $(\mathbf{w}, \boldsymbol{\lambda})$ solves Program $\text{DEE}'_{l(\epsilon)}$, then Equations (4.155) and (4.156) generate an ϵ-optimal solution to Program EE$'$.

4.3.4 Computational Issues

Based on the results shown in previous subsections, we outline a dual based computation procedure as follows:

Algorithm 4.4: (Dual Perturbation Method – Entropy Constraints)

Step 1: Given a small number $\epsilon > 0$, a dual feasible vector $(\mathbf{w}^+, \boldsymbol{\lambda}^+)$ with $\lambda_i^+ > 0$, for $i = 1, ..., m$, and a dual upper bound V of $V(\mathbf{w}, \boldsymbol{\lambda})$;
Step 2: Choose a perturbation vector $\boldsymbol{l}\ (\epsilon) = (l(\epsilon)_1, ..., l(\epsilon)_n)$ according to Equation (4.154).
Step 3: Find an optimal solution (w_j^i, λ_i) for the perturbed dual program $\text{DEE}'_{l(\epsilon)}$.
Step 4: Compute an $\epsilon-$optimal solution $\mathbf{x}$ according to Equations (4.155) and (4.156). More precisely, x_j^0, $j = 1, 2, ..., n$, is an $\epsilon-$optimal solution to Program EE and

$$0 \le G_0(\mathbf{x}) - V(\mathbf{w}, \boldsymbol{\lambda}) \le \epsilon.$$

Step 5: Stop.

Several observations can be made here:

Observation 4.11: Notice that Program DEE'_l is equivalent to the following:

Program $\overline{\text{DEE}'_l}$:

$$\max\ \bar{V}(\bar{\mathbf{w}}, \boldsymbol{\lambda}) = -\ln\left(\sum_{j=1}^{n} p_j^0 \exp(-\sum_{i=1}^{m} w_j^i)\right)$$
$$-\sum_{i=1}^{m} \lambda_i \left[\ln\left(\sum_{j=1}^{n} p_j^i \exp(\frac{w_j^i}{\lambda_i})\right) + e_i\right] \quad (4.157)$$
$$\text{s.t.} \quad w_j^i \epsilon R, \quad i = 1, 2, ..., m, \quad j = 1, 2, ..., n, \quad (4.158)$$
$$\lambda_i \geq l_i, \quad i = 1, 2, ..., m, \quad (4.159)$$

which has only simple lower bound constraints.

If w_j^i and λ_i, $i = 1, ..., m$ and $j = 1, ..., n$, form a solution to Program $\overline{\text{DEE}'_l}$, then it is easy to see that w_j^0, w_j^i and λ_i form a solution to Program DEE'_l where

$$w_j^0 = -\sum_{i=1}^{m} w_j^i. \quad (4.160)$$

Observation 4.12: In Step 3, we can either solve Program DEE'_l or $\overline{\text{DEE}'_l}$. The former is a differentiable concave program with simple linear constraints while the latter is a differentiable nonconcave (and nonconvex) program with simple lower bound constraints. Compared to the primal program P, both of these two programs are relatively simple and attractive. We may use a commercial nonlinear optimizer, e.g., MINOS, to solve this dual problem. The additional customization of computer codes is minimal.

According to our experience, it is advisable to solve Program DEE'_l. Although it has linear constraints, the convexity property of its objective function will reduce the computational effort.

Observation 4.13: $(\mathbf{w}^+, \boldsymbol{\lambda}^+)$ in Step 1 can be easily chosen, since there are only lower bound constraints in Program $\overline{\text{DEE}'}$.

Observation 4.14: Usually a dual bound V can be estimated easily. If it is not provided, we can always solve a Program DEE$'_l$ with $l_i = \lambda_i^+$ first, with arbitrary λ_i^+, and find a dual feasible solution. Then, we can use Equations (4.144) and (4.156) to find a corresponding primal solution $\mathbf{x}^*$ and dual upper bound $V = G_0(\mathbf{x}^*)$. By the definition of $\mathbf{x}$, we know that $\mathbf{x}^0 = \mathbf{q}$. Hence, $\mathbf{x}^0$ obtained in Step 4 is an ϵ-optimal solution to Program EE.

To demonstrate the computation procedure, we provide two examples and use MINOS [20] for solving the dual program in these examples.

Example 4.1: In this example, Program EE has 10 variables and a single entropy constraint. The *a priori* probability distribution associated with the objective function is:

$$\mathbf{p}^0 = (0.01,0.19,0.05,0.15,0.10,0.10,0.10,0.10,0.05,0.15).$$

That associated with the entropy constraint is:

$$\mathbf{p}^1 = (0.05,0.05,0.05,0.05,0.05,0.15,0.15,0.15,0.15,0.15).$$

The error tolerance for the single constraint is given by

$$e_1 = 0.0002,$$

and the error bound for the duality gap is given by

$$\epsilon = 0.00001.$$

The Dual Program DEE$'_l$ can be easily verified to be the following:

$$\begin{aligned}
\max \quad & V(\mathbf{w}, \boldsymbol{\lambda}) = -\ln\Big(\sum_{j=1}^{10} p_j^0 \exp(w_j^0)\Big) - \lambda_1 \left[\ln\Big(\sum_{j=1}^{10} p_j^1 \exp\Big(\frac{w_j^1}{\lambda_1}\Big)\Big) + e_1 \right] \\
\text{s.t.} \quad & [\mathbf{I}_{10}, \mathbf{I}_{10}]\mathbf{w} = \mathbf{0}, \\
& \lambda_1 \geq l_1 > 0,
\end{aligned}$$

where $\mathbf{I}_{10}$ is the identity matrix as defined in (4.115).

First we need a dual upper bound V and a dual feasible solution. Since neither is provided, following Observation 4.14, we pick $l_1 = 0.1$ and solve Program DEE$'_l$ via MINOS. This gives us a dual solution $(\mathbf{w}^+, \boldsymbol{\lambda}^+)$ whose components are:

1	1.169496	2	−1.690627	3	−0.393854	4	−1.461006
5	1.067153	6	0.000000	7	0.000000	8	0.000000
9	0.673298	10	0.393854	11	−1.169496	12	1.690627
13	0.393854	14	1.461006	15	1.067153	16	0.000000
17	0.000000	18	0.000000	19	−0.673298	20	0.393854,

$\lambda_1^+ = 33.921$.

Its dual objective value $V(\mathbf{w}) = 0.25764$.

Now, we use Equations (4.155) and (4.156) to find a corresponding primal solution $\mathbf{x}$ of Program EE′. Notice that $\mathbf{x}$ has 20 components, instead of 10 in Program EE, and the values are:

1	0.04811	2	0.05234	3	0.05038	4	0.05199
5	0.05139	6	0.14940	7	0.14940	8	0.14940
9	0.14646	10	0.15114	11	0.04811	12	0.05234
13	0.05038	14	0.05199	15	0.05139	16	0.14940
17	0.14940	18	0.14940	19	0.14646	20	0.15114.

We also have primal objective value $G_0(\mathbf{x}) = 0.25764$. Furthermore, $G_1(\mathbf{x}) = 0$. Since the primal objective value is exactly equal to the dual objective value $V(\mathbf{w})$, we have generated an optimal solution to Program EE with $q_j = x_j$, $j = 1, ..., 10$, and objective value 0.25764, in one step.

Example 4.2: In this example, Program EE has 10 variables and 2 entropy constraints. The *a priori* probability distribution associated with the objective function is:

$$\mathbf{p}^0 = (0.01,0.19,0.05,0.15,0.10,0.10,0.10,0.10,0.05,0.15).$$

That associated with the first entropy constraint is:

$$\mathbf{p}^1 = (0.05,0.05,0.05,0.05,0.05,0.15,0.15,0.15,0.15,0.15).$$

Its counterpart associated with the second entropy constraint is:

$$\mathbf{p}^2 = (0.05,0.05,0.05,0.05,0.05,0.16,0.14,0.15,0.15,0.15).$$

Let the required error tolerances for the constraints 1 and 2 be

$e_1 = 0.0002$ and $e_2 = 0.0002$,

and the given error bound for the duality gap be

$\epsilon = 0.00001.$

The Dual Program DEE'_l is

$$\max \quad V(\mathbf{w}, \boldsymbol{\lambda}) = -\ln(\sum_{j=1}^{10} p_j^0 \exp(w_j^0)) - \lambda_1 \left[\ln(\sum_{j=1}^{10} p_j^1 \exp(\frac{w_j^1}{\lambda_1})) + e_1\right]$$
$$-\lambda_2 \left[\ln(\sum_{j=1}^{10} p_j^2 \exp(\frac{w_j^2}{\lambda_2})) + e_2\right]$$
$$\text{s.t.} \quad [\mathbf{I}_{10}, \mathbf{I}_{10}, \mathbf{I}_{10}]\mathbf{w} = \mathbf{0},$$
$$\lambda_1 \geq l_1 > 0,$$
$$\lambda_2 \geq l_2 > 0.$$

Step 1: First, we need a dual upper bound V and a dual feasible solution. Since neither is provided, following Observation 4.14, this time we pick $l_1 = 100.0$ and $l_2 = 100.0$ and solve Program DEE'_l via MINOS. A dual solution $(\mathbf{w}^+, \boldsymbol{\lambda}^+)$ is obtained with the following components:

1	1.347289	2	−1.582501	3	−0.254142	4	−1.347289
5	−0.943841	6	0.181414	7	0.114981	8	0.149306
9	0.839004	10	−0.254142	11	0.000000	12	1.464895
13	0.800715	14	1.347289	15	1.145565	16	3.809863
17	−2.833490	18	0.598991	19	0.254142	20	0.800715
21	−1.347289	22	0.117606	23	−0.546573	24	0.000000
25	−0.201724	26	−3.991278	27	2.718509	28	−0.748297
29	−1.093147	30	−0.546573,				

$\lambda_1^+ = 100.0,\ \lambda_2^+ = 100.0.$

Similar to Example 4.1, we use the dual feasible solution $(\mathbf{w}^+, \boldsymbol{\lambda}^+)$ to find an upper bound $V = 0.26912$ and $V(\mathbf{w}^+, \boldsymbol{\lambda}^+) = 0.26371$. Notice that the duality gap is 0.00541, which is not within the given $\epsilon = 0.00001$. Hence, we go to the next step.

Step 2: Equation (4.154) provides a perturbation vector

$$\boldsymbol{l}(\epsilon) = (0.184111, 0.18411).$$

Step 3: The solution (w_j^i, λ_i) to the perturbed dual program $\mathrm{DEE}'_{l(\epsilon)}$ is obtained via MINOS. The components are:

1	5.791906	2	2.881831	3	4.201251	4	3.115461
5	3.516194	6	4.634242	7	4.567501	8	4.601984
9	5.287042	10	4.201251	11	−3.811377	12	−2.372975
13	−3.025143	14	−2.488454	15	−2.686530	16	−1.872991
17	−4.666637	18	−3.223218	19	−3.561831	20	−3.025143
21	−1.980529	22	−0.508856	23	−1.176109	24	−0.627007
25	−0.829664	26	−2.761251	27	0.099136	28	−1.378766
29	−1.725211	30	−1.176109,				

$\lambda_1 = 41.858074$, $\lambda_2 = 42.826271$.

The dual objective value of this solution is $V(\mathbf{w}, \boldsymbol{\lambda}) = 0.26590$. Moreover, substituting this in Equations (4.155) and (4.156) produces the following primal solution $\mathbf{x}$ with 30 components:

1	0.04923	2	0.05095	3	0.05017	4	0.05081
5	0.05057	6	0.15470	7	0.14471	8	0.14979
9	0.14858	10	0.15050	11	0.04923	12	0.05095
13	0.05017	14	0.05081	15	0.05057	16	0.15470
17	0.14471	18	0.14979	19	0.14858	20	0.15050
21	0.04923	22	0.05095	23	0.05017	24	0.05081
25	0.05057	26	0.15470	27	0.14471	28	0.14979
29	0.14858	30	0.15050.				

Based on this solution, we have $G_0(\mathbf{x}) = 0.26590$, $G_1(\mathbf{x}) = 0$, and $G_2(\mathbf{x}) = 0$. Notice that the primal objective value is exactly equal to the dual objective value $V(\mathbf{w})$. Therefore, we have generated an optimal solution, $q_j = x_j$, $j = 1, ..., 10$, with objective value 0.26590.

4.3.5 Concluding Remarks

In this section, we have provided a geometric programming based dual approach to solving the minimum cross-entropy problems with entropy-type constraints. Essentially, this approach converts the nonlinearly constrained problem to a dual problem, which is a maximization problem with a concave objective function subject to linear constraints. The dual problem is easily solvable by using any general purpose optimization codes. Finding a solution to the original Program EE can be easily done through a simple conversion formula provided in this section. The numerical results show the practicality of this approach.

4.4 ENTROPY OPTIMIZATION WITH CONVEX CONSTRAINTS

The minimum cross-entropy problem with convex constraints to be studied in this section has the following form:

Program EC:

$$\min \quad g(\mathbf{q}) = \sum_{j=1}^{n} q_j \ln(q_j/p_j) \tag{4.161}$$

$$\text{s.t.} \quad g_i(\mathbf{q}) = h_i(\mathbf{A}_i\mathbf{q}) + \mathbf{b}_i^T\mathbf{q} + c_i \leq 0, \quad i = 1, 2, ..., m, \tag{4.162}$$

$$\sum_{j=1}^{n} q_j = 1, \tag{4.163}$$

$$\mathbf{q} \geq \mathbf{0}, \tag{4.164}$$

where $\mathbf{A}_i$ is a $k_i \times n$ matrix with full row-rank, $\mathbf{b}_i$ an n-dimensional column vector, c_i a constant, and h_i a convex function on R^{k_i}, for $i = 1, 2, \ldots, m$. We assume that $\mathbf{p} > \mathbf{0}$ is a known probability distribution.

For the sake of simplicity, we further assume in our derivation that (1) h_i is differentiable, and (2) h_i is co-finite, i.e., it is a closed proper convex function whose epi-graph contains no non-vertical half-lines [23]. The first assumption allows us to use the gradient notation in the derivation. For a general non-differentiable h_i, we only have to replace the gradient function by the subgradient set for the derivation without affecting validity. The second assumption ensures that the conjugate function h_i^* of h_i is well-defined on the whole space of R^{k_i}. For a nonco-finite h_i, it is required that the corresponding dual variables belong to the effective domain of h_i^*, and this requirement has to be included as a constraint in the dual program.

The cross-entropy optimization problems with linear constraints ([4] and Chapter 3), quadratic constraints ([12, 26], and Section 4.2), and entropic constraints ([2, 8], and Section 4.3) have been studied before. In these cases, it has been observed that a dual program of the cross-entropy minimization problem has trivial constraints and, hence, enjoys some computational advantages. The dual program can be either a Lagrangian dual ([2, 26], and Subsection 4.1.1) or a geometric dual ([12, 8], and Sections 4.2 and 4.3). Teboulle [24] extended the Lagrangian approach to obtain a dual program for the case with convex constraints. But neither the computational issues nor the duality properties were explicitly addressed.

The purpose of this section is twofold. First, we show that the geometric dual approach, used in Sections 4.2 and 4.3, can also be generalized to handle convex constraints. By using one simple geometric inequality and the conjugate inequality, the strong duality theorem and a dual-to-primal conversion formula can be explicitly derived. Second, we discuss how to combine the geometric dual with the dual perturbation method [10, 7, 12] to gain computational advantages.

Section 4.4 is organized as follows. In Subsection 4.4.1, a simple geometric inequality and the conjugate inequality are used to derive a geometric dual with concave objective and linear constraints. Then we prove the "strong duality theorem" for the minimum cross-entropy problem with convex constraints (Subsection 4.4.2). Contrary to previous research, we notice that a "partial" geometric dual program, with open lower bound constraints, is enough to accomplish the numerical advantages of the dual problem. A dual-to-primal conversion mechanism is studied in Subsection 4.4.3. The conversion formula constructs an "ϵ-optimal" solution to the original problem. A computational procedure that allows us to use a general purpose optimizer like MINOS [20] to solve the problem is proposed in Subsection 4.4.4. The effectiveness of the computational procedure is demonstrated via an example in Subsection 4.4.5. Concluding remarks are given in Subsection 4.4.6.

4.4.1 Geometric Dual

To derive a geometric dual for Program EC, we adhere to the following standard notation, which has been used extensively in the geometric programming literature [6, 12, 8]:

$$[0] = \{1, 2, ..., n\}, \qquad]0[= n, \tag{4.165}$$

$$[i] = \{\,]i-1[+1, ...,]i-1[+k_i\}, \quad]i[=]i-1[+k_i+1, \quad \text{for } i = 1, ..., m, \tag{4.166}$$

$$]m[= k. \tag{4.167}$$

With $\mathbf{I}_n$ denoting the $n \times n$ identity matrix, we can define a $k \times n$ matrix $\mathbf{M}$ by its transpose as follows:

$$\mathbf{M}^T \equiv (\mathbf{I}_n, \mathbf{A}_1^T, \mathbf{b}_1, ..., \mathbf{A}_m^T, \mathbf{b}_m). \tag{4.168}$$

Moreover, we let X denote the column space of matrix $\mathbf{M}$. Hence each $\mathbf{x} \in X$ is a k-dimensional column vector that can be expressed in the form of the Cartesian product $\mathbf{x} = \{\mathbf{x}^0\} \times_{i=1}^{m} \{\mathbf{x}^i,\ x_{]i[}\}$, where $\mathbf{x}^0 = (x_1, ..., x_{]0[})^T$ and $\mathbf{x}^i = (x_{]i-1[+1}, ..., x_{]i-1[+k_i})^T$, for $i = 1, ...m$. It is a simple routine to check that Program EC is equivalent to the following "primal" geometric program:

Program EC′ :

$$\min \quad G(\mathbf{x}) = \sum_{j \in [0]} x_j \ln(x_j/p_j) \tag{4.169}$$

$$\text{s.t.} \quad G_i(\mathbf{x}) = h_i(\mathbf{x}^i) + x_{]i[} + c_i \le 0, \quad i = 1, 2, ..., m, \tag{4.170}$$

$$\sum_{j \in [0]} x_j = 1, \tag{4.171}$$

$$x_j \ge 0, j \in [0], \tag{4.172}$$

$$\mathbf{x} \in X. \tag{4.173}$$

In order to construct a geometric dual problem, we define, corresponding to the primal vector $\mathbf{x}$, a dual vector $\mathbf{w} = \{\mathbf{w}_0\} \times_{i=1}^{m} \{\mathbf{w}_i,\ w_{]i[}\}$, where $\mathbf{w}_0 = (w_1, ..., w_{]0[})^T$ and $\mathbf{w}_i = (w_{]i-1[+1}, ..., w_{]i-1[+k_i})^T$, for $i = 1, ...m$. In this way,

$\mathbf{w}$ is also a k-dimensional column vector. Now, let us consider the arithmetic-geometric inequality used in [6]:

$$\sum_{j\epsilon[0]} p_j \exp(w_j) \;\geq\; \prod_{j\epsilon[0]} \{p_j \exp(w_j)/x_j\}^{x_j}, \tag{4.174}$$

which is valid for any $w_j \in R$ and $x_j > 0$ with $\sum_{j\epsilon[0]} x_j \;=\; 1.$

Taking the logarithm on both sides and simplifying it, we have

$$\sum_{j\epsilon[0]} x_j w_j \;\leq\; \sum_{j\epsilon[0]} x_j \;\ln(x_j/p_j) \;+\; \ln(\sum_{j\epsilon[0]} p_j \exp\,(w_j)). \tag{4.175}$$

With the convention of $0 \times \ln 0 = 0$, the above inequality is true for all $w_j \in R$, $x_j \geq 0,\ j \in [0]$, and $\sum_{j\epsilon[0]} x_j \;=\; 1$. The equality holds if and only if there exists a constant K such that $p_j \exp(w_j) = K \times x_j$, for $j \in [0]$.

Another inequality we consider is the conjugate inequality used in [23]:

$$\sum_{j\epsilon[i]} x_j z_j \;\leq\; h_i(\mathbf{x}^i) \;+\; h_i^*(\mathbf{z}^i), \quad \text{for} \quad \mathbf{x}^i, \mathbf{z}^i \in R^{k_i}, \tag{4.176}$$

where $h_i^*(\mathbf{z}^i)$ is the conjugate function of $h_i(\mathbf{x}^i)$ defined by

$$h_i^*(\mathbf{z}^i) \;= \sup_{\mathbf{x}^i} \;\{\; \langle \mathbf{x}^i, \mathbf{z}^i \rangle - h_i(\mathbf{x}^i) \;\}. \tag{4.177}$$

The notation $\langle \mathbf{x}^i, \mathbf{z}^i \rangle$ denotes the scalar product defined by $\sum_{j\epsilon[i]} x_j z_j$.

Since h_i^* is well-defined on R^{k_i}, we can replace z_j by $\frac{w_j}{w_{]i[}}$. Multiplying both sides by $w_{]i[}$, and adding $x_{]i[}w_{]i[}$ to both sides gives

$$\sum_{j\epsilon[i]} x_j w_j + x_{]i[}w_{]i[} \;\leq\; w_{]i[}\{h_i(\mathbf{x}^i) + x_{]i[} + c_i\} + \{w_{]i[}\; h_i^*(\frac{\mathbf{w}_i}{w_{]i[}}) - c_i w_{]i[}\}, \tag{4.178}$$

which holds for all $w_{]i[} > 0$ and becomes an equality if and only if

$$\frac{\mathbf{w}_i}{w_{]i[}} \;=\; \nabla h_i(\mathbf{x}^i), \quad \text{for} \quad i = 1, 2, ..., m,$$

or

$$\mathbf{x}^i \;=\; \nabla h_i^*(\frac{\mathbf{w}_i}{w_{]i[}}), \quad \text{for} \quad i = 1, 2, ..., m,$$

where ∇ is the gradient operator.

Combining Inequalities (4.175) and (4.178), for $i = 1, 2, ..., m$, we obtain

$$\sum_{j=1}^{k} x_j w_j \leq \sum_{j \epsilon [0]} x_j \ln(x_j/p_j) + \sum_{i=1}^{m} w_{]i[}\{h_i(\mathbf{x}^i) + x_{]i[} + c_i\}$$

$$+ \ln(\sum_{j \epsilon [0]} p_j \exp(w_j)) + \sum_{i=1}^{m}\{w_{]i[}\ h_i^*(\frac{\mathbf{w}_i}{w_{]i[}}) - c_i w_{]i[}\}. \tag{4.179}$$

If $\mathbf{x}$ is primal feasible, $w_{]i[} > 0$, and $\mathbf{M}^T\mathbf{w} = \mathbf{0}$, then $\sum_{j=1}^{k} x_j w_j = 0$ and $w_{]i[}\{h_i(\mathbf{x}^i) + x_{]i[} + c_i\} \leq 0$, for $i = 1, ..., m$. Consequently,

$$-\ln(\sum_{j \epsilon [0]} p_j \exp(w_j)) - \sum_{i=1}^{m}\{ w_{]i[} h_i^*(\frac{\mathbf{w}_i}{w_{]i[}}) - c_i w_{]i[} \} \leq \sum_{j \epsilon [0]} x_j \ln(\frac{x_j}{p_j}). \tag{4.180}$$

Based on this inequality, we can define a "dual geometric program" DEC$'$ for Program EC$'$ as follows:

Program DEC$'$:

$$\min \quad V(\mathbf{w}) = -\ln(\sum_{j \epsilon [0]} p_j \exp(w_j)) - \sum_{i=1}^{m}\{ w_{]i[} h_i^*(\frac{\mathbf{w}_i}{w_{]i[}}) - c_i w_{]i[}\} \tag{4.181}$$

$$\text{s.t.} \quad \mathbf{M}^T\mathbf{w} = \mathbf{0}, \tag{4.182}$$

$$w_{]i[} > 0, \quad i = 1, 2, ..., m. \tag{4.183}$$

Several interesting observations can be made here:

Observation 4.15: Note that $\mathbf{w} \in R^k$ and $\mathbf{M}$, $[i]$, $]i[$, c_i are defined as before. We know that Program DEC$'$ is a concave optimization problem with k variables, n linear constraints, and m trivial positivity constraints. However, if we define $w_j = y_j w_{]i[} =$, for $j \in [i]$ and $i = 1, ..., m$, and $\mathbf{y}^i = (y_{]i-1[+1}, ..., y_{]i-1[+k_i})^T$, then $\mathbf{y}^i \in R^{k_i}$ and constraint (4.182) becomes $\mathbf{w}_0 = -\sum_{i=1}^{m} w_{]i[}\{\mathbf{A}_i^T\mathbf{y}^i + \mathbf{b}_i\}$. Plugging this relation into the objective function of Program DEC$'$, we have an optimization problem with $\sum_{i=1}^{m} k_i$ free variables in $\mathbf{y} = \times_{i=1}^{m}\mathbf{y}^i$ and m positive variables $w_{]i[}$, $i = 1, 2, ..., m$. However, this substitution may make the dual objective function non-concave and, hence, computationally disadvantageous.

Observation 4.16: Program DEC$'$ is only a "partial dual" to Program EC$'$ because we require $w_{]i[} > 0$, instead of $w_{]i[} \geq 0$ as commonly required in the literature [12, 8, 24, 26]. This greatly reduces the difficulties of defining $w_{]i[}h_i^*(\frac{\mathbf{w}_i}{w_{]i[}})$ at $w_{]i[} = 0$ and avoids the nondifferentiability problems in computation. This point will be re-addressed in the next subsection.

Observation 4.17: Although Program DEC$'$ is only a partial dual, we can easily derive the following "weak duality theorem" as an immediate consequence of Inequality (4.180).

Theorem 4.16. If $\mathbf{x}$ is a primal feasible solution of Program EC$'$ and $\mathbf{w}$ is a dual feasible solution of Program DEC$'$, then $V(\mathbf{w}) \leq G(\mathbf{x})$.

Observation 4.18: Consider the vector $\hat{\mathbf{w}} \in R^k$ such that $\hat{\mathbf{w}}_0 = -\sum_{i=1}^m \mathbf{b}_i$, $\hat{\mathbf{w}}_i = \mathbf{0}$, and $\hat{w}_{]i[} = 1$, for $i = 1, ..., m$. It is clear that $\hat{\mathbf{w}}$ satisfies Constraints (4.182) and (4.183). Hence we have the following result:

Theorem 4.17. Program DEC$'$ always has a strictly interior feasible solution.

This property also allows us to explore the possibility of using the newly developed "interior-point methods" to solve our problem [16].

Observation 4.19: The following result is a direct consequence of Observations 4.17 and 4.18.

Theorem 4.18. If Program EC$'$ is consistent, then Program DEC$'$ has a finite optimum.

4.4.2 Strong Duality Theorem

Although the dual Program DEC$'$ involves many more variables, it has much simpler constraints. Compared to the primal program EC$'$, this makes the dual approach computationally attractive. However, as we observed before, the dual problem is nondifferentiable at some of its boundary points where $w_{]i[} = 0$. This would cause some computational difficulties as described in previous sections. Hence we would like to stay away from the boundary area with $w_{]i[} = 0$ as far as possible. On the other hand, if we restrict ourselves too far away, we may miss optimal solutions. Therefore, a delicate balance is needed.

Fortunately, the recently developed "controlled perturbation" technique [10, 8] can be extended to solve our problem. Basically, for any given tolerance $\epsilon > 0$, we can find a perturbation vector $\boldsymbol{l} = (l_1, ..., l_m)^T > 0$ such that a "perturbed dual" program provides a dual ϵ-optimal solution to Program EC$'$. The perturbed dual program is in the following form:

Program DEC$'_l$:

$$\min \quad V(\mathbf{w}) = -\ln(\sum_{j\epsilon[0]} p_j \exp(w_j)) - \sum_{i=1}^{m}\{ w_{]i[} h_i^*(\frac{\mathbf{w}_i}{w_{]i[}}) - c_i w_{]i[}\} \tag{4.184}$$

$$\text{s.t.} \quad \mathbf{M}^T\mathbf{w} = \mathbf{0}, \tag{4.185}$$

$$w_{]i[} \geq l_i, \quad i = 1, 2, ..., m. \tag{4.186}$$

Recall the way we chose $\hat{\mathbf{w}}$ in Observation 4.18. As long as we set $0 < l_i < 1$, for $i = 1, ..., m$, the corresponding Program DEC$'_l$ always has an interior feasible solution. Moreover, Program DEC$'_l$ is a concave program with differentiable objective function over a linearly constrained feasible domain. As we observed in the two previous sections, this makes the dual approach very favorable, since a general purpose optimizer like MINOS [20] can be used for solving it without much difficulty. Again, we have to first make sure that the gap between Program EC$'$ and Program DEC$'_l$ vanishes. The following theorem will lead us to this goal:

Theorem 4.19. If Program DEC$'$ has a finite optimum, then Program EC$'$ is consistent. Moreover, for any given $\epsilon > 0$, we can choose an appropriate perturbation vector $\mathbf{l}(\epsilon) > \mathbf{0}$ such that the perturbed dual program DEC$'_{l(\epsilon)}$ has an optimal solution $\mathbf{w}$ and the Program EC$'$ has a feasible solution $\mathbf{x}$ satisfying the condition $0 \leq G(\mathbf{x}) - V(\mathbf{w}) \leq \epsilon$.

Proof. Let $\mathbf{w}^+$ be a feasible solution of Program DEC$'$ with $w^+_{]i[} > 0$, for $i = 1, ..., m$. Consider Program DEC$'_l$ with $l_i < w^+_{]i[}$, for $i = 1, ...m$. Since

Program DEC′ has a finite optimum, so does Program DEC$'_l$. The Lagrangian of Program DEC$'_l$ is defined as

$$L = V(\mathbf{w}) + \sum_{i=1}^{m} \lambda_i (w_{]i[} - l_i) + \gamma^T \mathbf{M}^T \mathbf{w}, \tag{4.187}$$

where γ (a column vector of size n) and λ_i, $i = 1, 2, ..., m$, are Lagrange multipliers of the constraints (4.185) and (4.186), respectively.

Because Program DEC$'_l$ is differentiable and concave, there exist solutions to the following K.K.T. conditions:

$$\partial L / \partial w_j = -(p_j \exp(w_j)) / \sum_{k \in [0]} (p_k \exp(w_k)) + \gamma^T \mathbf{M}_j = 0, \quad \text{for } j \in [0], \tag{4.188}$$

$$\partial L / \partial w_j = -w_{]i[} \langle \nabla h_i^* (\frac{\mathbf{w}_i}{w_{]i[}}), \frac{\mathbf{e}_j^i}{w_{]i[}} \rangle + \gamma^T \mathbf{M}_j = 0, \quad \text{for } j \in [i], \quad i = 1, 2, ..., m, \tag{4.189}$$

$$\partial L / \partial w_{]i[} = c_i + \langle \nabla h_i^* (\frac{\mathbf{w}_i}{w_{]i[}}), \frac{\mathbf{w}_i}{w_{]i[}} \rangle - h_i^* (\frac{\mathbf{w}_i}{w_{]i[}}) + \lambda_i + \gamma^T \mathbf{M}_{]i[} = 0, \ i = 1, 2, ..., m, \tag{4.190}$$

$$\lambda_i (w_{]i[} - l_i) = 0, \quad i = 1, 2, ..., m, \tag{4.191}$$

$$\lambda_i \geq 0, \quad i = 1, 2, ..., m, \tag{4.192}$$

plus conditions (4.185) and (4.186), where $\nabla h_i^* (\frac{\mathbf{w}_i}{w_{]i[}})$ is the gradient of h_i^* with respect to $\mathbf{w}_i$ at $\frac{\mathbf{w}_i}{w_{]i[}}$, $\langle ., . \rangle$ is the product of two vectors, $\mathbf{e}_j^i$ is an k_i dimensional vector with 1 at its jth element and 0 elsewhere, and $\mathbf{M}_j$ (a column vector of size n) is the jth column of matrix $\mathbf{M}^T$.

Define

$$x_j = \gamma^T \mathbf{M}_j, \quad \text{for } j = 1, ..., k. \tag{4.193}$$

By definition, we know $\mathbf{x} \in X$, the column space of $\mathbf{M}$. Moreover, condition (4.188) implies that $x_j \geq 0$ and $\sum_{j \in [0]} x_j = 1$ and condition (4.189) implies that $\mathbf{x}^i = \nabla h_i^*(\frac{\mathbf{w}_i}{w_{]i[}})$, for $i = 1, ..., m$. Therefore, condition (4.190) becomes

$$\langle \mathbf{x}^i, \frac{\mathbf{w}_i}{w_{]i[}} \rangle - h_i^*(\frac{\mathbf{w}_i}{w_{]i[}}) + x_{]i[} + c_i + \lambda_i = 0, \quad i = 1, 2, ..., m. \tag{4.194}$$

Furthermore, the conjugate inequality, i.e., Inequality (4.178), turns out to be an equality. Hence, $\langle \mathbf{x}^i, \frac{\mathbf{w}_i}{w_{]i[}} \rangle = h_i(\mathbf{x}^i) + h_i^*(\frac{\mathbf{w}_i}{w_{]i[}})$ and (4.194) reduces to

$$h_i(\mathbf{x}^i) + x_{]i[} + c_i + \lambda_i = 0, \quad i = 1, 2, ..., m. \tag{4.195}$$

Since $\lambda_i \geq 0$, we know that $\mathbf{x}$ satisfies Equation (4.170). Consequently, Program EC′ has a feasible solution $\mathbf{x}$. To prove the second part of this theorem, let us consider the right-hand side of the Lagrangian defined by Equation (4.187). Since condition (4.191) holds and $\mathbf{x}$ and $\mathbf{w}$ are complementary, we have

$$L = V(\mathbf{w}). \tag{4.196}$$

Moreover, substituting $V(\mathbf{w})$ defined by (4.184) in the right-hand side of Equation (4.187) and rearranging terms, we get

$$L = -\ln(\sum_{j \in [0]} p_j \exp(w_j)) + \sum_{j \in [0]} \gamma^T \mathbf{M}_j w_j + \sum_{i=1}^{m} w_{]i[}\{c_i - h_i^*(\frac{\mathbf{w}_i}{w_{]i[}}) + \lambda_i + \gamma^T \mathbf{M}_{]i[}\}$$
$$+ \sum_{i=1}^{m} \sum_{j \in [i]} \gamma^T \mathbf{M}_j w_j - \sum_{i=1}^{m} \lambda_i l_i.$$

Equations (4.193) and (4.196) imply

$$V(\mathbf{w}) = -\ln(\sum_{j \in [0]} p_j \exp(w_j)) + \sum_{j \in [0]} x_j w_j + \sum_{i=1}^{m} w_{]i[}\{c_i - h_i^*(\frac{\mathbf{w}_i}{w_{]i[}}) + \lambda_i + x_{]i[}\}$$
$$+ \sum_{i=1}^{m} \sum_{j \in [i]} x_j w_j - \sum_{i=1}^{m} \lambda_i l_i. \tag{4.197}$$

By Equation (4.187), we know $p_j \exp(w_j) = x_j \times \sum_{j \in [0]}(p_j \exp(w_j))$, for each $j \in [0]$. But $\sum_{j \in [0]}(p_j \exp(w_j))$ is a constant. Therefore, Inequality (4.175) becomes an equality, and

$$\sum_{j \in [0]} x_j w_j = \ln(\sum_{j \in [0]} p_j \exp w_j) + \sum_{j \in [0]} x_j \ln(x_j / p_j). \tag{4.198}$$

In addition, by Equation (4.194), we see that

$$c_i \; - \; h_i^*(\frac{\mathbf{w}_i}{w_{]i[}}) \; + \; \lambda_i \; + \; x_{]i[} \; = \; - \sum_{j \epsilon [i]} x_j w_j \; / \; w_{]i[}, \tag{4.199}$$

and

$$w_{]i[} \; \{ \; c_i \; - \; h_i^*(\frac{\mathbf{w}_i}{w_{]i[}}) \; + \; \lambda_i \; + \; x_{]i[} \; \} \; = \; - \sum_{j \epsilon [i]} x_j w_j \; , \tag{4.200}$$

for $i = 1, ..., m$. Equation (4.197), after substitution using Equations (4.198) and (4.200), and Equation (4.169) imply

$$V(\mathbf{w}) \; = \; G(\mathbf{x}) \; - \; \sum_{i=1}^{m} \lambda_i l_i. \tag{4.201}$$

By Theorem 4.16, we further have

$$0 \; \leq \; G(\mathbf{x}) \; - \; V(\mathbf{w}) \; = \; \sum_{i=1}^{m} \lambda_i l_i. \tag{4.202}$$

Now, consider Inequalities (4.175) and (4.178) with $\mathbf{w}$ replaced by $\mathbf{w}^+$. By summing these inequalities and using condition (4.195), for $i = 1, ..., m$, we have

$$0 \; \leq \; -\sum_{i=1}^{m} \lambda_i w_{]i[}^+ \; - V(\mathbf{w}^+) + G(\mathbf{x}). \tag{4.203}$$

Equation (4.201) and Inequality (4.203) imply

$$0 \; \leq \; \sum_{i=1}^{m} \lambda_i (w_{]i[}^+ \; - \; l_i) \; \leq \; V(\mathbf{w}) - V(\mathbf{w}^+) \; \leq \; V - V(\mathbf{w}^+), \tag{4.204}$$

where V is an upper bound for the dual program, which exists by assumption.

Now, for any given $\epsilon > 0$, if we choose l_i such that

$$l_i \; = \; l_i(\epsilon) \; = \; \epsilon \delta w_{]i[}^+ / (V - V(\mathbf{w}^+) + \epsilon), \tag{4.205}$$

for any $0 < \delta < 1$ and $i = 1, 2, ..., m$, then there are two possible cases for consideration:

Case 1: If $V \; - \; V(\mathbf{w}^+) \; = \; 0$, then since $l_i \; = \; \delta w_{]i[}^+$ and $w_{]i[}^+ - l_i = (1-\delta) w_{]i[}^+ > 0$, λ_i in (4.204) has to be 0 for every i. Consequently, we have $G(\mathbf{x}) - V(\mathbf{w}) = 0$ by Equation (4.202).

Case 2: If $V - V(\mathbf{w}^+) > 0$, by (4.205), we obtain $w_{]i[}^+ = [(V - V(\mathbf{w}^+) + \epsilon)/\epsilon\delta]l_i$. Plugging this into (4.204) produces $\sum_{i=1}^m \lambda_i[((V - V(\mathbf{w}^+))/\delta\epsilon) + (1/\delta) - 1]l_i \leq V - V(\mathbf{w}^+)$. This implies that $\sum_{i=1}^m \lambda_i l_i \leq \delta\epsilon < \epsilon$. Therefore, by Equation (4.202), we have $0 \leq G(\mathbf{x}) - V(\mathbf{w}) < \epsilon$. This completes the proof.

□

The result of Theorem 4.19 can be further strengthened to prove the following "strong duality theorem" for the minimum cross-entropy problem with convex constraints.

Theorem 4.20. Program EC$'$ is consistent if and only if its partial dual Program DEC$'$ has a finite optimum value. Moreover, both programs attain a common optimum value in this case.

Proof. By Theorems 4.18 and 4.19, Program EC$'$ has a feasible solution if and only if Program DEC$'$ has a finite optimum value. Moreover, for any given $\epsilon > 0$, we can find feasible $\mathbf{x}$ and $\mathbf{w}$ such that $0 \leq G(\mathbf{x}) - V(\mathbf{w}) \leq \epsilon$. Hence we have $V(\mathbf{w}) \leq V^* \leq G^* \leq G(\mathbf{x})$, where V^* and G^* are optimum values of DEC$'$ and EC$'$, respectively. Consequently, $0 \leq G^* - V^* \leq G(\mathbf{x}) - V(\mathbf{w}) \leq \epsilon$, for any $\epsilon > 0$. Therefore $G^* = V^*$.

□

Once we have shown that there is no duality gap between Programs EC$'$ and DEC$'$, the gap between Programs EC and DEC$'_{\boldsymbol{l}}$ can be controlled to any tolerance level for a properly chosen $\boldsymbol{l}$. This is shown as follows.

Theorem 4.21. Given any $\epsilon > 0$, if $\boldsymbol{l}(\epsilon)$ is chosen according to Equation (4.205) and γ is a vector of Lagrange multipliers associated with the constraint set $\mathbf{M}^T\mathbf{w} = \mathbf{0}$ of Program DEC$'_{\boldsymbol{l}(\epsilon)}$, then γ is a feasible solution to Program EC and $0 \leq g(\gamma) - V(\mathbf{w}) \leq \epsilon$.

Proof. Since the first n columns of matrix $\mathbf{M}$ form an identity matrix $\mathbf{I}_n$, Equation (4.193) implies that $x_j = \gamma_j$, for $j \in [0]$, and $G(\mathbf{x}) = g(\gamma)$. By Equation (4.188), we know $\gamma_j \geq 0$ and $\sum_{j \in [0]} \gamma_j = 1$. Since $\lambda_i \geq 0$, Equation

(4.195), after substitution using Equations (4.193) and (4.168), implies that $g_i(\gamma) = h_i(\mathbf{A}_i\gamma) + \mathbf{b}_i^T\gamma + c_i \leq 0$, for $i = 1, 2, ..., m$. Hence γ is a solution to Program EC. Theorem 4.19 implies that $0 \leq g(\gamma) - V(\mathbf{w}) \leq \epsilon$.

□

4.4.3 Dual-to-Primal Conversion Formula

Now, if $\mathbf{w}$ solves Program $\text{DEC}'_{l(\epsilon)}$ as stated in Theorem 4.21, we would like to express the primal solution γ in terms of $\mathbf{w}$. To do so, we first define

$$\mu = \ln(\sum_{j \epsilon [0]} p_j \exp{(w_j)}). \tag{4.206}$$

The constraint set $\mathbf{M}^T\mathbf{w} = \mathbf{0}$ implies

$$\mathbf{w}_0 = -\sum_{i=1}^{m}\{\mathbf{A}_i^T\mathbf{w}_i + \mathbf{b}_i w_{]i[}\}. \tag{4.207}$$

Hence we have

$$w_j = -\sum_{i=1}^{m}\{\mathbf{A}_{i,j}^T\mathbf{w}_i + b_{i,j} w_{]i[}\}, \quad \text{for } j \in [0], \tag{4.208}$$

where $\mathbf{A}_{i,j}^T$ is the jth row of $\mathbf{A}_i^T$ and $b_{i,j}$ is the jth component of $\mathbf{b}_i$.

Equations (4.193) and (4.188) imply $p_j \exp{(w_j)} = \gamma_j \exp\ \mu$, for $j \in [0]$. Now, substitution of w_j in this expression using Equation (4.208) yields

$$\gamma_j = p_j \exp(-\mu - \sum_{i=1}^{m}\{\mathbf{A}_{i,j}^T\mathbf{w}_i + b_{i,j} w_{]i[}\}), \quad \text{for } j \in [0]. \tag{4.209}$$

Therefore, Equations (4.206) and (4.209) provide a dual-to-primal conversion formula.

4.4.4 Computational Issues

Although the perturbed dual program $\text{DEC}'_{l(\epsilon)}$ has more variables, it possesses a differentiable concave objective function and relatively simple linear constraints. Compared to the original Program EC, this makes a dual approach computationally attractive. The theoretical insights obtained in previous subsections can now be put together to outline a dual based computation procedure as follows:

Algorithm 4.5: (Dual Perturbation Method – Convex Constraints)

Step 1 : Given a small number $\epsilon > 0$, a dual feasible vector $\mathbf{w}^+$ with $w^+_{]i[} > 0$, for $i = 1, ..., m$, and a dual upper bound V of $V(\mathbf{w})$;

Step 2 : Choose a perturbation vector $\boldsymbol{l}\ (\epsilon) \ = \ (l(\epsilon)_1, ..., l(\epsilon)_m)$ according to Equation (4.205).

Step 3 : Find an optimal solution $\mathbf{w}$ for the perturbed dual program $\mathrm{DEC}'_{l(\epsilon)}$.

Step 4 : Generate an ϵ–optimal solution γ according to Equations (4.206) and (4.209). Then, we have

$$0 \ \leq \ g(\gamma) \ - \ V(\mathbf{w}) \ \leq \ \epsilon.$$

Step 5 : Stop.

Several observations can be made here.

Observation 4.20: In Step 1, $\mathbf{w}^+$ can be easily obtained, as we derived in the proof of Theorem 4.17. In this case

$\mathbf{w}_0{}^+ \ = \ -\sum_{i=1}^m \mathbf{b}_i, \ w^+{}_{]i-1[+1} \ = \ ... \ = \ w^+{}_{]i-1[+k_i} \ = \ 0$, and $w^+_{]i[} \ = \ 1$, for $i = 1, ..., m$.

Observation 4.21: A dual bound V can usually be estimated before solving the problem. If it is not provided, we can always choose $0 < \delta < 1$ and solve Program DEC'_l with $l_i \ = \ \delta w^+_{]i[}$. Then, use Equations (4.206) and (4.209) to find a corresponding primal solution γ^*. In this way, $V \ \leq \ g(\gamma^*)$ is ensured.

Observation 4.22: The real computation work is done in Step 3. Since Program $\mathrm{DEC}'_{l(\epsilon)}$ is nicely structured, we can either use a general optimizer like MINOS to solve the problem with minimum customization efforts, or we can explore the possibility of applying the newly developed interior-point methods. Implementation of this computational procedure with MINOS for solving minimum cross-entropy problems with entropic constraints was discussed in Section 4.3. Developing interior-point methods for solving the minimum cross-entropy models with convex constraints using the dual program constructed in this section could be a worthy subject for research.

4.4.5 Computational Results

In this subsection, we use one example to illustrate the computational procedure described in this section.

Example 4.3:

In this example Program EC has 4 variables and two constraints. The first constraint is a quadratic constraint while the second is an entropic constraint. The *a priori* probability distribution associated with the objective function is:

$$\mathbf{p}^0 = (0.20,\ 0.10,\ 0.50,\ 0.20).$$

The quadratic constraint is given by

$$g_1(\mathbf{q}) \;=\; 1/2(\mathbf{q}^T\mathbf{A}^T\mathbf{A}\mathbf{q}) + \mathbf{b}^T\mathbf{q} + c_1 \;\le\; 0, \tag{4.211}$$

where

$$\begin{aligned}\mathbf{A}^T &= \begin{bmatrix} 1.0 & 0.0 & 2.0 & 3.0 \\ 0.0 & 2.0 & -1.0 & 1.0 \end{bmatrix},\\ \mathbf{b}^T &= (-1.0, 1.0, 1.0, -2.0), \text{and}\\ c_1 &= -20.0.\end{aligned}$$

The entropic constraint is given by

$$g_2(\mathbf{q}) \;= \sum_{j=1}^{4} q_j \ln(q_j/p_j^1) + c_2 \le\; 0, \tag{4.212}$$

where

$$\mathbf{p}^1 = (0.10,\ 0.15,\ 0.45,\ 0.30) \text{ and } c_2 = \text{-}0.0002.$$

The error bound for the duality gap is assumed to be $\epsilon = 0.0001$.

The Dual Program DEC_l' can be shown to be as follows:

$$\begin{aligned}\max V(\mathbf{w}) \ = \ & - \ \ln(\sum_{j=1}^{4} p_j^0 \exp(w_j)) \ - \ \frac{1}{2}\sum_{j=5}^{6}(w_j^2/w_7) \\ & - \ c_1 w_7 - w_{12}\ln(\sum_{j=8}^{11} p_j^1 \exp(w_j/w_{12})) - c_2 w_{12} \\ \text{s.t.} \quad & \left[\mathbf{I}_4, \ \mathbf{A}^T, \ \mathbf{b}^T, \ \mathbf{I}_4, \ \mathbf{0}\right] \mathbf{w} \ = \ \mathbf{0}, \\ & w_7 \ \geq \ l_1, \\ & w_{12} \ \geq \ l_2,\end{aligned}$$

where $\mathbf{I}_4$ is the identity matrix as defined in Equation (4.168).

First, we need a dual upper bound V and a dual feasible solution. Since neither is provided, following Step 1, we choose $\mathbf{w}^+$ to be

1	1.000000	2	−1.000000	3	−1.000000	4	2.000000
5	0.000000	6	0.000000	7	1.000000	8	0.000000
9	0.000000	10	0.000000	11	0.000000	12	1.000000.

To find an upper bound for the dual, we solve Program DEC$'_l$ using MINOS, with $l_1 = 0.5w_7^+$ and $l_2 = 0.5w_{12}^+$. The upper bound is found to be $V = 0.06244$. With this value, the value of $V(\mathbf{w}^+)$, and ϵ, Equation (4.205) yields $l_1 = 0.000005$ and $l_2 = 0.000005$.

With these values, the solution of Program DEC$'_{l(\epsilon)}$ is:

1	−1.521794	2	−0.485636	3	−0.967423	4	−0.485637
5	0.000009	6	0.000001	7	0.000005	8	1.521789
9	0.485630	10	0.967401	11	0.485619	12	16.590445.

By Equations (4.206) and (4.209), we obtain the primal solution $\gamma = (0.10439, 0.14710, 0.45431, 0.29420)$. It can be easily verified that the value of the dual objective function is $V(y) = 0.0588239$, the value of the primal objective function is $g(\gamma) = 0.0589113$, the value of the first constraint is $g_1(\gamma) = -18.28572082$, and the value of the second constraint is $g_2(\gamma) = 0.0$.

Moreover, the duality gap is 0.0000874, which is within the specified error bound $\epsilon = 0.0001$.

4.4.6 Concluding Remarks

In this section, we have generalized the geometric dual approach presented in the two previous sections to solving the minimum cross-entropy optimization problems with convex constraints. We used only one simple geometric inequality and the conjugate inequality to derive the related duality theory. The concept of dual perturbational method has led us to a computational procedure that would allow us to use a general-purpose optimizer to solve efficiently the minimum cross-entropy problems with general convex constraints.

REFERENCES

[1] Ben-Tal A., and Charnes, A., "A Dual Optimization Framework for Some Problems of Information Theory and Statistics," *Problems of Control and Information Theory*, Vol. 8, 1979, pp. 387–401.

[2] Ben-Tal, A., Teboulle, M., and Charnes, A., "The Role of Duality in Optimization Problems involving Entropy Functionals with Applications to Information Theory," *Journal of Optimization Theory and Applications*, Vol. 58, 1988, pp. 209–223.

[3] Boltzmann, L., *Vorlesungen uber Gastheorie*, J. A. Barth, Leipzig, 1896.

[4] Censor, Y., "On Linearly Constrained Entropy Maximization," *Linear Algebra and Its Applications*, Vol. 80, 1986, pp. 191–195.

[5] Charnes, A., Cooper, W.W., and Seiford, L., "Extremal Principles and Optimization Qualities for Khinchin-Kullback-Leibler Estimation," *Mathematische Operationsforschung und Statistik, Series Optimization*, Vol. 9, 1978, pp. 21–29.

[6] Duffin, R.J., Peterson, E.L., and Zener, C., *Geometric Programming - Theory And Applications*, John Wiley, New York, 1967.

[7] Fang, S.-C., Peterson, E.L., and Rajasekera, J.R., "Controlled Dual Perturbations for Posynomial Programs," *European Journal of Operational Research*, Vol. 35, 1988, pp. 111–117.

[8] Fang, S.-C., Peterson, E.L., and Rajasekera, J.R., "Minimum Cross-Entropy Analysis with Entropy-Type Constraints," *Journal of Computational and Applied Mathematics*, Vol. 39, 1992, pp. 165–178.

[9] Fang, S.-C., and Rajasekera, J.R., "Minimum Cross-Entropy Analysis with Convex Constraints," *Information and Computation*, Vol. 116, 1995, pp. 304–311.

[10] Fang, S.-C., and Rajasekera, J.R., "Controlled Perturbations for Quadratically Constrained Quadratic Programs" *Mathematical Programming*, Vol. 36, 1986, pp. 276–289.

[11] Fang, S.-C., and Rajasekera, J.R., "Controlled Dual Perturbations for l_p-Programming," *Zeitschrift fur Operations Research*, Vol. 30 1986, pp. A29–A42.

[12] Fang, S.-C., and Rajasekera, J.R., "Quadratically Constrained Minimum Cross-Entropy Analysis," *Mathematical Programming*, Vol. 44, 1989, pp. 85-96.

[13] Jefferson, T.R., and Scott, C.H., "The Analysis of Flexible Entropy Models of Spatial Interaction Using Geometric Programming," *Proceedings of the OR Society of New Zealand*, Vol. 1, 1978, pp. 39–47.

[14] Jefferson, T.R., and Scott, C.H., "Entropy and Geometric Programming in Transportation Planning," *Opsearch*, Vol. 15, 1978, pp. 22–34.

[15] Ji, J., and Potra, F.A., "Interior-Point Algorithm for Quadratically Constrained Entropy Minimization Problems," *Journal of Optimization Theory and Applications*, Vol. 77, 1993, pp. 79–95.

[16] Kortanek, K.O., Potra, F., and Ye, Y., "On Some Efficient Interior Point Methods for Nonlinear Convex Programming," *Linear Algebra and Its Applications*, Vol. 152, 1991, pp. 169–189.

[17] Kullback, S., *Information and Statistics*, John Wiley, New York, 1959.

[18] Kullback, S., and Libler, R.A., "On Information and Sufficiency," *Annals of Mathematical Statistics*, Vol. 22, 1951, pp. 79–86.

[19] Mehrotra, S., and Sun, J., "An Algorithm for Convex Quadratic Programming That Requires $O(n^{3.5}L)$ Arithmetic Operations," *Mathematics of Operations Research*, Vol. 15, 1990, pp. 342–363.

[20] Murtagh, B.A., and Saunders, M.A., *MINOS 5.1 User's Guide*, Technical Report SOL 83-20R, Stanford University, Stanford, California, 1983.

[21] Potra, F., and Ye, Y., "A Quadratically Convergent Polynomial Algorithm for Solving Entropy Optimization Problems," *SIAM Journal on Optimization*, Vol. 3, 1993, pp. 843–860.

[22] Rajasekera, J.R., and Fang, S.-C., "On the Convex Programming Approach to Linear Programming," *Operations Research Letters*, Vol. 10, 1991, pp. 309–312.

[23] Rockafellar, R.T., *Convex Analysis*, Princeton University Press, Princeton, New Jersey, 1970.

[24] Teboulle, M., "A Simple Duality Proof for Quadratically Constrained Entropy Functionals and Extension to Convex Constraints," *SIAM Journal on Applied Mathematics*, Vol. 49, 1989, pp. 1845–1850.

[25] Ye, Y., "Interior Algorithms for Linear, Quadratic, and Linearly Constrained Programming," Ph.D. Thesis, Department of Engineering-Economic Systems, Stanford University, Stanford, California, 1987.

[26] Zhang, J., and Brockett, P.L., "Quadratically Constrained Information Theoretic Analysis," *SIAM Journal on Applied Mathematics*, Vol. 47, 1987, pp. 871–885.

5

ENTROPIC PERTURBATION APPROACH TO MATHEMATICAL PROGRAMMING

The barrier and penalty function methods for solving mathematical programming problems have been widely used for both theoretical and computational purposes. In a penalty approach, any point outside of the feasible region is assigned a penalty while, in a barrier approach, those feasible solutions near the boundary of the feasible region are subject to a penalty. Both approaches are designed to prevent the search process for an optimal solution from wondering away from the feasible region. They can be considered as an objective-perturbation approach. This chapter studies the objective-perturbation approach by using the entropic function, $\sum_j x_j \ln x_j$, for solving four classes of problems, namely, linear programming problems in Karmarkar's form, linear programming programs in standard form, convex quadratic programming problems, and linear and convex quadratic semi-infinite programming problems.

5.1 LINEAR PROGRAMMING: KARMARKAR-FORM

Since N. Karmarkar introduced his projective scaling algorithm [24] in 1984, the study of interior-point methods for solving linear programs has received much attention. Good reviews can be found in [20, 10, 36]. Among the key research directions is the framework of first incorporating a logarithmic function in the linear objective function and then solving the Karush-Kuhn-Tucker (K.K.T.) conditions by Newton's method [18, 25, 34]. It has been shown that many so-called primal methods, dual methods, and primal-dual methods of the interior-point approach can be derived from this unifying framework [38, 37]. Early work on using the entropic perturbation function to solve linear programming problems can be found in Erlander [7] and Eriksson [6]. Their methods involve solving a sequence of perturbed problems with decreasing perturbation parameters until a solution is found to satisfy a prespecified duality gap.

In this section, we are interested in obtaining, by solving only one unconstrained convex program, an ϵ-optimal solution to a linear program in Karmarkar's standard form, where an ϵ-optimal solution is defined to be a feasible solution whose objective value differs from the true optimum by at most ϵ. This can be achieved by first perturbing the objective function of the linear program by an entropic function, with an appropriate perturbation parameter, and then solving the perturbed program. The optimal solution of the perturbed program turns out to be an ϵ-optimal solution to the original linear program. Rather than solving the perturbed program directly, one can solve its unconstrained dual. The origin of this approach goes back to Fang [8], where the generalized geometric programming approach [31] was used to derive an unconstrained convex dual program.

After introducing the idea of perturbing the objective function of a Karmarkar linear program, Subsection 5.1.1 shows that an unconstrained convex dual of the perturbed program can be easily derived from a simple geometric inequality without borrowing results from the generalized geometric programming or performing the conventional Lagrangian derivation. Subsection 5.1.2 proves the strong duality theorem. Subsection 5.1.3 determines an upper bound for the perturbation parameter that ensures the ϵ-optimality. Subsection 5.1.4 improves the upper bound with a larger value. With the perturbation parameters being set within the upper bound, an ϵ-optimal solution to the original linear program can be obtained by solving the unconstrained convex dual. Subsection 5.1.5 provides a conventional perspective involving the K.K.T. conditions of the perturbed program. Subsection 5.1.6 concludes this section.

5.1.1 Entropic Perturbation for Karmarkar's Form

Consider the following linear program in Karmarkar's standard form [24]:

Program KL:

$$\min \quad \mathbf{c}^T\mathbf{x} \tag{5.1}$$

$$\text{s.t.} \quad \mathbf{A}\mathbf{x} = \mathbf{0}, \tag{5.2}$$

$$\mathbf{e}^T\mathbf{x} = 1, \tag{5.3}$$

$$\mathbf{x} \geq \mathbf{0}, \tag{5.4}$$

where $\mathbf{c}$ and $\mathbf{x}$ are n-dimensional column vectors, $\mathbf{A}$ is an $m \times n$ matrix whose (i, j)th entry is a_{ij}, $\mathbf{0}$ is the zero vector of appropriate dimension, and $\mathbf{e}$ is an n-dimensional vector of 1's. We assume that matrix $\mathbf{A}$ has full row-rank and Program KL has an interior feasible solution $\mathbf{x} > \mathbf{0}$.

A linear dual of Program KL can be written in the following form:

Program DKL:

$$\max \quad w_{m+1} \tag{5.5}$$

$$\text{s.t.} \quad \sum_{i=1}^{m} a_{ij}w_i + w_{m+1} \leq c_j, \quad j = 1, 2, .., , n, \tag{5.6}$$

$$w_i \in R, \quad i = 1, 2, ..., m + 1. \tag{5.7}$$

Note that Program DKL is always feasible because, for any $w_1, \ldots, w_m, w_{m+1} \in R$ such that

$$w_{m+1} \equiv \min_{j=1,..,n}(c_j - \sum_{i=1}^{m} a_{ij}w_i), \tag{5.8}$$

the vector $(w_1, w_2, ..., w_{m+1})^T \in R^{m+1}$ is a feasible solution.

Note that Equations (5.3) and (5.4) indicate that the decision vector $\mathbf{x}$ of Program KL can be viewed as a probability distribution, while Equation (5.2) specifies m expectations to be satisfied. Our objective is to find such a probability distribution with the minimum expected cost with respect to a given $\mathbf{c}$-vector. Based on the maximum entropy principle, among all feasible probability distributions, nature favors the one with maximum entropy. Therefore, instead of solving Program KL directly, we may consider incorporating an entropic perturbation into the objective function of Program KL. With the convention of $0 \ln 0 = 0$, we consider the following convex programming problem :

Program KL(μ):

$$\min \quad \mathbf{c}^T\mathbf{x} + \mu \sum_{j=1}^{n} x_j \ln x_j \tag{5.9}$$

$$\text{s.t.} \quad \mathbf{A}\mathbf{x} = \mathbf{0}, \tag{5.10}$$

$$\mathbf{e}^T\mathbf{x} = 1, \tag{5.11}$$

$$\mathbf{x} \geq \mathbf{0}. \tag{5.12}$$

Now consider the following well-known "geometric-arithmetic inequality" [4]:

$$\sum_{j=1}^{n} x_j z_j \geq \prod_{j=1}^{n} {z_j}^{x_j}, \tag{5.13}$$

where $z_j > 0$ and $x_j \geq 0$, $j = 1, 2, \ldots, n$, such that $\sum_{j=1}^{n} x_j = 1$. This inequality becomes an equality if and only if there exists a constant $C > 0$ such that

$$z_j = C, \ \forall \ j \text{ with } x_j > 0. \tag{5.14}$$

By imposing an extra condition of $\mathbf{x} > \mathbf{0}$ and substituting z_j with e^{y_j}/x_j, where $y_j \in R$, we have

$$\sum_{j=1}^{n} e^{y_j} \geq \prod_{j=1}^{n} \{e^{y_j}/x_j\}^{x_j}. \tag{5.15}$$

Again, this inequality becomes an equality if and only if

$$x_j = \lambda e^{y_j}, \quad j = 1, 2, ..., n, \tag{5.16}$$

for some constant $\lambda > 0$. We further expand Inequality (5.15) by substituting

$$y_j = (\sum_{i=1}^{m} a_{ij} w_i - c_j)/\mu, \quad j = 1, 2, ..., n, \tag{5.17}$$

where $\mathbf{w} \equiv (w_1, w_2, \ldots, w_m)^T \in R^m$ and $\mu > 0$. Taking the logarithm on both sides results in

$$\sum_{j=1}^{n}(\sum_{i=1}^{m} a_{ij}w_i - c_j)x_j \leq \mu \sum_{j=1}^{n} x_j \ln x_j + \mu \ln \left\{ \sum_{j=1}^{n} \exp[(\sum_{i=1}^{m} a_{ij}w_i - c_j)/\mu] \right\}, \tag{5.18}$$

which holds for any $\mu > 0$, $\mathbf{w} \in R^m$ and $\mathbf{x} \in R^n$ such that $\mathbf{x} > \mathbf{0}$ and $\sum_{j=1}^{n} x_j = 1$. Moreover, Inequality (5.18) becomes an equality if and only if

$$x_j = \lambda \exp[(\sum_{i=1}^{m} a_{ij}w_i - c_j)/\mu], \quad j = 1, 2, ..., n. \tag{5.19}$$

Assume that $\mathbf{x}$ also satisfies Constraint (5.2) of Program KL. Then, $\sum_{j=1}^{n} a_{ij}x_j = 0$, $i = 1, 2, ..., m$, and

$$\sum_{j=1}^{n}(\sum_{i=1}^{m} a_{ij}w_i - c_j)x_j = \sum_{i=1}^{m} w_i(\sum_{j=1}^{n} a_{ij}x_j) - \sum_{j=1}^{n} c_j x_j = -\sum_{j=1}^{n} c_j x_j. \tag{5.20}$$

Hence, after rearrangement, (5.18) reduces to

$$-\mu \ln \left\{ \sum_{j=1}^{n} \exp[(\sum_{i=1}^{m} a_{ij}w_i - c_j)/\mu] \right\} \leq \mathbf{c}^T\mathbf{x} + \mu \sum_{j=1}^{n} x_j \ln x_j, \tag{5.21}$$

which holds for any $\mathbf{w} \in R^m$ and any $\mathbf{x} \in R^n$ that satisfies Constraints (5.2) - (5.4). The equality holds if and only if the Condition (5.19) is satisfied. This leads to the following unconstrained convex dual program:

Program DKL(μ):

$$\max_{\mathbf{w} \in R^m} \left\{ d_\mu(\mathbf{w}) \equiv -\mu \ln \left[\sum_{j=1}^{n} \exp[(\sum_{i=1}^{m} a_{ij}w_i - c_j)/\mu] \right] \right\}. \tag{5.22}$$

Note that Inequality (5.21) naturally links Program KL(μ) and Program DKL(μ) into a primal-dual pair of convex programs, as originally obtained via geometric programming in [8]. This duality can also be obtained via the conventional Lagrangian derivation. Inequality (5.21) also establishes the so-called "weak duality theorem." In the next subsection, we shall strengthen this result as the "strong duality theorem."

5.1.2 Strong Duality Theorem

Let us examine the objective function of Program DKL(μ) in Equation (5.22). When $\mathbf{A}$ has full row-rank then $d_\mu(\mathbf{w})$ is a strictly concave function of $\mathbf{w}$. Also, under the assumption that Program KL has an interior feasible solution, it can be shown that $d_\mu(\mathbf{w})$ has a unique maximum solution $\mathbf{w}^*$ [41]. We state and prove this result as follows:

Lemma 5.1. If matrix $\mathbf{A}$ has full row-rank and Program KL has an interior feasible solution, then Program DKL(μ) has a unique maximum solution.

Proof. If Program DKL(μ) attains an optimal solution, then the solution is unique because $d_\mu(\mathbf{w})$ is strictly concave under the assumption that matrix A has full row-rank. Consequently, it suffices to focus on the existence issue.

When Program KL is feasible, the weak duality property implies that there exist $\theta < \infty$ and $\{\mathbf{w}^k \in R^m \mid k = 1, 2...\}$ such that, as $k \to \infty$,

$$d_\mu(\mathbf{w}^k) \to \theta \equiv \sup_{\mathbf{w} \in R^m} d_\mu(\mathbf{w}). \tag{5.23}$$

If the sequence $\{\mathbf{w}^k\}$ is bounded, then any cluster point $\mathbf{w}^*$ satisfies $d_\mu(\mathbf{w}^*) = \theta$. To show the boundedness of $\{\mathbf{w}^k\}$, let us consider

$$\mathbf{y}^k = \mathbf{A}^T\mathbf{w}^k - \mathbf{c} \tag{5.24}$$

or, equivalently,

$$y_j^k = \sum_{i=1}^{m} a_{ij} w_i^k - c_j, \; j = 1, 2, \ldots, n. \tag{5.25}$$

When matrix $\mathbf{A}$ has full row-rank, the boundedness of $\{\mathbf{y}^k\}$ implies the boundedness of $\{\mathbf{w}^k\}$ and, equivalently, the unboundedness of $\{\mathbf{w}^k\}$ implies the unboundedness of $\{\mathbf{y}^k\}$.

Suppose that $\bar{\mathbf{x}} > 0$ is an interior feasible solution of Program KL. Then,

$$\bar{\mathbf{x}}^T \mathbf{y}^k = \bar{\mathbf{x}}^T \mathbf{A}^T \mathbf{w}^k - \bar{\mathbf{x}}^T \mathbf{c} = -\bar{\mathbf{x}}^T \mathbf{c} \geq \min_{1 \leq j \leq n} \{-c_j\}. \tag{5.26}$$

If $\{\mathbf{w}^k\}$ is unbounded, so is $\{\mathbf{y}^k\}$. Hence there exists an index $j_0 \in \{1, 2, \ldots, n\}$ and a subsequence $\{\mathbf{y}^{k_l}\}$ such that either $y_{j_0}^{k_l} \to +\infty$ or $y_{j_0}^{k_l} \to -\infty$.

For the first case, we have $y_{j_0}^{k_l} = \sum_{i=1}^{m} a_{ij_0} w_i^{k_l} - c_{j_0} \to +\infty$. Since $\mu > 0$, this leads to a contradiction to Equation (5.23). For the second case, to balance out Equation (5.26), there must exist a different index $j_1 \in \{1, 2, \ldots, n\}$, such that $y_{j_1}^{k_l} \to +\infty$. This again leads to a contradiction. Hence we can conclude that $\{\mathbf{w}^k\}$ is bounded and the proof is complete.

□

Lemma 5.1 established the existence and uniqueness of the maximum solution of Program DKL(μ). We now establish the absence of duality gap. The differentiability of $d_\mu(\mathbf{w})$ is clear. By taking derivatives of $d_\mu(\mathbf{w})$ at $\mathbf{w}^*$, we have, for $i = 1, ..., m$,

$$\left\{ \sum_{j=1}^{n} \exp[(\sum_{l=1}^{m} a_{lj} w_l^* - c_j)/\mu] \right\}^{-1} \sum_{j=1}^{n} a_{ij} \exp[(\sum_{l=1}^{m} a_{lj} w_l^* - c_j)/\mu] = 0. \tag{5.27}$$

By taking the second-order derivatives, it can be shown that $\mathbf{w}^*$ really achieves the maximum of $d_\mu(\mathbf{w})$ over $\mathbf{w} \in R^n$.

Let us define the n-vector $\mathbf{x}^*$ as follows, for $j = 1, ..., n$,

$$x_j^* = \left\{ \sum_{k=1}^{n} \exp[(\sum_{i=1}^{m} a_{ik} w_i^* - c_k)/\mu] \right\}^{-1} \exp[(\sum_{i=1}^{m} a_{ij} w_i^* - c_j)/\mu]. \quad (5.28)$$

Then, Equation (5.27) implies that $\mathbf{x}^*$ satisfies Constraint (5.2) and Equation (5.28) implies that $\mathbf{x}^*$ satisfies Constraints (5.3) and (5.4) of Program KL. Hence $\mathbf{x}^*$ is a feasible solution to Program KL. Moreover, each x_j^* satisfies the condition specified in (5.19) and, hence, (5.21) becomes an equality with $\mathbf{x}$ and $\mathbf{w}$ replaced by $\mathbf{x}^*$ and $\mathbf{w}^*$, respectively. We summarize the results as follows:

Theorem 5.1. Suppose that Program KL has a nonempty interior and matrix $\mathbf{A}$ has full row-rank. Let $\mu > 0$ and $\mathbf{w}^*$ be the unique maximal solution of the concave function $d_\mu(\mathbf{w})$. If $\mathbf{x}^*$ is defined by Equation (5.28), then $\mathbf{x}^*$ is the optimal solution to Program KL(μ) and the duality gap vanishes, i.e.,

$$d_\mu(\mathbf{w}^*) = -\mu \ln \left\{ \sum_{j=1}^{n} \exp[(\sum_{i=1}^{m} a_{ij} w_i^* - c_j)/\mu] \right\} = \mathbf{c}^T \mathbf{x}^* + \mu \sum_{j=1}^{n} x_j^* \ln x_j^*. \quad (5.29)$$

5.1.3 ϵ-Optimality

As we mentioned before, by defining

$$w_{m+1}^* = \min_{j=1,..,n} \left\{ c_j - \sum_{i=1}^{m} a_{ij} w_i^* \right\}, \quad (5.30)$$

it can be easily seen that $\mathbf{w}^* = (w_1^*, w_2^*, ..., w_{m+1}^*)$ is a feasible solution to the dual program DKL. Without loss of generality, we assume that the minimum of the right-hand side of Equation (5.30) occurs at $j = 1$ and

$$w_{m+1}^* = c_1 - \sum_{i=1}^{m} a_{i1} w_i^*. \quad (5.31)$$

By taking the logarithm of x_1^*, as defined in Equation (5.28), and multiplying the result by μ, we have

$$\mu \ln x_1^* = (\sum_{i=1}^{m} a_{i1} w_i^* - c_1) - \mu \ln \left\{ \sum_{j=1}^{n} \exp[(\sum_{i=1}^{m} a_{ij} w_i^* - c_j)/\mu] \right\}. \tag{5.32}$$

Hence, by Equations (5.29) and (5.31),

$$\mu \ln x_1^* = -w_{m+1}^* + \mathbf{c}^T \mathbf{x}^* + \mu \sum_{j=1}^{n} x_j^* \ln x_j^*. \tag{5.33}$$

Moreover, note that the weak duality theory of linear programming assures that $0 \leq \mathbf{c}^T \mathbf{x}^* - w_{m+1}^*$. Thus, we have

$$\begin{aligned} 0 \leq \mathbf{c}^T \mathbf{x}^* - w_{m+1}^* &= \mu \ln x_1^* - \mu \sum_{j=1}^{n} x_j^* \ln x_j^* && (5.34) \\ &= \mu \sum_{j=1}^{n} x_j^* \ln x_1^* - \mu \sum_{j=1}^{n} x_j^* \ln x_j^* \\ &= \mu \sum_{j=1}^{n} \ln (x_1^*/x_j^*)^{x_j^*}. && (5.35) \end{aligned}$$

Since $x_j^* > 0$, for $j = 1, 2, \ldots, n$, and $\sum_{j=1}^{n} x_j^* = 1$, we have, by considering the geometric inequality again,

$$\sum_{j=1}^{n} x_1^* \geq \prod_{j=1}^{n} (x_1^*/x_j^*)^{x_j^*}. \tag{5.36}$$

Since $1 \geq x_1^*$, we obtain

$$n \geq \prod_{j=1}^{n} (x_1^*/x_j^*)^{x_j^*}. \tag{5.37}$$

Therefore,

$$\sum_{j=1}^{n} \ln (x_1^*/x_j^*)^{x_j^*} \leq \ln n \tag{5.38}$$

and Equation (5.35) reduces to

$$0 \leq \mathbf{c}^T \mathbf{x}^* - w^*_{m+1} \leq \mu \ln n. \tag{5.39}$$

Now for any tolerance level $\epsilon > 0$, if $\mu = \epsilon / \ln n$, then $0 \leq \mathbf{c}^T \mathbf{x}^* - w^*_{m+1} \leq \epsilon$. Therefore, we have the following result:

Theorem 5.2. Suppose that matrix $\mathbf{A}$ has full row-rank and Program KL has a nonempty interior. Let $\epsilon > 0$ be any given tolerance level and $\mu = \epsilon / \ln n$ be the perturbation parameter. If $\mathbf{w}^*$ is the unique maximum of $d_\mu(\mathbf{w})$ and $\mathbf{x}^*$ is defined by Equation (5.28), then

$$0 \leq \mathbf{c}^T \mathbf{x}^* - w^*_{m+1} \leq \epsilon \tag{5.40}$$

and $(\mathbf{x}^*, \mathbf{w}^*)$ becomes an ϵ-optimal primal-dual solution pair.

5.1.4 A Tighter Bound for ϵ-Optimality

Theorem 5.2 actually assures that any perturbation parameter with a value less than $\epsilon / \ln n$ would produce an ϵ-optimal solution to Program KL. Since minute perturbation values may lead to numerical difficulty for computer implementation, we can benefit from an upper bound that is larger than $\epsilon / \ln n$. This is the goal of this subsection.

When $n \geq 3$, the upper bound for the perturbation parameter μ can be further improved as follows [41]. Let us begin with Equation (5.34), which can be written as

$$(1/\mu)(\mathbf{c}^T \mathbf{x}^* - w^*_{m+1}) = (1 - x^*_1) \ln x^*_1 - \sum_{j=2}^{n} x^*_j \ln x^*_j. \tag{5.41}$$

Our objective is to find a tighter upper bound for the right-hand side of Equation (5.41). Since the upper bound turns out to be an implicit function of n, the total number of variables, we will approximate it by some explicit functions of n.

It can be verified that, given x^*_1, the optimal value of the problem

$$\max\{-\sum_{j=2}^{n} x_j \ln x_j \mid \sum_{j=2}^{n} x_j = 1 - x_1^*, x_j > 0\} \tag{5.42}$$

is $-(1 - x_1^*) \ln[(1 - x_1^*)/(n - 1)]$. Hence, we focus on finding the optimal value of the following problem:

$$\max_{s \in (0,1)} f(s) \equiv (1 - s) \ln s - (1 - s) \ln[(1 - s)/(n - 1)]. \tag{5.43}$$

The first- and second-order derivatives of $f(s)$ are:

$$f'(s) = \frac{1}{s} - \ln s + \ln[\frac{(1 - s)}{(n - 1)}], \tag{5.44}$$

$$f''(s) = -\frac{1}{s^2} - \frac{1}{s} - \frac{1}{(1 - s)}. \tag{5.45}$$

Since $f''(s) < 0$ for $s \in (0, 1)$, we know that $f(s)$ is strictly concave and $f'(s)$ is strictly decreasing in $(0, 1)$. Moreover, $f'(s) = 0$ has a unique solution $s_n^* \in (1/n, 1)$ because $f'(1/n) > 0$ and $f'(1) \to -\infty$ from the left. Also, s_n^* is the optimal solution of $f(s)$ in $(0, 1)$. Since $f'(s_n^*) = 0$, Equation (5.43) reduces to

$$f(s_n^*) = 1/s_n^* - 1. \tag{5.46}$$

In order to find s_n^*, we set $f'(s) = 0$ and obtain

$$(1 - s) \exp(1/s) = s(n - 1). \tag{5.47}$$

The variable s can be solved through a numerical search, and, as will be demonstrated later via an example, this upper bound could be tight. Notice that although s is a function of n, it is not explicitly defined. Therefore, we substitute

$$s = [\ln(n - 1) - \ln \ln(n - 1) + r]^{-1} \tag{5.48}$$

in Equation (5.47) and obtain

$$(1 - e^{-r}) \ln(n - 1) + r = 1 + \ln \ln(n - 1). \tag{5.49}$$

It can be shown that Equation (5.49) has a unique solution, denoted by r_n. Therefore, by Equation (5.48), we have

$$s_n^* = [\ln(n-1) - \ln\ln(n-1) + r_n]^{-1}. \tag{5.50}$$

By Equation (5.46), we also have

$$\max_{s\in(0,1)} f(s) = f(s_n^*) = \ln(n-1) - \ln\ln(n-1) + r_n - 1. \tag{5.51}$$

We now try to find a lower bound and an upper bound for r_n. Clearly, $r_n \neq 0$. Since $1 - e^{-r} < r$, for $r \neq 0$, replacing $(1 - e^{-r})$ by r in Equation (5.49) results in

$$r\ln(n-1) + r > 1 + \ln\ln(n-1). \tag{5.52}$$

Hence

$$r_n > \underline{r}_n \equiv [1 + \ln(n-1)]^{-1}[1 + \ln\ln(n-1)]. \tag{5.53}$$

Similarly, by replacing r by $(1 - e^{-r})$ in the second term on the left-hand side of Equation (5.49), we have

$$r_n < \bar{r}_n \equiv \ln[1 + \ln(n-1)] - \ln[\ln(n-1) - \ln\ln(n-1)]. \tag{5.54}$$

Clearly, $\underline{r}_n > 0$. Now we show $\bar{r}_n < 1$. Note that $\bar{r}_n < 1$ is equivalent to

$$(e-1)\ln(n-1) - e\ln\ln(n-1) - 1 > 0, \tag{5.55}$$

which follows from the fact that

$$\inf_{t\in(0,\infty)} g(t) > 0, \tag{5.56}$$

with $g(t) \equiv (e-1)t - e\ln t - 1$.

We summarize this analysis by the following theorem:

Theorem 5.3. Assume that Program KL has a nonempty interior and the matrix $\mathbf{A}$ has full row-rank. Let $\mathbf{x}^*$ and w_{m+1}^* be defined by Equations (5.28) and (5.30), respectively. When $n \geq 3$, we have

$$0 \leq \mathbf{c}^T\mathbf{x}^* - w_{m+1}^* \leq \mu[\ln(n-1) - \ln\ln(n-1) + r_n - 1], \tag{5.57}$$

where r_n solves Equation (5.49) and $r_n \in (\underline{r}_n, \bar{r}_n) \subset (0,1)$ with $\underline{r}_n$ and $\bar{r}_n$ defined by Equations (5.53) and (5.54), respectively.

Several observations can be made here:

Observation 5.1: Both $\underline{r}_n$ and $\bar{r}_n$ approach zero as n tends to infinity.

Observation 5.2: The error bound specified in Equation (5.57) could be a tight one. Consider the following example with $n \geq 3$:

$$\begin{aligned} \min \quad & x_1 \\ s.t. \quad & (1/2a)x_1 - (1/2b)(x_2 + x_3 + \ldots + x_n) = 0, \\ & x_1 + x_2 + \ldots + x_n = 1, \\ & x_j \geq 0, \quad j = 1, 2, \ldots, n, \end{aligned}$$

where $a,\ b > 0$ with $a + b = 1$.

Its feasible domain is $\{\mathbf{x} \in R^n \mid x_1 = a,\ x_2 + x_3 + \ldots + x_n = 1 - a,\ x_j \geq 0,\ j \geq 2\}$. Note that $x_1 = a$ holds for all feasible solutions. It can be shown that the optimal solution to the corresponding Program KL(μ), with $\mu > 0$, is $x_1^* = a$ and $x_j^* = (1-a)/(n-1),\ j = 2, \ldots, n$. Substituting the solution in Equation (5.34), we have

$$x_1^* - w_2^* = \mu\{(1-a)\ln a - (1-a)\ln[(1-a)/(n-1)]\}. \tag{5.58}$$

When the right-hand side of Equation (5.58) is viewed as a function of a, it is identical to the objective function defined in Equation (5.43). Note that a is a parameter of the the problem. If a is set to s_n^*, then, by the very definition of s_n^*, the upper bound specified in Equation (5.57) is tight.

Observation 5.3: The upper bound given in Theorem 5.3 has significant theoretical value, because it is tighter. To obtain a practical upper bound, simply replace r_n by $\bar{r}_n$ in Equation (5.57). For $n \geq 3$, this leads to

$$0 \leq \mathbf{c}^T \mathbf{x}^* - w_{m+1}^* \leq \mu B_n, \tag{5.59}$$

Table 5.1. Dependence of μ on the problem size n with $\epsilon = 0.0001$.

$size(n)$	$\mu = \epsilon/B_n$	$\mu = \epsilon/\ln n$	$\mu = \epsilon e/n$
100	.00003745	.00002171	.00000271
1000	.00002253	.00001447	.000000271
10000	.00001570	.00001086	.0000000271
100000	.00001192	.00000866	.00000000271

where

$$B_n = \begin{array}{l} [\ln(n-1) - \ln\ln(n-1)] + \ln[1 + \ln(n-1)] \\ - \ln[\ln(n-1) - \ln\ln(n-1)] - 1. \end{array} \tag{5.60}$$

Corollary 5.1. Assume that matrix $\mathbf{A}$ has full row-rank and Program KL has a nonempty interior. Let $\epsilon > 0$ be any tolerance level and $\mu = \epsilon/B_n$ be the perturbation parameter, where B_n is defined by Equation (5.60). If $\mathbf{w}^*$ is the unique maximum of $d_\mu(\mathbf{w})$ and $\mathbf{x}^*$ is defined by Equation (5.28), then

$$0 \leq \mathbf{c}^T\mathbf{x}^* - w^*_{m+1} \leq \epsilon \tag{5.61}$$

and $(\mathbf{x}^*, \mathbf{w}^*)$ becomes an ϵ-optimal primal-dual solution pair.

Although $B_n \sim \ln n$ for sufficiently large n, the two values still differ when n is of suitably large size. Therefore, we can use Equation (5.59) to derive the perturbation parameter μ. Table 5.1 shows the difference with $\epsilon = 0.0001$.

Since $d_\mu(\mathbf{w})$ is a concave function, maximizing $d_\mu(\mathbf{w})$ becomes an unconstrained convex minimization problem. Once μ is determined according to Corollary 5.1, we can use unconstrained optimization techniques to find an ϵ-optimal solution to the original linear program. In the next subsection, we briefly describe an alternative perspective for solving Program KL(μ).

5.1.5 The Karush-Kuhn-Tucker Perspective

To find an ϵ-optimal solution for Program KL, we first compute $\mu = \bar{\mu} = \epsilon/B_n$, and solve the following problem:

Program KL($\bar{\mu}$):

$$\min \quad \mathbf{c}^T\mathbf{x} + \bar{\mu} \sum_{j=1}^{n} x_j \ \ln \ x_j \tag{5.62}$$

$$\text{s.t.} \quad \mathbf{A}\mathbf{x} = \mathbf{0}, \tag{5.63}$$

$$\mathbf{e}^T\mathbf{x} = 1, \tag{5.64}$$

$$\mathbf{x} \geq \mathbf{0}. \tag{5.65}$$

If **A** has full row-rank and Program KL has an interior feasible solution, then the unique pair of optimal primal-dual solution to Program KL($\bar{\mu}$) is also part of the unique solution to the K.K.T. conditions associated with Program KL($\bar{\mu}$). Since $\mathbf{x} > \mathbf{0}$ is ensured by the K.K.T. conditions, these conditions can be simplified as follows:

$$\mathbf{c} + \bar{\mu}(\mathbf{e} + \ln \mathbf{x}) - \mathbf{A}^T\mathbf{w} - \gamma\mathbf{e} = \mathbf{0}, \tag{5.66}$$

$$\mathbf{A}\mathbf{x} = \mathbf{0}, \tag{5.67}$$

$$\mathbf{e}^T\mathbf{x} = 1, \tag{5.68}$$

where $\ln \mathbf{x} = (\ln x_1, \ldots, \ln x_n)^T$.

Now, consider a related system of equations:

$$-\bar{\mu} \ln \mathbf{x} + \mathbf{A}^T\mathbf{w} = \mathbf{c}, \tag{5.69}$$

$$\mathbf{A}\mathbf{x} = \mathbf{0}, \tag{5.70}$$

for $\mathbf{x} \in R^n_+$ and $\mathbf{w} \in R^m$. If $(\tilde{\mathbf{x}}, \tilde{\mathbf{w}})$ solves Equations (5.69) and (5.70), then it can be shown that

$$\bar{\mathbf{x}} = \tilde{\mathbf{x}}/(\mathbf{e}^T\tilde{\mathbf{x}}), \tag{5.71}$$

$$\bar{\mathbf{w}} = \tilde{\mathbf{w}}, \tag{5.72}$$

$$\gamma = \bar{\mu}(1 - \ln(\mathbf{e}^T\tilde{\mathbf{x}})) \tag{5.73}$$

is a solution to the K.K.T. conditions of Program KL($\bar{\mu}$). Therefore we have the following result:

Theorem 5.4. Suppose that Program KL has a nonempty interior. Let $\epsilon > 0$ and $\bar{\mu} \equiv \epsilon/B_n$. If $(\tilde{\mathbf{x}}, \tilde{\mathbf{w}})$ solves Equations (5.69) and (5.70), then $\bar{\mathbf{x}} = \tilde{\mathbf{x}}/(\mathbf{e}^T\tilde{\mathbf{x}})$

and $\bar{\mathbf{w}} = \tilde{\mathbf{w}}$ form a primal-dual pair of optimal solutions for Programs KL($\bar{\mu}$) and DKL($\bar{\mu}$), and $\bar{\mathbf{x}}$ is an ϵ-optimal solution to Program KL. If matrix $\mathbf{A}$ has full row-rank, then this pair is unique.

Note that Equations (5.69) and (5.70) form a system of $(n+m)$ variables in $(n+m)$ equations. To solve such a system, we can apply some standard numerical methods such as the Newton's method. Also note that, by Equation (5.69), we have

$$\mathbf{x} = \exp[(\mathbf{A}^T\mathbf{w} - \mathbf{c})/\bar{\mu}]. \tag{5.74}$$

Hence, Equation (5.70) requires that

$$\mathbf{A}\exp[(\mathbf{A}^T\mathbf{w} - \mathbf{c})/\bar{\mu}] = \mathbf{0}. \tag{5.75}$$

This is equivalent to the first-order condition (5.27) of maximizing $d_{\bar{\mu}}(\mathbf{w})$. Although Equation (5.75) is a system of m convex equations in m variables, solving Equations (5.69) and (5.70) may still have some advantages in numerical stability.

5.1.6 Concluding Remarks

We have shown how to derive an unconstrained convex program with a controlled perturbation parameter for solving linear programming problems based on a simple geometric inequality. For a problem of n variables with a given tolerance level $\epsilon > 0$, we can set $\mu = \epsilon/B_n$ for ϵ-optimality. When large-scale problems are solved via the unconstrained convex programming approach, a bigger value of μ that still guarantees the ϵ-optimality will reduce the potential difficulty in arithmetic overflow. Some computational results based on the theory described here can be found in Wang [40]. Further numerical aspects will be discussed in the next section.

5.2 LINEAR PROGRAMMING: STANDARD-FORM

In this section, we focus on linear programming problems in standard form. By using only one simple inequality "$\ln z \leq z-1, \forall\, z > 0$," we extend the unconstrained convex programming approach introduced in the previous section to solving standard-form linear programs. Moreover, based on the curved-search methods proposed by Ben-Tal et al. [2], we design a quadratically convergent global algorithm for solving linear programming problems. Related convergence results and computational experience are also included.

The unconstrained convex programming approach will be introduced in Subsection 5.2.1. Issues related to ϵ-optimality will be discussed in Subsection 5.2.2. Then we propose a curved-search algorithm in Subsection 5.2.3 and prove its convergence properties in Subsection 5.2.4. The implementation issues and computational experience will be reported in Subsections 5.2.5 and 5.2.6, respectively. Then we make concluding remarks in the final subsection.

5.2.1 Entropic Perturbation for Standard Form LP

Consider the following (primal) linear program in its standard form:

Program L:

$$\min \quad \mathbf{c}^T\mathbf{x} \tag{5.76}$$

$$\text{s.t.} \quad \mathbf{A}\mathbf{x} = \mathbf{b}, \tag{5.77}$$

$$\mathbf{x} \geq \mathbf{0}, \tag{5.78}$$

where $\mathbf{c}$ and $\mathbf{x}$ are n-dimensional column vectors, $\mathbf{A}$ is an $m \times n$ $(m \leq n)$ matrix, $\mathbf{b}$ is an m-dimensional column vector, and $\mathbf{0}$ is the n-dimensional zero vector.

A linear dual of Program L is given as follows:

Program DL:

$$\max \quad \mathbf{b}^T\mathbf{w} \tag{5.79}$$

$$\text{s.t.} \quad \mathbf{A}^T\mathbf{w} \leq \mathbf{c}, \tag{5.80}$$

$$\mathbf{w} \in R^m. \tag{5.81}$$

Instead of solving Program L directly, following Subsection 5.1.1, we consider the following convex program with a parameter $\mu > 0$:

Program L(μ):

$$\min \quad \mathbf{c}^T\mathbf{x} + \mu \sum_{j=1}^{n} x_j \ln x_j \tag{5.82}$$

$$\text{s.t.} \quad \mathbf{A}\mathbf{x} = \mathbf{b}, \tag{5.83}$$

$$\mathbf{x} \geq \mathbf{0}. \tag{5.84}$$

Note that, with the convention of $0 \ln 0 = 0$, the entropic function $x \ln x$ is a strictly convex function well-defined on $[0, \infty)$, with a unique minimum value of $-1/e$ at $x = 1/e$, where $e = 2.718....$

As assumed for most interior-point methods, we assume that Program L has an interior feasible solution $\mathbf{x} > \mathbf{0}$. We will refer to this assumption as the Interior-Point Assumption. Under this assumption, Program L(μ) is feasible for all $\mu > 0$. Moreover, since $0 \ln 0 = 0$, $c_j x_j + x_j \ln x_j \rightarrow \infty$ as $x_j \rightarrow \infty$, and $x_j \ln x_j$ is strictly convex over its domain, for each j, we know that Program L(μ) achieves a finite minimum at a unique point $\mathbf{x}_\mu^* \in R^n$ for each $\mu > 0$. More interestingly, as discussed in [8], if Program L is assumed to have a bounded feasible domain, then, as $\mu \rightarrow 0$, the optimal solution of Program L(μ) approaches an optimal solution of Program L. We will refer to this assumption as the Bounded Feasible Domain Assumption.

Superficially, Program L(μ) seems to be more complicated than Program L because of the involvement of the nonlinear entropic function. In reality, this nonlinearity is the key to deriving an unconstrained dual convex program. To achieve this goal, consider the following simple inequality:

$$\ln z \leq z - 1, \ \forall\, z > 0. \tag{5.85}$$

This inequality can be verified by the graph of $\ln z$. Moreover, Inequality (5.85) becomes an equality if and only if $z = 1$.

Now, for any $\mu > 0$, $w_i \in R$, $i = 1, \ldots, m$, and $x_j > 0$, $j = 1, \ldots, n$, we define

$$z_j = \frac{\exp\{[(\sum_{i=1}^m a_{ij} w_i - c_j)/\mu] - 1\}}{x_j}, \quad j = 1, \ldots, n. \tag{5.86}$$

Since $x_j > 0$, $z_j > 0$ is implied. By (5.85), we have

$$[(\sum_{i=1}^m a_{ij} w_i - c_j)/\mu] - 1 - \ln x_j \le \frac{\exp\{[(\sum_{i=1}^m a_{ij} w_i - c_j)/\mu] - 1\}}{x_j} - 1. \tag{5.87}$$

Consequently,

$$x_j[(\sum_{i=1}^m a_{ij} w_i - c_j)/\mu] - \exp\{[(\sum_{i=1}^m a_{ij} w_i - c_j)/\mu] - 1\} \le x_j \ln x_j \ . \tag{5.88}$$

By first multiplying both sides by μ and then summing over j, we have

$$\sum_{j=1}^n x_j(\sum_{i=1}^m a_{ij} w_i) - \mu \sum_{j=1}^n \exp\{[(\sum_{i=1}^m a_{ij} w_i - c_j)/\mu] - 1\} \le \sum_{j=1}^n c_j x_j + \mu \sum_{j=1}^n x_j \ln x_j. \tag{5.89}$$

If x_j, $j = 1, \ldots, n$, satisfies $\sum_{j=1}^n a_{ij} x_j = b_i$, for $i = 1, 2, \ldots, m$, (i.e., $\mathbf{Ax} = \mathbf{b}$), then

$$\sum_{j=1}^n x_j(\sum_{i=1}^m a_{ij} w_i) = \sum_{i=1}^m (\sum_{j=1}^n a_{ij} x_j) w_i = \sum_{i=1}^m b_i w_i \ . \tag{5.90}$$

Therefore, for any $\mathbf{x} > \mathbf{0}$ such that $\mathbf{Ax} = \mathbf{b}$, we have

$$\sum_{i=1}^m b_i w_i - \mu \sum_{j=1}^n \exp\{[(\sum_{i=1}^m a_{ij} w_i - c_j)/\mu] - 1\} \le \sum_{j=1}^n c_j x_j + \mu \sum_{j=1}^n x_j \ln x_j. \tag{5.91}$$

Recall that the right-hand side of (5.91) is the objective function of Program L(μ). Hence we define a geometric dual program DL(μ) as follows:

Program DL(μ):

$$\max_{\mathbf{w}\in R^m} \left\{ d_\mu(\mathbf{w}) \equiv \sum_{i=1}^{m} b_i w_i - \mu \sum_{j=1}^{n} \exp\{[(\sum_{i=1}^{m} a_{ij} w_i - c_j)/\mu] - 1\} \right\}. \quad (5.92)$$

Note that Program DL(μ) is an unconstrained problem and the sum in each of the n exponents in the second term of its objective function is simply the amount of violation of the corresponding constraint of Program DL. However, the nonlinear term of $d_\mu(\mathbf{w})$ is neither a penalty function nor a barrier function in the traditional sense, since the traditional penalty function does not impose penalty on any feasible solution and the traditional barrier function ensures that the objective value approaches infinity at the boundary of feasible region. More importantly, if Program DL(μ) attains a finite optimum at $\mathbf{w}^*(\mu)$, then $\mathbf{w}^*(\mu)$ becomes a feasible solution to Program DL as μ approaches 0. Actually, it solves Program DL as μ approaches 0.

Several observations can be made here.

Observation 5.4: Since Inequality (5.91) holds for any $\mathbf{w} \in R^m$ and $\mathbf{x} \in R^n$ such that $\mathbf{Ax} = \mathbf{b}$ and $\mathbf{x} > \mathbf{0}$, by denoting the minimum objective value of Program L(μ) by min (L(μ)) and the maximum objective value of Program DL(μ) by max (DL(μ)), we have the following "weak duality theorem":

Theorem 5.5. min (L(μ)) $\geq$ max (DL(μ)).

Observation 5.5: Recall that Inequality (5.85) becomes an equality if and only if $z = 1$. Hence Inequality (5.91) becomes an equality if and only if

$$z_j = \frac{\exp\{[(\sum_{i=1}^{m} a_{ij} w_i - c_j)/\mu] - 1\}}{x_j} = 1 \quad (5.93)$$

or, equivalently,

$$x_j = \exp\{[(\sum_{i=1}^{m} a_{ij} w_i - c_j)/\mu] - 1\}. \quad (5.94)$$

Consequently, we have the following theorem:

Theorem 5.6. Given that $\mathbf{w}^* \in R^m$ and $\mathbf{x}^* \in R^n$ such that $\mathbf{Ax}^* = \mathbf{b}$ and $\mathbf{x}^* > \mathbf{0}$, if

$$x_j^* = \exp\{[(\sum_{i=1}^{m} a_{ij} w_i^* - c_j)/\mu] - 1\}, \quad j = 1, \ldots, n, \tag{5.95}$$

then $\mathbf{x}^*$ is an optimal solution to Program L(μ) and $\mathbf{w}^*$ is an optimal solution to Program DL(μ). Moreover, min (L(μ)) = max (DL(μ)).

Observation 5.6: The (k_1, k_2)th element of the Hessian matrix of function $d_\mu(\mathbf{w})$ is given by

$$\frac{\partial^2 d_\mu(\mathbf{w})}{\partial w_{k_1} \partial w_{k_2}} = -\frac{1}{\mu} \sum_{j=1}^{n} \exp\{[(\sum_{i=1}^{m} a_{ij} w_i - c_j)/\mu] - 1\} a_{k_1 j} a_{k_2 j}. \tag{5.96}$$

Therefore, the Hessian matrix can be written as $\mathbf{AD_r}(\mathbf{w})\mathbf{A}^T$, where $\mathbf{D_r}(\mathbf{w})$ is an $n \times n$ diagonal matrix with

$$r_j(\mathbf{w}) = -\frac{1}{\mu} \exp\{[(\sum_{i=1}^{m} a_{ij} w_i - c_j)/\mu] - 1\} < 0 \tag{5.97}$$

being its jth diagonal element. By matrix theory, the Hessian matrix must be nonsingular and negative definite as long as $\mathbf{A}$ has full row-rank. In this case, we know that $d_\mu(\mathbf{w})$ is strictly concave, and we have the following result:

Theorem 5.7. If the constraint matrix $\mathbf{A}$ in Program L has full row-rank, then Program DL(μ) has a strictly concave objective function $d_\mu(\mathbf{w})$.

Observation 5.7: Under the Interior-Point Assumption, Program L (hence Program L(μ)) has an interior feasible solution. From convex analysis (Fenchel's Theorem) [31, 35], we know that there is no duality gap between the pair of Programs L(μ) and DL(μ). Recall that Program L(μ) always achieves a finite optimum as long as $\mu > 0$. Therefore, if $\mathbf{A}$ has full row-rank, then $d_\mu(\mathbf{w})$ is strictly concave and Program DL(μ) must also achieve a finite optimum at a unique maximizer $\mathbf{w}^*(\mu) \in R^m$. Now, since $d_\mu(\mathbf{w})$ is continuously differentiable, the first-order optimality conditions hold at $\mathbf{w}^*(\mu)$. In other words, by setting $\nabla d_\mu(\mathbf{w}^*(\mu)) = \mathbf{0}$, we have

$$\sum_{j=1}^{n} \exp\{[(\sum_{i=1}^{m} a_{ij} w_i^*(\mu) - c_j)/\mu] - 1\} a_{kj} = b_k, \quad k = 1, 2, \ldots, m. \tag{5.98}$$

If we further define $\mathbf{x}^*(\mu) \in R^n$ according to Equation (5.95), then the above equation becomes $\mathbf{A}\mathbf{x}^*(\mu) = \mathbf{b}$. Hence Theorem 5.6 further implies the following "strong duality theorem":

Theorem 5.8. If Program L has an interior feasible solution and a full row-rank constraint matrix $\mathbf{A}$, then Program DL(μ) has a unique optimal solution $\mathbf{w}^*(\mu) \in R^m$. In this case, formula (5.95) provides a dual-to-primal conversion which defines the optimal solution $\mathbf{x}^*(\mu)$ of Program L(μ). Moreover, $\min(\mathrm{L}(\mu)) = \max(\mathrm{DL}(\mu))$.

Note that the above strong duality theorem can also be viewed as a special case of the Fenchel duality theorem in separable convex programming, but it is derived in an explicit manner which tackles our problem directly. Throughout Section 5.2, we make the Interior-Point Assumption and assume that matrix $\mathbf{A}$ has full row-rank. Besides, $\mathbf{x}^*(\mu)$ and $\mathbf{w}^*(\mu)$ will be used to denote the unique optimal solutions of Program L(μ) and DL(μ), respectively.

5.2.2 An ϵ-Optimal Solution

Since Program L(μ) and Program L have the same feasible domain, any solution to Program L(μ) must be feasible to Program L. For any given $\epsilon > 0$, our objective is to find a sufficiently small $\mu > 0$ such that $\mathbf{x}^*(\mu)$ is an ϵ-optimal solution to Program L.

To achieve our goal, we first study the behavior of $\mathbf{x}^*(\mu)$ as μ becomes smaller. For $\mu_1 > \mu_2 > 0$, since $\mathbf{x}^*(\mu_1)$ is an optimal solution to Program L(μ_1), we have

$$\sum_{j=1}^{n} c_j x_j^*(\mu_1) + \mu_1 \sum_{j=1}^{n} x_j^*(\mu_1) \ln x_j^*(\mu_1) \leq \sum_{j=1}^{n} c_j x_j^*(\mu_2) + \mu_1 \sum_{j=1}^{n} x_j^*(\mu_2) \ln x_j^*(\mu_2). \tag{5.99}$$

Similarly, since $\mathbf{x}^*(\mu_2)$ solves Program L(μ_2), we see

$$\sum_{j=1}^{n} c_j x_j^*(\mu_2) + \mu_2 \sum_{j=1}^{n} x_j^*(\mu_2) \ln x_j^*(\mu_2) \leq \sum_{j=1}^{n} c_j x_j^*(\mu_1) + \mu_2 \sum_{j=1}^{n} x_j^*(\mu_1) \ln x_j^*(\mu_1). \tag{5.100}$$

Multiplying Inequality (5.100) by -1 and adding the resulting inequality to (5.99) results in

$$(\mu_1 - \mu_2) \sum_{j=1}^{n} x_j^*(\mu_1) \ln x_j^*(\mu_1) \leq (\mu_1 - \mu_2) \sum_{j=1}^{n} x_j^*(\mu_2) \ln x_j^*(\mu_2). \tag{5.101}$$

Since $\mu_1 > \mu_2$, we know that

$$\sum_{j=1}^{n} x_j^*(\mu_1) \ln x_j^*(\mu_1) \leq \sum_{j=1}^{n} x_j^*(\mu_2) \ln x_j^*(\mu_2). \tag{5.102}$$

After rearranging terms in (5.100) and using (5.102), we further have

$$\begin{aligned} 0 \quad &\leq \quad \mu_2 \left[\sum_{j=1}^{n} x_j^*(\mu_2) \ln x_j^*(\mu_2) - \sum_{j=1}^{n} x_j^*(\mu_1) \ln x_j^*(\mu_1) \right] \\ &\leq \quad \sum_{j=1}^{n} c_j x_j^*(\mu_1) - \sum_{j=1}^{n} c_j x_j^*(\mu_2). \end{aligned}$$

Rearranging terms in (5.99), we obtain

$$\begin{aligned} &\sum_{j=1}^{n} c_j x_j^*(\mu_1) - \sum_{j=1}^{n} c_j x_j^*(\mu_2) \\ &\leq \mu_1 \left[\sum_{j=1}^{n} x_j^*(\mu_2) \ln x_j^*(\mu_2) - \sum_{j=1}^{n} x_j^*(\mu_1) \ln x_j^*(\mu_1) \right]. \end{aligned}$$

Combining the two inequalities above together gives the following theorem:

Theorem 5.9. As μ decreases to 0, $\mathbf{c}^T\mathbf{x}^*(\mu)$ decreases monotonically. Moreover, for $\mu_1 > \mu_2 > 0$,

$$\begin{aligned} 0 &\le \mu_2 \left[\sum_{j=1}^n x_j^*(\mu_2) \ln x_j^*(\mu_2) - \sum_{j=1}^n x_j^*(\mu_1) \ln x_j^*(\mu_1) \right] \\ &\le \mathbf{c}^T\mathbf{x}^*(\mu_1) - \mathbf{c}^T\mathbf{x}^*(\mu_2) \\ &\le \mu_1 \left[\sum_{j=1}^n x_j^*(\mu_2) \ln x_j^*(\mu_2) - \sum_{j=1}^n x_j^*(\mu_1) \ln x_j^*(\mu_1) \right]. \end{aligned} \tag{5.103}$$

Now, in addition to the Interior-Point Assumption, we add the Bounded Feasible Domain Assumption. Under this circumstance, it is not difficult to see that Program L has an optimal solution $\mathbf{x}^*$ and $\mathbf{x}^*(\mu) \to \mathbf{x}^*$ as $\mu \to 0$. Moreover, there exists a large positive number $M > 0$ such that the primal feasible domain is contained in the spheroid centered at the origin with a radius of M. Thus, for any primal feasible solution $\mathbf{x}$, we have

$$| x_j \ln x_j | \le \tau \equiv \max\{1/e, | M \ln M |\}, \text{ for } j = 1, \ldots, n. \tag{5.104}$$

Consequently, by taking $\mu_2 \to 0$ in (5.103), we have

$$| \mathbf{c}^T\mathbf{x}^*(\mu_1) - \mathbf{c}^T\mathbf{x}^* | \le \mu_1(n\tau + n\tau) = 2n\tau\mu_1. \tag{5.105}$$

Therefore, for any given $\epsilon > 0$, if we choose

$$\mu_1 = \epsilon/2n\tau, \tag{5.106}$$

then $|\mathbf{c}^T\mathbf{x}^*(\mu_1) - \mathbf{c}^T\mathbf{x}^*| \le \epsilon$. In other words, $\mathbf{x}^*(\mu_1)$ is an ϵ-optimal solution.

In summary, we have the following result:

Theorem 5.10. Under the Interior-Point and Bounded Feasible Domain Assumptions, if $\mu > 0$ is chosen according to Equation (5.106), then the optimal solution to Program L(μ) is an ϵ-optimal solution to Program L.

A couple of observations can be made here.

Observation 5.8: For a linear program with bounded feasible domain, the variables can be rescaled so that the program is equivalent to a linear programming problem with all variables bounded above by 1. (See den Hertog et al. [22] for details.) In this case, since $-1/e \leq x \ln x \leq 0$ for $0 \leq x \leq 1$, we see that $\tau = 1/e$ and $\mu = \epsilon e/2n$ are good enough for any linear program scaled according to den Hertog et al.

Observation 5.9: Note that the magnitude of the second bracket of Inequality (5.103) is proportional to the difference of entropy evaluated at two different points, one of which is to be used as an approximation of the other. The bound provided in Theorem 5.10 is calculated based on the triangular inequality and hence may be too conservative. In other words, for a practical implementation, μ may be chosen to be much bigger than the value specified by Equation (5.106).

5.2.3 A Curved-Search Algorithm

We now outline the unconstrained convex programming approach to solving standard-form linear programs under the Interior-point and Bounded Feasible Domain Assumptions:
Step 1: Given $\epsilon > 0$, compute μ according to formula (5.106).
Step 2: Solve Program DL(μ) by unconstrained convex optimization techniques for an optimal solution $\mathbf{w}^*(\mu)$.
Step 3: Compute the optimal solution $\mathbf{x}^*(\mu)$ of Program L(μ) according to formula (5.95). Then $\mathbf{x}^*(\mu)$ is an ϵ-optimal solution to Program L.

It is clear that Step 2 is the major source of required computation. In this subsection, we customize the "curved-search method" proposed by Ben-Tal et al. [2] to achieve global convergence with a quadratic rate of local convergence.

Instead of solving Program DL(μ) directly, we solve the following equivalent convex minimization problem:

Program DL$'$(μ) :

$$\min_{\mathbf{w} \in R^m} \{f(\mathbf{w}) \equiv -d_\mu(\mathbf{w})\}. \tag{5.107}$$

As discussed before, when matrix $\mathbf{A}$ is of full row-rank, $f(\mathbf{w})$ is a strictly convex and twice continuously differentiable function over R^m with

$$\frac{\partial f}{\partial w_{k_1}} = \sum_{j=1}^{n} \exp\{[(\sum_{i=1}^{m} a_{ij} w_i^* - c_j)/\mu] - 1\} a_{k_1 j} - b_{k_1} \tag{5.108}$$

and

$$\frac{\partial^2 f}{\partial w_{k_1} \partial w_{k_2}} = \frac{1}{\mu} \sum_{j=1}^{n} \exp\{[(\sum_{i=1}^{m} a_{ij} w_i - c_j)/\mu] - 1\} a_{k_1 j} a_{k_2 j}. \tag{5.109}$$

The Hessian matrix $\mathbf{H}$ is positive definite and can be written as $\mathbf{A}\mathbf{D_s}(\mathbf{w})\mathbf{A}^T$, where $\mathbf{D_s}(\mathbf{w})$ is an $n \times n$ diagonal matrix with $s_j(\mathbf{w})$ as its jth diagonal element such that, for $j = 1, \ldots, n$,

$$s_j(\mathbf{w}) = -r_j(\mathbf{w}) = \frac{1}{\mu} \exp\{[(\sum_{i=1}^{m} a_{ij} w_i - c_j)/\mu] - 1\} > 0. \tag{5.110}$$

Based on the curved-search method described in Subsection 3.2.1 [2, 11], we propose the following CS-LPE (Curved Search for Linear Programs with Entropic perturbation) algorithm for solving Program DL$'(\mu)$:

Algorithm 5.1 (The CS-LPE Algorithm):

Initialization:

Choose sufficiently small real numbers $\zeta > 0$, $\delta > 0$ and $\gamma > 0$.
Choose a sequence α_k such that $\overline{\alpha} \geq \alpha_k \geq 0$ with $\overline{\alpha} > 0$.
Choose a sequence β_k such that $\overline{\beta} \geq \beta_k \geq \underline{\beta}$ with $\overline{\beta} \geq \underline{\beta} > 0$.
Set $k = 0$ and select any starting solution $\mathbf{w}_0 \in R^m$.

Iteration:

Compute $\mathbf{g}_k \equiv \nabla f(\mathbf{w}_k)$ with its k_1th element defined by Equation (5.108).
Compute $\mathbf{H}_k \equiv \nabla^2 f(\mathbf{w}_k)$ with its (k_1, k_2)th element defined by Equation (5.109).
If $\|\mathbf{g}_k\| \leq \zeta$, step. Otherwise, proceed.
Solve $\mathbf{H}_k \mathbf{v}_k = \mathbf{g}_k$ for $\mathbf{v}_k$.
Compute $G_k \equiv \mathbf{g}_k^T \mathbf{v}_k$, $\gamma_k \equiv \frac{|G_k|}{\|\mathbf{v}_k\| \|\mathbf{g}_k\|^2}$ and $\delta_k \equiv (\det \mathbf{H}_k)^2$.

If $\delta_k > \delta$ and $\gamma_k > \gamma$, then
compute $\mathbf{d}_k \equiv -\beta_k \frac{\|\mathbf{g}_k\|^2}{G_k} \mathbf{v}_k$ and $\mathbf{z}_k \equiv -\alpha_k \|\mathbf{g}_k\| \mathbf{g}_k$.
Otherwise, set $\mathbf{d}_k = -\mathbf{g}_k$ and $\mathbf{z}_k = \mathbf{0}$.
Compute a step size $t_k \in \arg\min_{t>0} f(\mathbf{w}_k + t\mathbf{d}_k + \frac{1}{2}t^2 \mathbf{z}_k)$.
Iterate $\mathbf{w}_{k+1} = \mathbf{w}_k + t_k \mathbf{d}_k + \frac{1}{2}t_k^2 \mathbf{z}_k$.
Reset $k \leftarrow k + 1$ for the next iteration.

5.2.4 Convergence Results

In this subsection, we show that, under some mild conditions, the CS-LPE algorithm indeed produces a sequence of iterates that converge to the unique optimal solution of Program $\mathrm{DL}'(\mu)$ with a quadratic rate of convergence. Actually, this result is expected, since the curved-search method is a combination of the signed Newton method (which achieves a quadratic rate of local convergence) and the steepest descent method (which converges globally).

The next two theorems state these facts:

Theorem 5.11. If Program L has an interior feasible solution and a full row-rank constraint matrix **A**, then the CS-LPE algorithm either stops after a finite number of steps or generates an infinite sequence $\{\mathbf{w}_k \in R^m | k = 1, 2, \ldots.\}$ such that (i) $f(\mathbf{w}_{k+1}) < f(\mathbf{w}_k)$ for each k; (ii) $\{\mathbf{w}_k \in R^m | k \in N\}$ has at least a cluster point $\mathbf{w}^c$ in the level set $L_0 \equiv \{\mathbf{w} \in R^m | f(\mathbf{w}) \leq f(\mathbf{w}_0)\}$; (iii) for each such cluster point $\mathbf{w}^c$, $|\nabla f(\mathbf{w}^c)| \leq \zeta$.

Proof. Recall that the objective function f of Program $\mathrm{DL}'(\mu)$ is twice continuously differentiable. According to Theorem 3.1 of [2], if we can further show that the level set L_0 is compact, then the proof is complete. Under current assumptions, by Theorem 5.8, we know that f has a unique minimizer $\mathbf{w}^* \in R^m$ with a finite minimum value. Therefore, each level set of f, including L_0, is compact.

□

Note that, under the assumptions of Theorem 5.11, since f has a unique optimal solution $\mathbf{w}^*$ satisfying the first-order optimality conditions, the infinite sequence generated by the CS-LPE algorithm, with $\zeta = 0$, must converge to the unique cluster point $\mathbf{w}^c = \mathbf{w}^*$.

Theorem 5.12. Under the assumptions of Theorem 5.11, the infinite sequence $\{\mathbf{w}_k \in R^m | k \in N\}$ generated by the CS-LPE algorithm converges to $\mathbf{w}^*$ at a quadratic rate.

Proof. Based on Theorem 5.4 of [2], to establish the quadratic rate of convergence for the CS-LPE algorithm, we need to show that the following four conditions are satisfied: (i) $\{\mathbf{w}_k \in R^m | k \in N\}$ converges to $\mathbf{w}^*$, (ii) $\nabla \mathrm{f}(\mathbf{w}^*) = 0$, (iii) $\nabla^2 f(\mathbf{w}^*)$ is positive definite, and (iv) $\mathbf{H}(\cdot) \equiv \nabla^2 f(\cdot)$ is locally Lipschitz-continuous in a neighborhood of $\mathbf{w}^*$.

The first three conditions are clearly met from our previous discussion. As to condition (iv), recall that $\mathbf{H} = \mathbf{A}\mathbf{D_s}(\mathbf{w})\mathbf{A}^T$ and $\mathbf{D_s}(\mathbf{w})$ is a diagonal matrix with $s_j(\mathbf{w})$ as its jth diagonal element. From Equation (5.110), we know that $s_j(\mathbf{w})$, $j = 1, 2, \ldots, n$, belongs to C^∞. Therefore, every element of $\mathbf{H}$ is a finite combination of C^∞ functions. Since differentiability implies Lipschitz continuity, we know that condition (iv) is satisfied too and the proof is complete.

□

5.2.5 Implementation Issues

Before we report our computational experience with the CS-LPE algorithm, we discuss four related implementation issues:

1. *Step size*: In each iteration of the CS-LPE algorithm, major calculations include solving $\mathbf{H}_k\mathbf{v}_k = \mathbf{g}_k$ for $\mathbf{v}_k$ and finding t_k by minimizing $f(\mathbf{w}_k + t\mathbf{d}_k + \frac{1}{2}t^2\mathbf{z}_k)$ over $t > 0$.

To calculate $\mathbf{v}_k$ in our implementation, we chose the Gaussian Elimination method because our main interest is in investigating the total number of iterations. Clearly, the more efficient Cholesky factorization method may be applied to solving this system of equations involving a positive definite matrix. For the same reason, we chose a commonly used line-search procedure, consisting of a search for a three-point pattern and the Golden-Section search [29], to determine the step size t_k.

2. *Numerical Overflow*: Consider Program $DL'(\mu)$ again. Evaluation of exponential functions is required in calculating f, ∇f, and $\nabla^2 f$. By examing the definition of f and Equations (5.108) and (5.109), it is clear that we have to overcome the possible arithmetic overflow problem, in particular when μ is very small.

To prevent the overflow problem, note that Program $DL'(\mu)$ is a minimization problem and the exponential terms contained in the gradient vector and the Hessian matrix are identical to those contained in the objective function. Therefore, if an initial solution $\mathbf{w}_0$ is properly chosen such that the initial objective value does not cause numerical overflow and hence can be represented as a floating point, then there is no overflow problem in evaluating the exponential functions subsequently. Technically, this can be achieved by choosing $\mathbf{w}_0$ to be feasible to Program DL, since for such a solution all the exponents involved in the objective function are strictly negative no matter how small μ is. Moreover, in searching for the three-point pattern, we need to start at a point that is close to the current solution. With these two precautionary steps, the overflow problem never occurred in our implementation.

3. *Gradient vector and Hessian matrix*: Since the exponential function belongs to the category of transcendental functions, a significant amount of computational effort is needed in evaluating f, ∇f, and $\nabla^2 f$. Realizing the fact that there are n exponential functions involved in the gradient vector and $m^2 n$ exponential functions involved in the Hessian matrix, one might think that the CS-LPE algorithm would require an excessive amount of computation. Fortunately, this is not the case. In our implementation, only n "seed exponential functions" need to be evaluated in each iteration. They are the diagonal elements $s_j(\mathbf{w})$, $j = 1, \ldots, n$, of the diagonal matrix $\mathbf{D_s}(\mathbf{w})$ defined by Equation (5.110). Once these n seeds are computed and stored, both the gradient vector and Hessian matrix can be computed via simple arithmetic operations. Therefore, no excessive computational effort is required. Also, because the total number of these seeds grows linearly with n, i.e., the number of primal variables, the computational time spent on these calculations becomes decreasingly significant, in terms of proportion, as the problem size increases.

4. *Matrix inversion:* Once the n seed exponential functions are calculated and stored, it takes $O(m^2 n)$ elementary operations to compute the Hessian matrix and $O(n^3)$ (or fewer) elementary operations for the matrix inversion. This is the real computational bottleneck for solving large problems. Since every element of the Hessian matrix $\mathbf{H}_k$ can be independently calculated, an implementation with parallel computation can certainly improve the performance of the CS-LPE algorithm. Moreover, the use of sparse matrix techniques can also lead to more efficient implementation for solving large problems. But this is beyond the scope of this book.

5.2.6 Computational Experience

In this subsection, we report our computational experience with the CS-LPE algorithm. The test problems are randomly generated with sizes ranging from 30 variables with 10 constraints to 4,000 variables with 1,000 constraints. All the test problems have a full row-rank constraint matrix $\mathbf{A}$ and satisfy both the Interior-Point and Bounded Feasible Domain assumptions.

We briefly describe how an $m \times n$ standard-form linear program is "randomly" generated for the test. A test linear program contains $m-1$ "regular" constraints and one bounding constraint. The regular constraints are randomly generated as inequality constraints first in the $(n-m)$-dimensional vector space and then converted to equality ones in the n-dimensional space. The coefficients on the left-hand side of the inequalities are randomly generated between -1 and 1. The right-hand side of each of these inequalities is chosen so that (i) the associated $(n-m)$-dimensional hyperplane intersects the line segment connecting (0,0,...,0) and (1,1,...,1) and (ii) the point of intersection is uniformly distributed over this line segment. The sign of the inequality is chosen so that the point (0.5,0.5,...,0.5) is feasible. The bounding constraint is set to be $\sum_{j=1}^{n} x_j \leq M$ for a large constant M. All elements of the cost vector $\mathbf{c}$ are randomly generated between 0 and 1. In this way, the m-dimensional zero vector $\mathbf{0}$ is feasible to Program DL and hence an initial solution $\mathbf{w}_0$ for the CS-LPE algorithm which avoids the otherwise potential problem of arithmetic overflow.

Table 5.2. CS-LPE Algorithm for Randomly-Generated Problems

size	*#prob.*	# it.	cpu%-Hess.	cpu%-inv.	cpu%-other
10×30	4	14.3	18%	11%	71%
25×100	4	35.5	36%	19%	45%
50×150	4	45.0	43%	32%	25%
100×300	4	57.0	47%	37%	16%
300×1000	4	64.3	54%	37%	9%
500×1500	4	80.3	55%	41%	4%
1000×4000	1	94.0	NA*	NA*	NA*

To demonstrate the computational procedure, regardless of problem size, we choose $\mu = 0.001$ in the tests. For purposes of simplicity, we always set $\alpha_k = \beta_k = 1$ and $\zeta = 10^{-5}$. The line search stops when the difference between the objective values of two consecutive test points is within 10^{-10}, and the CS-LPE algorithm stops when $||\mathbf{g}_k|| \leq \zeta = 10^{-5}$.

We ran the CS-LPE algorithm on a SUN SPARCII workstation for all test problems except the 1000×4000 problem, for which we used an Alliant FX8 with its parallel/vector processing capability.

Table 5.2 summarizes our computational experience. The six columns correspond to problem size, the number of test problems for that size, the average number of iterations for the test problems of that size, CPU time spent on Hessian calculations, CPU time spent on Hessian inversion, and CPU for the rest of the calculations. (NA indicates items not included for fair comparison because parallel/vector processing was invoked.)

Some important facts supplement Table 5.2:

(i) According to our experience, the CS-LPE algorithm always converges as it should.

(ii) In order to verify the quadratic rate of convergence, we calculated $r_k \equiv ||\mathbf{w}_k - \mathbf{w}^*|| \,/||\mathbf{w}_{k-1} - \mathbf{w}^*||^2$. In our experiment, r_k indeed converges up until the current solution is in an "immediate" neighborhood of the optimal solution $\mathbf{w}^*$. Then the ratio becomes sensitive to numerical accuracy. This observation together with the low number of iterations could confirm the quadratic rate of convergence of the CS-LPE algorithm.

(iii) By setting the initial solution at a point feasible to Program DL and choosing appropriate first step size in the three-point search, no arithmetic overflow problem ever occurred in our experiment.

(iv) The percentage of cpu-other decreases as problem size increases. This indicates that the computational effort spent in evaluating the exponential functions is relatively small in solving large problems.

(v) It is confirmed that the calculation of the Hessian matrix $\mathbf{H}$ and its inversion really dominates the computational complexity in solving large problems.

(vi) The number of total iterations grows slowly with the size of the problem.

(vii) For test problems with the same size, the number of iterations required varies only slightly.

5.2.7 Concluding Remarks

We have introduced an unconstrained convex programming approach for solving standard-form linear programs. An ϵ-optimal solution is obtained by solving an unconstrained dual convex program. The dual to primal conversion is effortless. We also customized a curved-search method for solving the unconstrained dual convex program. This algorithm converges globally with a quadratic rate of convergence.

Note that the direction of movement of the CS-LPE algorithm depends on its Hessian matrix $\mathbf{H} = \mathbf{A}\mathbf{D_s}(\mathbf{w})\mathbf{A}^T$, while the moving direction of all known interior-point algorithm [20] is determined by a symmetric positive definite matrix with exactly the same structure. This should lead to comparable computing time for one iteration among these different algorithms when applied to a common problem. The fact that the total number of iterations grows slowly with respect to the increase in problem size demonstrates the potential of the unconstrained convex programming approach for efficiently solving large-scale linear programming problems. Among other potential advantages, this approach does not require Phase-I iterations. Finally, perturbing the objective function of a standard-form linear program with an entropic function is equivalent to imposing an exponential penalty function on the objective function of the dual linear program [13].

5.3 CONVEX QUADRATIC PROGRAMMING

In this section, we extend the unconstrained convex programming approach discussed in the two previous sections to solve convex quadratic programming problems [12]. With the aid of two simple inequalities, we show that a convex quadratic programming problem can be suitably perturbed with an entropic function such that its geometric dual [31] turns out to be an unconstrained convex program. Furthermore, we show that, with a well-chosen perturbation parameter, the solution to the geometric dual program generates an ϵ-optimal solution to the original convex quadratic program. This new setting allows us to apply unconstrained convex optimization techniques [3, 17] to solving convex quadratic programming problems.

In this section, we introduce the unconstrained convex dual approach in Subsection 5.3.1. Then an ϵ-optimal primal solution is derived in Subsection 5.3.2. We report some computational results in Subsection 5.3.3 and conclude the section with Subsection 5.3.4.

5.3.1 An Unconstrained Convex Dual Approach

Consider the following convex quadratic programming problem [10]:

Program Q:

$$\min \quad f(\mathbf{x}) \equiv \frac{1}{2}\mathbf{x}^T\mathbf{D}\mathbf{x} + \mathbf{c}^T\mathbf{x} \tag{5.111}$$

$$\text{s.t.} \quad \mathbf{A}\mathbf{x} = \mathbf{b}, \tag{5.112}$$

$$\mathbf{x} \geq \mathbf{0}, \tag{5.113}$$

where $\mathbf{c}, \mathbf{x} \in R^n$, $\mathbf{b} \in R^m$, $\mathbf{D}$ is an $n \times n$ diagonal matrix with diagonal elements $d_j \geq 0$, for $j = 1, \ldots, n$, $\mathbf{A}$ is an $m \times n$ $(m \leq n)$ matrix, and $\mathbf{0}$ is the n-dimensional zero vector.

Note that any linearly constrained convex quadratic program can be converted into the "standard form" as defined by Program Q [10]. In this section, we focus on treating the standard-form convex quadratic programs. Similar to the approach taken in Sections 5.1 and 5.2 [8, 11, 32, 33], for any given $\mu > 0$, instead of solving Program Q directly, we consider a nonlinear program with an entropic perturbation, namely,

Program Q(μ):

$$\min \quad f_\mu(\mathbf{x}) \equiv \frac{1}{2}\mathbf{x}^T\mathbf{D}\mathbf{x} + \mathbf{c}^T\mathbf{x} + \mu\sum_{j=1}^{n} x_j \ln x_j \tag{5.114}$$

$$\text{s.t.} \quad \mathbf{A}\mathbf{x} = \mathbf{b}, \tag{5.115}$$

$$\mathbf{x} \geq \mathbf{0}. \tag{5.116}$$

Since Program Q(μ) and Program Q have the same feasible domain, any solution to Program Q(μ) must be feasible to Program Q. As in the treatment of all interior-point methods, we assume, in this section, that Program Q has an interior feasible solution $\mathbf{x} > \mathbf{0}$ (i.e., the Interior Point Assumption). Under this assumption, we see that Program Q(μ) is feasible for any given $\mu > 0$. Also note that the strict convexity of the entropic function and the convention of $0 \ln 0 = 0$ imply that Program Q(μ) achieves a finite minimum at a unique point $\mathbf{x}^*(\mu) \in R^n$ for $\mu > 0$. More interestingly, if Program Q has a bounded feasible domain (i.e., the Bounded Feasible Domain Assumption), then $\mathbf{x}^*(\mu)$ approaches an optimal solution of Program Q as μ approaches 0. Our objective is to find a sufficiently small $\mu > 0$ such that its optimal solution $\mathbf{x}^*(\mu)$ is an ϵ-optimal solution to Program Q, i.e., $|f(\mathbf{x}^*(\mu)) - f(\mathbf{x}^*)| \leq \epsilon$, where $\mathbf{x}^*$ solves Program Q.

As shown in the linear case, superficially, Program Q(μ) seems to be more complicated than Program Q because of the inclusion of the nonlinear entropic function. However, this nonlinearity is the key to deriving an unconstrained dual convex program. First, let us consider the following inequality:

$$\ln z \leq z - 1, \ \forall\, z > 0. \tag{5.117}$$

The above inequality becomes an equality if and only if $z = 1$. Now, for any given positive function $h_j(\mathbf{w})$ defined on $\mathbf{w} \in R^m$ and $x_j > 0$, we define

$$z \equiv \frac{h_j(\mathbf{w})}{x_j}. \tag{5.118}$$

In this way, Inequality (5.117) implies that, for $\mu > 0$,

$$\mu \ln \left(\frac{h_j(\mathbf{w})}{x_j} \right) \leq \frac{\mu h_j(\mathbf{w})}{x_j} - \mu. \tag{5.119}$$

Next, we consider the following simple inequality:

$$yz^2 \geq 0, \ \forall \ z \in R \text{ and } y \geq 0. \tag{5.120}$$

This time, with the positive function $h_j(\mathbf{w})$, any positive x_j, and the nonnegative diagonal element d_j of the given matrix $\mathbf{D}$, we define $z \equiv x_j - h_j(\mathbf{w})$ and $y \equiv \frac{d_j}{2x_j}$. Then Inequality (5.120) implies that

$$-\frac{d_j(x_j - h_j(\mathbf{w}))^2}{2x_j} \leq 0, \ \forall \ x_j > 0. \tag{5.121}$$

Combining (5.119) and (5.121), we have

$$-\frac{d_j x_j^2 + d_j h_j^2(\mathbf{w}) - 2d_j x_j h_j(\mathbf{w})}{2x_j} + \mu \ln \left(\frac{h_j(\mathbf{w})}{x_j} \right) \leq \frac{\mu h_j(\mathbf{w})}{x_j} - \mu \tag{5.122}$$

or, equivalently,

$$-\frac{d_j x_j^2 + d_j h_j^2(\mathbf{w})}{2x_j} + [d_j h_j(\mathbf{w}) + \mu \ln h_j(\mathbf{w})] - \mu \ln x_j \leq \frac{\mu h_j(\mathbf{w})}{x_j} - \mu. \tag{5.123}$$

Note that, for any $\mathbf{w} \in R^m$, $h_j(\mathbf{w}) > 0$ can be defined as the unique solution of the following equation:

$$d_j h_j(\mathbf{w}) + \mu \ln h_j(\mathbf{w}) = \sum_{i=1}^{m} a_{ij} w_i - c_j - \mu. \tag{5.124}$$

Hence Inequality (5.123) becomes

$$-\frac{d_j x_j^2 + d_j h_j^2(\mathbf{w})}{2x_j} + \sum_{i=1}^{m} a_{ij} w_i - c_j - \mu - \mu \ln x_j \leq \frac{\mu h_j(\mathbf{w})}{x_j} - \mu. \tag{5.125}$$

By multiplying both sides by $x_j > 0$ and summing over index j, we have

$$-\frac{1}{2}\sum_{j=1}^{n} d_j x_j^2 - \frac{1}{2}\sum_{j=1}^{n} d_j h_j^2(\mathbf{w}) + \sum_{i=1}^{m} w_i \sum_{j=1}^{n} a_{ij} x_j - \sum_{j=1}^{n} c_j x_j - \mu \sum_{j=1}^{n} x_j \ln x_j$$

$$\leq \mu \sum_{j=1}^{n} h_j(\mathbf{w}). \qquad (5.126)$$

Suppose that $\mathbf{x} > \mathbf{0}$ also satisfies (5.115). Then, after rearrangement, we get

$$\frac{1}{2}\sum_{j=1}^{n} d_j x_j^2 + \sum_{j=1}^{n} c_j x_j + \mu \sum_{j=1}^{n} x_j \ln x_j$$

$$\geq -\frac{1}{2}\sum_{j=1}^{n} d_j h_j^2(\mathbf{w}) - \mu \sum_{j=1}^{n} h_j(\mathbf{w}) + \sum_{i=1}^{m} b_i w_i. \qquad (5.127)$$

Note that the left-hand side of the above inequality is exactly the objective function of Program Q(μ). For $\mathbf{w} \in R^m$ and $\mu > 0$, Equation (5.124) defines a unique solution function $h_j(\mathbf{w}) > 0$. With this function $h_j(\mathbf{w})$, we are now ready to define an unconstrained dual program:

Program DQ(μ):

$$\max_{\mathbf{w} \in R^m} \left\{ d_\mu(\mathbf{w}) \equiv -\frac{1}{2}\sum_{j=1}^{n} d_j h_j^2(\mathbf{w}) - \mu \sum_{j=1}^{n} h_j(\mathbf{w}) + \sum_{i=1}^{m} b_i w_i \right\}. \qquad (5.128)$$

Several observations can be made here.

Observation 5.10: If we treat $h_j(\mathbf{w})$ as a variable, then Inequality (5.124) is a single-variable equation which could be solved by a straightforward one-dimensional search. Also note that, in the absence of the quadratic term in the primal objective function, $h_j(\mathbf{w})$ can be solved explicitly.

Observation 5.11: Since Inequality (5.127) holds for any $\mathbf{w} \in R^m$ and $\mathbf{x} \in R^n$ such that $\mathbf{Ax} = \mathbf{b}$ and $\mathbf{x} > \mathbf{0}$, by denoting the minimum objective value of Program Q(μ) by min (Q(μ)) and the maximum objective value of Program DQ(μ) by max (DQ(μ)), we have the following "weak duality theorem":

Theorem 5.13. $\min(Q(\mu)) \geq \max(DQ(\mu))$.

Observation 5.12: Recall that both Inequalities (5.119) and (5.121) become equalities if and only if $x_j = h_j(\mathbf{w})$. Hence, Inequality (5.127) becomes an equation if and only if $x_j = h_j(\mathbf{w})$. Consequently, we have the following theorem:

Theorem 5.14. Given $\mathbf{w}^* \in R^m$ and $\mathbf{x}^* \in R^n$ such that $\mathbf{A}\mathbf{x}^* = \mathbf{b}$ and $\mathbf{x}^* > \mathbf{0}$, if

$$d_j x_j^* + \mu \ln x_j^* = \sum_{i=1}^{m} a_{ij} w_i^* - c_j - \mu, \tag{5.129}$$

then $\mathbf{x}^*$ is an optimal solution to Program $Q(\mu)$ and $\mathbf{w}^*$ is an optimal solution to Program $DQ(\mu)$. Moreover, $\min(Q(\mu)) = \max(DQ(\mu))$.

We now investigate the gradient vector and the Hessian matrix of the dual objective function $d_\mu(\mathbf{w})$. The kth element of the gradient vector is given by

$$\frac{\partial d_\mu(\mathbf{w})}{\partial w_k} = -\frac{1}{2}\sum_{j=1}^{n} 2d_j h_j(\mathbf{w})\frac{\partial h_j(\mathbf{w})}{\partial w_k} - \mu \sum_{j=1}^{n} \frac{\partial h_j(\mathbf{w})}{\partial w_k} + b_k, \tag{5.130}$$

where $\frac{\partial h_j(\mathbf{w})}{\partial w_k}$ can be obtained as follows. First we let $x_j^* = h_j(\mathbf{w})$ and define

$$F_j(x_j^*, \mathbf{w}) \equiv d_j x_j^* + \mu \ln x_j^* - \sum_{i=1}^{m} a_{ij} w_i + c_j + \mu. \tag{5.131}$$

Then, from Equation (5.129) and by applying chain rule, we have

$$\frac{\partial F_j}{\partial x_j^*}\frac{\partial x_j^*}{\partial w_k} + \frac{\partial F_j}{\partial w_k} = 0. \tag{5.132}$$

Equivalently, we see that

$$\left(d_j + \frac{\mu}{x_j^*}\right)\frac{\partial h_j(\mathbf{w})}{\partial w_k} - a_{kj} = 0. \tag{5.133}$$

In other words,

$$\frac{\partial h_j(\mathbf{w})}{\partial w_k} = \frac{a_{kj}}{d_j + \frac{\mu}{x_j^*}} = \frac{a_{kj} h_j(\mathbf{w})}{d_j h_j(\mathbf{w}) + \mu}. \tag{5.134}$$

Plugging this into Equation (5.130), we finally have

$$\begin{aligned} \frac{\partial d_\mu(\mathbf{w})}{\partial w_k} &= -\frac{1}{2}\sum_{j=1}^{n} 2 d_j h_j(\mathbf{w}) \frac{a_{kj} h_j(\mathbf{w})}{d_j h_j(\mathbf{w}) + \mu} - \mu \sum_{j=1}^{n} \frac{a_{kj} h_j(\mathbf{w})}{d_j h_j(\mathbf{w}) + \mu} + b_k \\ &= -\sum_{j=1}^{n} a_{kj} \left(\frac{1}{d_j h_j(\mathbf{w}) + \mu} \right) [h_j(\mathbf{w})(d_j h_j(\mathbf{w}) + \mu)] + b_k \\ &= -\sum_{j=1}^{n} a_{kj} h_j(\mathbf{w}) + b_k. \end{aligned} \tag{5.135}$$

Moreover, the Hessian Matrix is given by

$$\begin{aligned} \frac{\partial^2 d_\mu(\mathbf{w})}{\partial w_{k_1} \partial w_{k_2}} &= \frac{\partial}{\partial w_{k_2}} \left[-\sum_{j=1}^{n} a_{k_1 j} h_j(\mathbf{w}) + b_{k_1} \right] \\ &= \sum_{j=1}^{n} -a_{k_1 j} \left[\frac{\partial}{\partial w_{k_2}} h_j(\mathbf{w}) \right] \\ &= \sum_{j=1}^{n} -a_{k_1 j} \frac{a_{k_2 j} h_j(\mathbf{w})}{d_j h_j(\mathbf{w}) + \mu} \\ &= \sum_{j=1}^{n} \left(-\frac{h_j(\mathbf{w})}{d_j h_j(\mathbf{w}) + \mu} \right) a_{k_1 j} a_{k_2 j}. \end{aligned} \tag{5.136}$$

Since $h_j(\mathbf{w}) > 0$, $d_j \geq 0$, and $\mu > 0$, we know that $-\frac{h_j(\mathbf{w})}{d_j h_j(\mathbf{w})+\mu} < 0$. Consequently, the Hessian matrix can be written as $\mathbf{A}\mathbf{D}_\mathbf{r}(\mathbf{w})\mathbf{A}^T$, where $\mathbf{D}_\mathbf{r}(\mathbf{w})$ is an $n \times n$ diagonal matrix with negative diagonal elements $r_j(\mathbf{w}) = -\frac{h_j(\mathbf{w})}{d_j h_j(\mathbf{w})+\mu}$. By matrix theory, the Hessian matrix must be nonsingular and negative definite as long as $\mathbf{A}$ has full row-rank. In this case, we know that $d_\mu(\mathbf{w})$ is strictly concave and we have the following result:

Theorem 5.15. If the constraint matrix $\mathbf{A}$ in Program Q has full row-rank, then Program DQ(μ) has a strictly concave objective function $d_\mu(\mathbf{w})$.

Observation 5.13: Under the Interior-Point Assumption, Program Q (hence Program Q(μ)) has an interior feasible solution. By Fenchel's Theorem of convex analysis [31, 35], we know that there is no duality gap between Programs Q(μ) and DQ(μ). Recall that Program Q(μ) always achieves a finite optimum as long as $\mu > 0$. Therefore, if $\mathbf{A}$ has full row-rank, then $d_\mu(\mathbf{w})$ is strictly concave and Program DQ(μ) must also achieve a finite optimum at a unique maximizer $\mathbf{w}^*(\mu) \in R^m$. Now, since $d_\mu(\mathbf{w})$ is continuously differentiable, the first-order optimality conditions hold at $\mathbf{w}^*(\mu)$. In other words, by setting $\nabla d_\mu(\mathbf{w}^*(\mu)) = \mathbf{0}$, we have

$$\sum_{j=1}^{n} a_{kj} h_j(\mathbf{w}^*(\mu)) = b_k, \quad k = 1, 2, \ldots, m. \tag{5.137}$$

When we define $x_j^*(\mu) = h_j(\mathbf{w}^*(\mu)) > 0$ according to (5.129), Equation (5.137) becomes $\mathbf{A}\mathbf{x}^*(\mu) = \mathbf{b}$. We further have the following "strong duality theorem."

Theorem 5.16. Given $\mu > 0$, if Program Q has an interior feasible solution and its constraint matrix $\mathbf{A}$ has full row-rank, then Program DQ(μ) has a unique optimal solution $\mathbf{w}^*(\mu) \in R^m$. In this case, formula (5.129) provides a dual-to-primal conversion which defines the optimal solution $\mathbf{x}^*(\mu)$ of Program Q(μ). Moreover, min(Q(μ)) = max(DQ(μ)).

Throughout this section, we hold the Interior-Point Assumption and assume that matrix $\mathbf{A}$ has full row-rank. In this section, $\mathbf{x}^*(\mu)$ and $\mathbf{w}^*(\mu)$ will be used to denote the unique optimal solutions of Q(μ) and DQ(μ), respectively.

5.3.2 An ϵ-Optimal Solution

Note that Programs Q(μ) and Q have the same feasible domain. For any given $\epsilon > 0$, our strategy is to find a sufficiently small $\mu > 0$ such that $\mathbf{x}^*(\mu)$ is an ϵ-optimal solution to Program Q. In order to achieve this goal, we first study the behavior of $\mathbf{x}^*(\mu)$ as μ becomes smaller.

Assume that $\mathbf{x}^*(\mu_1) \equiv (x_1^*(\mu_1), \ldots, x_n^*(\mu_1))^T$ solves Program Q(μ_1) and $\mathbf{x}^*(\mu_2) \equiv (x_1^*(\mu_2), \ldots, x_n^*(\mu_2))^T$ solves Program Q(μ_2), where $\mu_1 > \mu_2 > 0$. Then, since $\mathbf{x}^*(\mu_1)$ is optimal to Program Q(μ_1), we have

$$\frac{1}{2}\sum_{j=1}^{n} d_j(x_j^*(\mu_1))^2 + \sum_{j=1}^{n} c_j x_j^*(\mu_1) + \mu_1 \sum_{j=1}^{n} x_j^*(\mu_1)\ln x_j^*(\mu_1)$$

$$\leq \frac{1}{2}\sum_{j=1}^{n} d_j(x_j^*(\mu_2))^2 + \sum_{j=1}^{n} c_j x_j^*(\mu_2) + \mu_1 \sum_{j=1}^{n} x_j^*(\mu_2)\ln x_j^*(\mu_2). \qquad (5.138)$$

Similarly, since $\mathbf{x}^*(\mu_2)$ solves Program Q(μ_2), we see

$$\frac{1}{2}\sum_{j=1}^{n} d_j(x_j^*(\mu_2))^2 + \sum_{j=1}^{n} c_j x_j^*(\mu_2) + \mu_2 \sum_{j=1}^{n} x_j^*(\mu_2)\ln x_j^*(\mu_2)$$

$$\leq \frac{1}{2}\sum_{j=1}^{n} d_j(x_j^*(\mu_1))^2 + \sum_{j=1}^{n} c_j x_j^*(\mu_1) + \mu_2 \sum_{j=1}^{n} x_j^*(\mu_1)\ln x_j^*(\mu_1). \qquad (5.139)$$

By multiplying (5.139) by -1 and adding the resulting inequality to (5.138), we obtain

$$(\mu_1 - \mu_2)\sum_{j=1}^{n} x_j^*(\mu_1)\ln x_j^*(\mu_1) \leq (\mu_1 - \mu_2)\sum_{j=1}^{n} x_j^*(\mu_2)\ln x_j^*(\mu_2). \qquad (5.140)$$

Since $\mu_1 > \mu_2$, we have

$$\sum_{j=1}^{n} x_j^*(\mu_1)\ln x_j^*(\mu_1) \leq \sum_{j=1}^{n} x_j^*(\mu_2)\ln x_j^*(\mu_2). \qquad (5.141)$$

Rearranging terms in (5.139) and using (5.141), we further have

$$\begin{aligned} 0 &\leq \mu_2 \left(\sum_{j=1}^{n} x_j^*(\mu_2)\ln x_j^*(\mu_2) - \sum_{j=1}^{n} x_j^*(\mu_1)\ln x_j^*(\mu_1) \right) \\ &\leq \left(\frac{1}{2}\sum_{j=1}^{n} d_j(x_j^*(\mu_1))^2 + \sum_{j=1}^{n} c_j x_j^*(\mu_1) \right) \\ &\quad - \left(\frac{1}{2}\sum_{j=1}^{n} d_j(x_j^*(\mu_2))^2 + \sum_{j=1}^{n} c_j x_j^*(\mu_2) \right) \qquad (5.142) \end{aligned}$$

By rearranging terms in (5.138), we see that

$$\left[\frac{1}{2}\sum_{j=1}^{n} d_j(x_j^*(\mu_1))^2 + \sum_{j=1}^{n} c_j x_j^*(\mu_1)\right] - \left[\frac{1}{2}\sum_{j=1}^{n} d_j(x_j^*(\mu_2))^2 + \sum_{j=1}^{n} c_j x_j^*(\mu_2)\right]$$

$$\leq \mu_1 \left[\sum_{j=1}^{n} x_j^*(\mu_2)\ln x_j^*(\mu_2) - \sum_{j=1}^{n} x_j^*(\mu_1)\ln x_j^*(\mu_1)\right]. \qquad (5.143)$$

Combining the above two inequalities together gives the following result:

Theorem 5.17. As μ decreases to 0, $f(\mathbf{x}^*(\mu))$ decreases monotonically. Moreover, for $\mu_1 > \mu_2 > 0$, if $\mathbf{x}^*(\mu_1)$ and $\mathbf{x}^*(\mu_2)$ solves Programs Q(μ_1) and Q(μ_2), respectively, then

$$\begin{aligned} 0 &\leq \mu_2 \left[\sum_{j=1}^{n} x_j^*(\mu_2)\ln x_j^*(\mu_2) - \sum_{j=1}^{n} x_j^*(\mu_1)\ln x_j^*(\mu_1)\right] \\ &\leq f(\mathbf{x}^*(\mu_1)) - f(\mathbf{x}^*(\mu_2)) \\ &\leq \mu_1 \left[\sum_{j=1}^{n} x_j^*(\mu_2)\ln x_j^*(\mu_2) - \sum_{j=1}^{n} x_j^*(\mu_1)\ln x_j^*(\mu_1)\right]. \qquad (5.144) \end{aligned}$$

Observation 5.14: Now, in addition to the Interior-Point Assumption, we add the Bounded Feasible Domain Assumption. In this case, as we discussed before, Program Q has an optimal solution $\mathbf{x}^*$ and $\mathbf{x}^*(\mu) \to \mathbf{x}^*$ as $\mu \to 0$. Moreover, there exists a number $M > 0$ such that the primal feasible domain defined by (5.112) and (5.113) is contained in the spheroid centered at the origin with a radius of M. Thus, for any primal feasible solution $\mathbf{x}$, we have

$$|x_j \ln x_j| \leq \tau \equiv \max\{1/e, |M \ln M|\}, \text{ for } j = 1, \ldots, n, \qquad (5.145)$$

where e = 2.713.... Consequently, by forcing $\mu_2 \to 0$ in Theorem 5.17, we have

$$|f(\mathbf{x}^*(\mu_1)) - f(\mathbf{x}^*)| \leq \mu_1(n\tau + n\tau) = 2n\tau\mu_1. \qquad (5.146)$$

Therefore, for any given $\epsilon > 0$, if we choose

$$\mu_1 = \epsilon/2n\tau, \tag{5.147}$$

then $|f(\mathbf{x}^*(\mu_1)) - f(\mathbf{x}^*)| \leq \epsilon$. In other words, $\mathbf{x}^*(\mu_1)$ is an ϵ-optimal solution to Program Q. In summary, we have the following result:

Theorem 5.18. Under the Interior-Point and Bounded Feasible Domain Assumptions, if $\mu > 0$ is chosen according to Equation (5.147), then the optimal solution $\mathbf{x}^*(\mu)$ to Program Q(μ) is an ϵ-optimal solution to Program Q.

One more observation can be made here.

Observation 5.15: Note that the magnitude of the second bracket of Inequality (5.144) in Theorem 5.17 is proportional to the difference of the entropy function evaluated at two different points, one of which is to be used as an approximation of the other. The bound provided in Theorem 5.18 is calculated based on the triangular inequality and hence may be too conservative. For practical implementations, an ϵ-optimal solution may be obtained by using a much bigger μ than the value specified by (5.147).

5.3.3 Computational Experience

We now outline a computational procedure for solving standard-form convex quadratic programs with an interior solution and a bounded feasible domain as follows.

Step 1: Given $\epsilon > 0$, compute μ according to Equation (5.147).

Step 2: Evaluate $h_j(\mathbf{w})$ by solving Equation (5.124) and solve Program DQ (μ) by using any unconstrained convex programming techniques for an optimal solution $\mathbf{w}^*(\mu)$.

Step 3: Compute the optimal solution $\mathbf{x}^*(\mu)$ of Program Q(μ) according to Equation (5.129).

Step 4: Output $\mathbf{x}^*(\mu)$ as an ϵ-optimal solution to Program Q.

It is clear that Step 2 accounts for the major computation. Different unconstrained convex optimization techniques may lead to different performance. We suggest a closer investigation of this subject. Here we report some compu-

Table 5.3: QP and LP Computational Results

size	QP			LP		
	#(iter)	cpu(s)	cpu/iter	#(iter)	cpu(s)	cpu/iter
10×30	17	13	0.7	9	1.4	0.15
25×100	21	57	2.7	34	37.7	1.11
100×300	26	900	34.6	63	2305.7	36.60
300×1000	28	41285	1474.5	59	55748.2	944.88
500×1500	27	176848	6549.9	73	315312.5	4319.35

tational experience with the "curved-search method" described in Subsection 3.2.1 [2, 11]. We only compare the results of applying the curved-search method to both the randomly generated standard-form convex quadratic programs and linear programs. The linear programs are solved using exactly the same approach, only with the absence of the quadratic term in the objective function $f_\mu(\mathbf{x})$ of Program Q(μ). As pointed out in *Observation 5.10*, in the absence of this quadratic term, $h_j(\mathbf{w})$ can be solved explicitly. The computer program is written in FORTRAN and runs on Alliant FX8 with the parallelism capability suppressed.

With $\mu = 10^{-3}$, the results are summarized in Table 5.3.

Several facts can be read from Table 5.3: (i) The total number of iterations needed for solving the convex quadratic programs grows very slowly as the problem size becomes larger; (ii) As problem size becomes larger, each iteration of QP consumes more cpu-time; (iii) Compared to LP, QP takes fewer iterations but consumes more cpu-time in each iteration.

We also observed that the calculation of the implicit function $h_j(\mathbf{w})$ takes very little computational overhead for large problems. Moreover, the quadratic term in the dual objective of QP takes relatively little computational effort too. But the calculation of the Hessian matrix of QP involves more elementary operations than the LP case and, hence, QP consumes more cpu-time in each iteration. As for the slow growth of the total number of iterations needed by QP, it is because the quadratic terms appearing in the dual objective function make the unconstrained dual smoother, which the curved-search method can take advantage of.

5.3.4 Concluding Remarks

We have extended the unconstrained convex programming approach to solving convex quadratic programming problems. An ϵ-optimal solution is obtained by solving an unconstrained dual convex program. A dual-to-primal conversion formula is also provided. The computational results, with the use of the curved-search method, show that the number of iterations required for convergence is insensitive to the size of the problem. Moreover, the post-optimality analysis, including parametric programming and sensitivity analysis, can be more easily performed because of the unconstrained nature of this approach. Other unconstrained convex optimization techniques for solving the convex quadratic programming problems may be worth further investigation.

5.4 LINEAR AND CONVEX QUADRATIC SEMI-INFINITE PROGRAMMING

This section develops theory and algorithms for solving linear and convex quadratic semi-infinite programs. Since a linear semi-infinite program is a special case of convex quadratic semi-infinite programs, we develop the theory explicitly for the latter. For a good review on general semi-infinite programming, the reader is contained in Hettich and Kortanek [23].

Consider the following mathematical program with n variables and infinitely many constraints:

Program QSI:

$$\min \quad f(\mathbf{x}) = \frac{1}{2}\mathbf{x}^T\mathbf{D}\mathbf{x} + \sum_{j=1}^{n} c_j x_j \tag{5.148}$$

$$\text{s.t.} \quad \sum_{j=1}^{n} g_j(t)x_j \geq h(t), \quad \forall\, t \in T, \tag{5.149}$$

$$x_j \geq 0, \quad j = 1, 2, \cdots, n, \tag{5.150}$$

where $\mathbf{D}$ is an $n \times n$ diagonal matrix with diagonal elements $d_j \geq 0$, $j = 1, ..., n$, T is a compact metric space with an infinite cardinality, g_j, $j = 1, 2, ..., n$, and h are real-valued continuous functions defined on T. Note that when T reduces to a finite set, the problem becomes a finite convex quadratic programming problem with inequality constraints. Also note that T can be extended to a compact Hausdorff space [1, 26] without much difficulty. In general, when $\mathbf{D}$ is a symmetric positive semi-definite matrix, it is a standard procedure [10] to transform the problem into the form of Program QSI.

In case $\mathbf{D}$ is a zero matrix, then Program QSI becomes the following linear semi-infinite programming problem:

Program LSI:

$$\min \quad f(\mathbf{x}) = \sum_{j=1}^{n} c_j x_j \tag{5.151}$$

$$\text{s.t.} \quad \sum_{j=1}^{n} g_j(t) x_j \geq h(t), \quad \forall\, t \in T, \tag{5.152}$$

$$x_j \geq 0, \quad j = 1, 2, ..., n. \tag{5.153}$$

The theory and algorithms for solving Program QSI have been studied by many researchers [23, 19, 21, 26, 9]. According to [23], there exist three "basic" solution approaches, namely, the exchange methods (or cutting plane methods), discretization methods, and methods based on local reduction. All of these methods usually replace a semi-infinite programming problem, such as Program QSI or LSI, by one (for discretization methods) or a sequence of finite convex programming problems for approximation and, then, prove that the sequence of approximate solutions converges to an optimal solution of the original problem.

For Program QSI, a cutting-plane method can be developed as follows.

In the kth iteration with $T_k = \{t_1, t_2, \cdots, t_k\} \subset T$, consider the following convex quadratic program with inequality constraints:

Program QI^k:

$$\min \quad f(\mathbf{x}) = \frac{1}{2}\mathbf{x}^T \mathbf{D} \mathbf{x} + \sum_{j=1}^{n} c_j x_j \tag{5.154}$$

$$\text{s.t.} \qquad \sum_{j=1}^{n} g_j(t_i)x_j \geq h(t_i), \quad i = 1, \cdots, k, \tag{5.155}$$

$$x_j \geq 0, \quad j = 1, \cdots, n. \tag{5.156}$$

Then solve Program $\mathrm{QI^k}$ for an "exact" optimal solution $\mathbf{x}^k = (x_1^k, x_2^k, \cdots, x_n^k)^T$. Find

$$t_{k+1} \in \arg\min_{t \in T} \left\{ \sum_{j=1}^{n} g_j(t)x_j^k - h(t) \right\}. \tag{5.157}$$

If $\sum_{j=1}^{n} g_j(t_{k+1})x_j^k - h(t_{k+1}) \geq 0$, then $\mathbf{x}^k$ must be an optimal solution to Program QSI, because Program $\mathrm{QI^k}$ has a larger feasible domain than that of Program QSI. Otherwise, $t_{k+1} \notin T_k$. We set $T_{k+1} = T_k \bigcup \{t_{k+1}\}$ and continue the iterative process.

In this section, we study Program QSI through entropic perturbation and propose a path–following algorithm for solving Program QSI. A simpler version of the proposed algorithm can find an "ε–optimal solution" of Program QSI, for any given $\varepsilon > 0$.

In this section, we first introduce a perturbed quadratic semi–infinite programming problem with the help of an entropic function in Subsection 5.4.1. A solution procedure for this perturbed problem and a convergence proof are then given. In Subsection 5.4.2, the basic properties of entropic path-following are studied. Then, a path–following algorithm and its variants are proposed in Subsection 5.4.3. Some concluding remarks are made in Subsection 5.4.4.

5.4.1 Entropic Perturbation for QSIP

Given Program QSI and any perturbation parameter $\mu > 0$, consider the following perturbed quadratic semi–infinite program:

Program QSI(μ):

$$\min \qquad f_\mu(\mathbf{x}) = \frac{1}{2}\mathbf{x}^T \mathbf{D}\mathbf{x} + \sum_{j=1}^{n} c_j x_j + \mu \sum_{j=1}^{n} x_j \ln x_j \tag{5.158}$$

$$\text{s.t.} \quad \sum_{j=1}^{n} g_j(t)x_j \geq h(t), \quad \forall\, t \in T, \tag{5.159}$$

$$x_j \geq 0, \quad j = 1, \cdots, n. \tag{5.160}$$

As shown in the previous sections, the entropic perturbation approach can be used to solve the convex quadratic programs in a finite dimensional space. Here we further extend the approach for solving Program QSI.

Let us denote the common feasible domain of Programs QSI and QSI(μ) by

$$F \equiv \left\{ \mathbf{x} \in R^n \middle| \sum_{j=1}^{n} g_j(t)x_j \geq h(t), \forall\, t \in T, \ \text{ and } x_j \geq 0, \ j = 1, \cdots, n \right\},$$

and its interior by

$$F^0 \equiv \left\{ \mathbf{x} \in R^n \middle| \sum_{j=1}^{n} g_j(t)x_j > h(t), \forall\, t \in T, \ \text{ and } x_j > 0, \ j = 1, \cdots, n \right\}.$$

Since the functions $\frac{1}{2}\mathbf{x}^T\mathbf{D}\mathbf{x}, \mathbf{c}^T\mathbf{x}$ are convex and $x_j \ln x_j$ is strictly convex over $\mathbf{x} \geq \mathbf{0}$, the perturbed objective function $f_\mu(\mathbf{x})$ must be strictly convex and hence has a unique global minimizer. Therefore, Program QSI has a unique optimal solution as long as it is feasible, i.e., $F \neq \emptyset$. Similar to most treatment of interior–point approaches in the literature, we make an "Interior–point Assumption" in this section, i.e., $F^0 \neq \emptyset$.

For any finite subset $T_k = \{t_1, t_2, \cdots, t_k\} \subseteq T$, Program QSI($\mu$) can be approximated by the following entropy optimization problem with n variables and k explicit linear constraints:

Program QI^k(μ):

$$\min \quad f_\mu(\mathbf{x}) = \frac{1}{2}\mathbf{x}^T\mathbf{D}\mathbf{x} + \sum_{j=1}^{n} c_j x_j + \mu \sum_{j=1}^{n} x_j \ln x_j \tag{5.161}$$

$$\text{s.t.} \quad \sum_{j=1}^{n} g_j(t_i)x_j \geq h(t_i), \quad i = 1, \cdots, k, \tag{5.162}$$

$$x_j \geq 0, \quad j = 1, \cdots, n. \tag{5.163}$$

Some computationally efficient algorithms have recently become available for solving Program $\mathrm{QI}^{\mathrm{k}}(\mu)$ [12, 10]. Although Program $\mathrm{QI}^{\mathrm{k}}(\mu)$ has a larger feasible domain than Program QSI(μ), it has a unique finite optimal solution. Denote the optimal solution by $\mathbf{x}^k(\mu) = \left(x_1^k(\mu), \cdots, x_n^k(\mu)\right)^T$. If we define the following "constraint violation function":

$$v_{k+1}(t) \equiv h(t) - \sum_{j=1}^{n} g_j(t) x_j^k(\mu), \quad \forall\, t \in T, \tag{5.164}$$

then $v_{k+1}(t)$ is a continuous function over a compact metric space T. Hence a maximizer $t_{k+1} \in T$ can be found for $v_{k+1}(t)$ over T. Now, if $v_{k+1}(t_{k+1}) \leq 0$, then $\mathbf{x}^k(\mu)$ is also a feasible solution of Program QSI(μ) and hence $\mathbf{x}^k(\mu)$ is optimal for Program QSI(μ). Otherwise, we know $t_{k+1} \notin T_k$. We now have sufficient background to outline an algorithm for solving Program QSI(μ), for any given $\mu > 0$:

Algorithm 5.2: (for Program QSI(μ))

Step 1: Set $k = 1$; select any $t_1 \in T$, and set $T_1 = \{t_1\}$.

Step 2: Solve Program $\mathrm{QI}^{\mathrm{k}}(\mu)$ for an optimal solution $\mathbf{x}^k(\mu)$.

Step 3: Find a maximizer t_{k+1} of $v_{k+1}(t)$ over T with an optimum value

$$v_{k+1}(t_{k+1}) = h(t_{k+1}) - \sum_{j=1}^{n} g_j(t_{k+1}) x_j^k(\mu).$$

Step 4: If $v_{k+1}(t_{k+1}) \leq 0$, stop! $\mathbf{x}^k(\mu)$ is optimal to Program QSI(μ). Otherwise, set $T_{k+1} = T_k \cup \{t_{k+1}\}$, reset $k \leftarrow k+1$, and go to Step 2.

Under the Interior–point Assumption, it is easy to see that Algorithm 5.2 either terminates in finite iterations with an optimal solution or generates a sequence of points $\{\mathbf{x}^k(\mu) \mid k = 1, 2, \cdots\}$. We aim to show that $\{\mathbf{x}^k(\mu)\}$ converges to the optimal solution $\mathbf{x}^*(\mu)$ of Program QSI(μ). To simplify the convergence proof, we further make a "Bounded Feasible Domain Assumption". More precisely, we assume that there is a scalar $M > 0$ such that $||\mathbf{x}|| \leq M$ for every feasible solution $\mathbf{x}$ of Program $\mathrm{QI}^1(\mu)$. Consequently, there exists an

$\bar{M} > 0$ such that $\left|\sum_{j=1}^{n} x_j \ln x_j\right| \leq \bar{M}$ for every feasible solution $\mathbf{x}$ of Program $\text{QI}^1(\mu)$.

We now provide a convergence proof for Algorithm 5.2.

Theorem 5.19. Under the Interior–point and Bounded Feasible Domain Assumptions, if Algorithm 5.2 does not terminate in finite iterations, then $\{\mathbf{x}^k(\mu) \mid k = 1, 2, \cdots\}$ converges to the optimal solution $\mathbf{x}^*(\mu)$ of Program $\text{QSI}(\mu)$.

Proof. Since the feasible domain of Program $\text{QI}^k(\mu)$ contains that of Program $\text{QI}^{k+1}(\mu)$, for $k = 1, 2, \cdots$, we have

$$f_\mu(\mathbf{x}^1(\mu)) \leq f_\mu(\mathbf{x}^2(\mu)) \leq \cdots \leq f_\mu(\mathbf{x}^*(\mu)). \tag{5.165}$$

By denoting $\mathbf{x}^*(\mu)$ as the unique optimal solution of Program $\text{QSI}(\mu)$, we first show that every convergent subsequence $\{\mathbf{x}^{k_i}(\mu)\}$ of $\{\mathbf{x}^k(\mu)\}$ converges to $\mathbf{x}^*(\mu)$. If this is not true, denote that limit of the subsequence by $\mathbf{y}$. It is obvious that $\mathbf{y} \neq \mathbf{x}^*(\mu)$, $\mathbf{y} \geq \mathbf{0}$ and $f_\mu(\mathbf{x}^{k_i}(\mu)) \to f_\mu(\mathbf{y})$. From (5.165), we see that

$$f_\mu(\mathbf{y}) \leq f_\mu(\mathbf{x}^*(\mu)). \tag{5.166}$$

To show that $\mathbf{y} \in F$, consider

$$v(t) \equiv h(t) - \sum_{j=1}^{n} g_j(t) y_j, \forall\, t \in T, \tag{5.167}$$

and let $\bar{t} \in T$ be a maximizer of $v(t)$ over T. By the definition of $\mathbf{y}$, we know

$$v(t_{k_i+1}) \leq 0, \ \forall\ k_i, \tag{5.168}$$

where $t_{k_i+1} \in T$ is generated by Algorithm 5.2 for maximizing $v_{k_i+1}(t)$ over T. Since T is a compact metric space, there exists a subsequence $\{\mathbf{x}^{m_i}(\mu)\}$ of $\{\mathbf{x}^{k_i}(\mu)\}$ such that $\{t_{m_i+1}\}$ converges to a limit point t^*. Consequently, by (5.168), $v(t^*) \leq 0$.

Since t_{m_i+1} is the maximizer of $v_{m_i+1}(t)$ over T,

$$h(t_{m_i+1}) - \sum_{j=1}^{n} g_j(t_{m_i+1}) x_j^{m_i}(\mu) \geq h(\bar{t}) - \sum_{j=1}^{n} g_j(\bar{t}) x_j^{m_i}(\mu). \tag{5.169}$$

Moreover, since $\mathbf{x}^{m_i}(\mu)$ also converges to $\mathbf{y}$, we have

$$0 \geq v(t^*) \geq v(\bar{t}). \tag{5.170}$$

It follows that $\mathbf{y} \in F$ and hence

$$f_\mu(\mathbf{x}^*(\mu)) \leq f_\mu(\mathbf{y}). \tag{5.171}$$

By combining Inequalities (5.166) and (5.171), we see $\mathbf{y} \in F$ and $f_\mu(\mathbf{y}) = f_\mu(\mathbf{x}^*(\mu))$. This contradicts the fact that Program QSI(μ) has a unique optimal solution. Therefore, $\mathbf{y} = \mathbf{x}^*(\mu)$ and every convergent subsequence $\{\mathbf{x}^{k_i}(\mu)\}$ of $\{\mathbf{x}^k(\mu)\}$ converges to $\mathbf{x}^*(\mu)$.

Now, suppose that $\{\mathbf{x}^k(\mu)\}$ does not converge to $\mathbf{x}^*(\mu)$. Due to the boundedness of $\{\mathbf{x}^k(\mu)\}$, there exists a convergent subsequence $\{\mathbf{x}^{n_i}(\mu)\}$ which does not converge to $\mathbf{x}^*(\mu)$. This contradicts what we have just proven. Thus $\{\mathbf{x}^k(\mu)\}$ indeed converges to $\mathbf{x}^*(\mu)$.

□

5.4.2 A Path to Follow

With Theorem 5.19 in mind, we show in this subsection that $\{\mathbf{x}^*(\mu) \mid \mu > 0\}$ characterizes a continuous path in the interior of F which eventually leads to an optimal solution of Program QSI as μ decreases to zero.

Theorem 5.20. Suppose that the feasible region of Program QSI is bounded and has a nonempty interior. Then, $\{f(\mathbf{x}^*(\mu)) \mid \mu > 0\}$ is a monotonically nonincreasing sequence as the perturbation parameter μ decreases.

Proof. For $0 < \mu_2 < \mu_1$, since Program QSI(μ_1) and Program QSI(μ_2) both have unique optimal solutions and share the same feasible domain, we have

$$f(\mathbf{x}^*(\mu_1)) + \mu_1 \sum_{j=1}^{n} x_j^*(\mu_1) \ln x_j^*(\mu_1) \leq f(\mathbf{x}^*(\mu_2)) + \mu_1 \sum_{j=1}^{n} x_j^*(\mu_2) \ln x_j^*(\mu_2) \tag{5.172}$$

and

$$f(\mathbf{x}^*(\mu_2)) + \mu_2 \sum_{j=1}^{n} x_j^*(\mu_2) \ln x_j^*(\mu_2) \leq f(\mathbf{x}^*(\mu_1)) + \mu_2 \sum_{j=1}^{n} x_j^*(\mu_1) \ln x_j^*(\mu_1). \tag{5.173}$$

Adding Inequalities (5.172) and (5.173) together yields

$$(\mu_1 - \mu_2) \sum_{j=1}^{n} x_j^*(\mu_1) \ln x_j^*(\mu_1) \leq (\mu_1 - \mu_2) \sum_{j=1}^{n} x_j^*(\mu_2) \ln x_j^*(\mu_2). \tag{5.174}$$

Since $\mu_1 > \mu_2$, Inequality (5.174) becomes

$$\sum_{j=1}^{n} x_j^*(\mu_1) \ln x_j^*(\mu_1) \leq \sum_{j=1}^{n} x_j^*(\mu_2) \ln x_j^*(\mu_2). \tag{5.175}$$

Combining Equations (5.175) and (5.173) results in

$$f(\mathbf{x}^*(\mu_2)) \leq f(\mathbf{x}^*(\mu_1)), \text{ for } 0 < \mu_2 < \mu_1. \tag{5.176}$$

□

As a direct consequence of (5.172) and (5.176), the following result holds:

Corollary 5.2. For $0 < \mu_2 < \mu_1$,

$$0 \leq f(\mathbf{x}^*(\mu_1)) - f(\mathbf{x}^*(\mu_2)) \leq \mu_1 \Big[\sum_{j=1}^{n} x_j^*(\mu_2) \ln x_j^*(\mu_2) - \sum_{j=1}^{n} x_j^*(\mu_1) \ln x_j^*(\mu_1) \Big].$$

Theorem 5.21. Suppose that the feasible region of Program QSI is bounded and has a nonempty interior. The sequence $\{\mathbf{x}^*(\mu) \mid \mu > 0\}$ forms a continuous curve (with respect to μ) in F^0.

Proof. For any $\mu > 0$, $\mathbf{x}^*(\mu) \in F^0$. The only issue that remains is to show the continuity. Suppose that the curve is discontinuous at $\bar{\mu} > 0$, then there exists $\bar{\varepsilon} > 0$ such that no matter how small $\Delta\mu$ is, $\|\mathbf{x}^*(\bar{\mu}) - \mathbf{x}^*(\bar{\mu} + \Delta\mu)\| \geq \bar{\varepsilon}$. We now show that this is impossible by contradiction.

Note that, under the Bounded Feasible Domain Assumption, there exists an $\bar{M} > 0$ such that $\left|\sum_{j=1}^{n} x_j \ln x_j\right| \leq \bar{M}$, for every $\mathbf{x} \in F$. Therefore for any given $\varepsilon > 0$, if we choose $|\Delta\mu| = \frac{\varepsilon}{2\bar{M}}$, then

$$\left| f_{\bar{\mu}}(\mathbf{x}^*(\bar{\mu} + \Delta\mu)) - f_{\bar{\mu}+\Delta\mu}(\mathbf{x}^*(\bar{\mu} + \Delta\mu)) \right|$$

$$\begin{aligned} &= \left|\Delta\mu \sum_{j=1}^{n} x_j^*(\bar{\mu}+\Delta\mu)\ln x_j^*(\bar{\mu}+\Delta\mu)\right| \\ &\le |\Delta\mu| \times \bar{M} = \frac{\varepsilon}{2}. \end{aligned} \tag{5.177}$$

Consequently,

$$f_{\bar{\mu}}\big(\mathbf{x}^*(\bar{\mu}+\Delta\mu)\big) \le f_{\bar{\mu}+\Delta\mu}\big(\mathbf{x}^*(\bar{\mu}+\Delta\mu)\big) + \frac{\varepsilon}{2}. \tag{5.178}$$

Since $\mathbf{x}^*(\bar{\mu}+\Delta\mu)$ minimizes Program QSI($\bar{\mu}+\Delta\mu$), we have

$$f_{\bar{\mu}+\Delta\mu}\big(\mathbf{x}^*(\bar{\mu}+\Delta\mu)\big) \le f_{\bar{\mu}+\Delta\mu}\big(\mathbf{x}^*(\bar{\mu})\big). \tag{5.179}$$

Similar to (5.177), we further have

$$|f_{\bar{\mu}+\Delta\mu}\big(\mathbf{x}^*(\bar{\mu})\big) - f_{\bar{\mu}}\big(\mathbf{x}^*(\bar{\mu})\big)| \le \frac{\varepsilon}{2}. \tag{5.180}$$

Hence,

$$f_{\bar{\mu}+\Delta\mu}\big(\mathbf{x}^*(\bar{\mu})\big) \le f_{\bar{\mu}}\big(\mathbf{x}^*(\bar{\mu})\big) + \frac{\varepsilon}{2}. \tag{5.181}$$

Combining Inequalities (5.178), (5.179), and (5.181), we have

$$f_{\bar{\mu}}\big(\mathbf{x}^*(\bar{\mu}+\Delta\mu)\big) \le f_{\bar{\mu}}\big(\mathbf{x}^*(\bar{\mu})\big) + \varepsilon. \tag{5.182}$$

Note that $\mathbf{x}^*(\bar{\mu})$ minimizes Program QSI($\bar{\mu}$). Hence,

$$0 \le f_{\bar{\mu}}\big(\mathbf{x}^*(\bar{\mu}+\Delta\mu)\big) - f_{\bar{\mu}}\big(\mathbf{x}^*(\bar{\mu})\big) \le \varepsilon. \tag{5.183}$$

Since $\mathbf{x}^*(\bar{\mu})$ is the unique optimal solution of Program QSI($\bar{\mu}$), Inequality (5.183) further implies that, as $\varepsilon \to 0$, $|\Delta\mu| = \dfrac{\varepsilon}{2\bar{M}} \to 0$ and $\mathbf{x}^*(\bar{\mu}+\Delta\mu) \to \mathbf{x}^*(\bar{\mu})$. This certainly contradicts the existence of $\bar{\varepsilon} > 0$ such that $\|\mathbf{x}^*(\bar{\mu}) - \mathbf{x}^*(\bar{\mu}+\Delta\mu)\| \ge \bar{\varepsilon}$, no matter how small $\Delta\mu$ is.

□

Observe that under the Bounded Feasible Domain Assumption, for each $\mathbf{x} \in F$, $f_\mu(\mathbf{x}) \to \frac{1}{2}\mathbf{x}^T\mathbf{D}\mathbf{x} + \mathbf{c}^T\mathbf{x} = f(\mathbf{x})$ as $\mu \to 0$. Hence, it is expected that Program QSI(μ) becomes Program QSI when μ becomes zero. By following the continuous interior path $\{\mathbf{x}^*(\mu) \mid \mu > 0\}$ from large μ toward 0, we can eventually solve Program QSI.

Theorem 5.22. Under the Bounded Feasible Domain Assumption, if Program QSI has a unique optimal solution $\mathbf{x}^*$, then $\mathbf{x}^*(\mu_i) \rightarrow \mathbf{x}^*$ as μ_i decreases to zero.

Proof. Under the Bounded Feasible Domain Assumption, $\{\mathbf{x}^*(\mu_i) | \mu_i > 0\}$ has a convergent subsequence. We claim that every convergent subsequence of $\{\mathbf{x}^*(\mu_i)\}$ converges to $\mathbf{x}^*$. Otherwise, let $\{\mathbf{x}^*(\mu_{l_i})\}$ be a converging subsequence with a limit $\mathbf{s} \neq \mathbf{x}^*$. In this case, since $\mathbf{x}^*(\mu_{l_i})$ is feasible for Program QSI and $\mathbf{x}^*(\mu_{l_i}) \rightarrow \mathbf{s}$, we know that $\mathbf{s}$ is feasible for Program QSI. Consequently, by the uniqueness of the optimal solution $\mathbf{x}^*$, there exists an $\alpha > 0$ such that

$$f(\mathbf{s}) - f(\mathbf{x}^*) > \alpha. \tag{5.184}$$

Moreover, since $\mathbf{x}^*$ is feasible for Program QSI(μ_{l_i}), for each l_i, we have

$$f(\mathbf{x}^*(\mu_{l_i})) + \mu_{l_i} \sum_{j=1}^{n} x_j^*(\mu_{l_i}) \ln x_j^*(\mu_{l_i}) \leq f(\mathbf{x}^*) + \mu_{l_i} \sum_{j=1}^{n} x_j^* \ln x_j^*. \tag{5.185}$$

Consequently,

$$f(\mathbf{x}^*(\mu_{l_i})) - f(\mathbf{x}^*) \leq \mu_{l_i} \Big[\sum_{j=1}^{n} x_j^* \ln x_j^* - \sum_{j=1}^{n} x_j^*(\mu_{l_i}) \ln x_j^*(\mu_{l_i})\Big]. \tag{5.186}$$

Note that

$$f(\mathbf{x}^*(\mu_{l_i})) - f(\mathbf{x}^*) = f(\mathbf{x}^*(\mu_{l_i})) - f(\mathbf{s}) + f(\mathbf{s}) - f(\mathbf{x}^*) \tag{5.187}$$

and recall that $\mathbf{x}^*(\mu_{l_i}) \rightarrow \mathbf{s}$. For a sufficiently small $\varepsilon > 0$ with $\alpha > \varepsilon$, there exists an integer $N > 0$ such that, for $l_i \geq N$,

$$-\varepsilon < f(\mathbf{x}^*(\mu_{l_i})) - f(\mathbf{s}) < \varepsilon. \tag{5.188}$$

Thus, for $l_i \geq N$, combining Inequalities (5.184) and (5.188) results in

$$f(\mathbf{x}^*(\mu_{l_i})) - f(\mathbf{x}^*) > \alpha - \varepsilon > 0. \tag{5.189}$$

Together with Inequality (5.186), for $l_i \geq N$, we have

$$\begin{aligned} 0 &< \alpha - \varepsilon < f(\mathbf{x}^*(\mu_{l_i})) - f(\mathbf{x}^*) \\ &\leq \mu_{l_i} \Big[\sum_{j=1}^{n} x_j^* \ln x_j^* - \sum_{j=1}^{n} x_j^*(\mu_{l_i}) \ln x_j^*(\mu_{l_i})\Big]. \end{aligned} \tag{5.190}$$

Since $\{\mu_i\}$ decreases to zero, $\mu_{l_i} \to 0$ as $l_i \to \infty$. Under the Bounded Feasible Domain Assumption, the right–most term in Inequality (5.190), i.e., the difference between the two sums, approaches zero as $l_i \to \infty$. This leads to a contradiction. Hence we know $\mathbf{s} = \mathbf{x}^*$. Since $\{\mathbf{x}^*(\mu_i)\}$ is a bounded sequence, if it does not converge to $\mathbf{x}^*$, then there exists a convergent subsequence which does not converge to $\mathbf{x}^*$. This contradicts what we have just proven. Thus $\mathbf{x}^*(\mu_i) \to \mathbf{x}^*$ as μ_i decreases to zero.

□

Note that, in Theorem 5.22, if Program QSI has multiple optimal solutions, from the proof given before, we know that $\{\mathbf{x}^*(\mu_i) | \mu_i > 0\}$ has a subsequence which converges to an optimal solution of Program QSI. Also note that, from Inequality (5.186), Corollary 5.2 can be further refined as follows:

Corollary 5.3. For $\mu > 0$,

$$0 \leq f(\mathbf{x}^*(\mu)) - f(\mathbf{x}^*) \leq \mu \Big[\sum_{j=1}^{n} x_j^* \ln x_j^* - \sum_{j=1}^{n} x_j^*(\mu) \ln x_j^*(\mu) \Big]. \tag{5.191}$$

5.4.3 Path–Following Algorithms for QSIP

We first focus on the concept of "ε–optimality". Given any $\varepsilon > 0$, as defined earlier in this chapter and in [8, 11], a feasible solution $\mathbf{x} \in F$ is called an "ε–optimal solution" of Program QSI, if $|f(\mathbf{x}) - f(\mathbf{x}^*)| \leq \varepsilon$, where $\mathbf{x}^*$ is an optimal solution of Program QSI. Under the Bounded Feasible Domain Assumption, we let $\bar{M} > 0$ such that $\left| \sum_{j=1}^{n} x_j \ln x_j \right| \leq \bar{M}$, for each $\mathbf{x} \in F$. If we choose $\mu = \dfrac{\varepsilon}{2\bar{M}} > 0$, then Corollary 5.3 implies that

$$0 \leq f(\mathbf{x}^*(\mu)) - f(\mathbf{x}^*) \leq \mu \times 2\bar{M} = \varepsilon.$$

Since Program QSI(μ) and Program QSI have the same feasible domain F, $\mathbf{x}^*(\mu)$ becomes an ε–optimal solution of Program QSI, if $0 < \mu \leq \frac{\varepsilon}{2\bar{M}}$. Hence, given any $\varepsilon > 0$ and the knowledge of $\bar{M} > 0$, the following algorithm finds an ε–optimal solution of Program QSI under the Interior–point and Bounded Feasible Domain Assumptions.

Algorithm 5.3:

Step 1: Set $i = 1$; choose any $0 < \theta < 1$ and $\mu_1 > 0$.

Step 2: Solve Program QSI(μ_i) by Algorithm 5.2 for an optimal solution $\mathbf{x}^*(\mu_i)$.

Step 3: If $\mu_i \leq \frac{\varepsilon}{2M}$, Stop! $\mathbf{x}^*(\mu_i)$ is an ε–optimal solution of Program QSI.

Otherwise, set $\mu_{i+1} = (1-\theta)\mu_i$, reset $i \leftarrow i+1$, and go to Step 2.

Since $\mu_{i+1} = (1-\theta)^i\mu_1$, for $i = 1, 2, \cdots$, and $\theta \in (0,1)$, Algorithm 5.3 always terminates with an ε–optimal solution of Program QSI, for any given $\varepsilon > 0$. By Theorem 5.20, we know that $\{f(\mathbf{x}^*(\mu)) | \mu > 0\}$ is a monotonically decreasing sequence as μ reduces to zero. Moreover, from Theorem 5.22 and related discussion, we know that every convergent subsequence of $\{\mathbf{x}^*(\mu)\}$ converges to an optimal solution of Program QSI as μ decreases to zero. Therefore Algorithm 5.3 can be modified as follows to find an optimal solution of Program QSI.

Algorithm 5.4:

Step 1: Set $i = 1$; select sufficiently small $\delta_1 > 0, \delta_2 > 0$, $0 < \theta < 1$, and $\mu_1 > 0$.

Step 2: Solve Program QSI(μ_i) by Algorithm 5.2 for an optimal solution $\mathbf{x}^*(\mu_i)$.

Step 3: For $i > 1$, if $||\mathbf{x}^*(\mu_i) - \mathbf{x}^*(\mu_{i-1})|| < \delta_1$ and $\mu_i < \delta_2$, then stop and output $\mathbf{x}^* = \mathbf{x}^*(\mu_i)$. Otherwise, go to Step 4.

Step 4: Set $\mu_{i+1} = (1-\theta)\mu_i$; reset $i \leftarrow i+1$; go to Step 2.

Note that the key purpose of Step 3 is to identify a limit point of $\{\mathbf{x}^*(\mu_i)\}$. Any procedure with such capability can be considered as a replacement for Step 3.

5.4.4 Concluding Remarks

1. In this section, the entropic perturbation approach has been extended to an entropic path-following scheme for solving convex quadratic semi-infinite programming problems.

2. Under the Interior-point and Bounded Feasible Domain Assumptions, since $\mathbf{x}^*(\mu) \in F^0$, $\forall\ \mu > 0$, the proposed path–following scheme approaches $\mathbf{x}^*$ from the interior of the feasible domain of Program QSI. This algorithm works on Program QSI directly without any discretization. Its convergence can be proved under proper conditions.

3. When the diagonal matrix D is a zero matrix, Program QSI reduces to a linear semi-infinite programming problem. In this case, the algorithms proposed in this section become new interior-point methods for linear semi-infinite programming problems, which are different from those in the existing literature [15, 16].

4. Related implementation issues and rather extensive computational experiments on Algorithm 5.2 and its variations can be found in [9, 28, 27, 14]. The computational results, although not included here, clearly confirmed that the proposed algorithms are capable of generating high-quality solutions to convex quadratic and linear semi-infinite programming problems. In particular, experiments showed that as the problem size grows, the required computational effort grows at a much slower pace. Therefore, the proposed algorithms are good candidates for solving large-scale problems.

REFERENCES

[1] Anderson, E. J., and Nash, P., *Linear Programming in Infinite-Dimensional Spaces*, John Wiley, Chichester, New York, 1987.

[2] Ben-Tal, A., Melman, A., and Zowe, J., "Curved Search Methods for Unconstrained Optimization," *Optimization*, Vol. 21, 1990, pp. 669–695.

[3] Dennis, J.E., and Schnabel, R.B., *Numerical Methods for Unconstrained Optimization and Nonlinear Equations*, Prentice Hall, Englewood Cliffs,

New Jersey, 1983.

[4] Duffin, R.J., Peterson, E.L., and Zener, C., *Geometric Programming : Theory And Applications*, John Wiley, New York, 1967.

[5] Erickson, J.R., "Algorithms for Entropy and Mathematical Programming," Ph.D. Thesis, Department of Mathematics, Linkoping University, Sweden, 1981.

[6] Erickson, J.R., "An Iterative Primal-Dual Algorithm for Linear Programming," Report LitH-MAT-R-1985-10, Department of Mathematics, Linkoping University, Linkoping, Sweden, 1985.

[7] Erlander, S., "Entropy in Linear Programming," *Mathematical Programming*, Vol. 21, 1981, pp. 137–151.

[8] Fang, S.-C., "A New Unconstrained Convex Programming Approach to Linear Programming," *Zeitschrift fur Operations Research*, Vol. 36, 1992, pp. 149–161.

[9] Fang, S.-C., Lin, C.-J., and Wu, S.-Y., "On Solving Convex Quadratic Semi-Infinite Programming Problems," *Optimization*, Vol. 31, 1994, pp. 107–125.

[10] Fang, S.-C., and Puthenpura, S., *Linear Optimization and Extensions: Theory and Algorithms*, Prentice Hall, Englewood Cliffs, New Jersey, 1993.

[11] Fang, S.-C., and Tsao, H.-S.J., "Linear Programming with Entropic Perturbation," *Zeitschrift fur Operations Research*, Vol. 37, 1993, pp. 171–186.

[12] Fang, S.-C., and Tsao, H.-S.J., "An Unconstrained Convex Programming Approach to Solving Convex Quadratic Programming Problems," *Optimization*, Vol. 27, 1993, pp. 235–243.

[13] Fang, S.-C., and Tsao, H.-S.J., "On the Entropic Perturbation and Exponential Penalty Methods for Linear Programming," *Journal of Optimization Theory and Applications*, Vol. 89, 1996, pp. 461–466.

[14] Fang, S.-C., and Wu, S.-Y., "An Inexact Approach to Solving Linear Semi-Infinite Programming Problems," *Optimization*, Vol. 28, 1994, pp. 291–299.

[15] Ferris, M.C., and Philpott, A.B., "An Interior Point Algorithm for Semi-Infinite Linear Programming," *Mathematical Programming*, Vol. 43, 1989, pp. 257–276.

[16] Ferris, M.C., and Philpott, A.B., "On the Affine Scaling and Semi-Infinite Programming," *Mathematical Programming*, Vol. 56, 1992, pp. 361–364.

[17] Fiacco, A.V., and McCormick, G.P., *Nonlinear Programming: Sequential Unconstrained Minimization Techniques*, John Wiley, New York, 1968.

[18] Gill, P.E., Murray, W., Saunders, M.A., Tomlin, J.A., and Wright, M.H., "On Projected Barrier Methods for Linear Programming and an Equivalence to Karmarkar's Projective Method," *Mathematical Programming*, Vol. 36, 1986, pp. 183–209.

[19] Glashoff, K., and Gustafson, S.A., *Linear Optimization and Approximation*, Springer-Verlag, New York, 1982.

[20] Goldfarb, D., and Todd, M.J., "Linear Programming," Cornell University, School of OR and IE, Tech. Report No. 777, 1988.

[21] Gustafson, S.A., and Kortanek, K.O., "Numerical Treatment of a Class of Semi–Infinite Programming Problems," *Naval Research Logistics Quarterly*, Vol. 20, 1973, pp. 473–504.

[22] Den Hertog, D., Roos, C., and Terlaky, T., "Inverse Barrier Methods for Linear Programming," Report of the Faculty of Technical Mathematics and Informatics, No. 91-27, Delft University of Technology, The Netherlands, 1991.

[23] Hettich, R., and Kortanek, K.O., "Semi-Infinite Programming : Theory, Method and Applications," *SIAM Review*, Vol. 35, 1993, pp. 380–429.

[24] Karmarkar, N., "A New Polynomial Time Algorithm for Linear Programming," *Combinatorica*, Vol. 4, 1984, pp. 373–395.

[25] Lagarias, J.C., and Bayer, D.A., "Karmarkar's Linear Programming Method and Newton's Method," *Bell Laboratories Technical Report*, 11218-870810-22TM, AT&T Bell Laboratories, Murray Hill, New Jersey, Aug 10, 1987.

[26] Lai, H.C., and Wu, S.Y., "On Linear Semi-Infinite Programming Problems, an Algorithm," *Numerical Functional Analysis and Optimization*, Vol. 13, 1992, pp. 287–304.

[27] Lin, C.J., Fang, S.-C., and Wu, S.Y., "An Unconstrained Convex Programming Approach to Linear Semi-Infinite Programming," OR Technical Report No. 296, North Carolina State University, Raleigh, North Carolina, 1994, to appear in *SIAM Journal on Optimization.*

[28] Lin, C.-J., Yang, E.K., Fang, S.-C., and Wu, S.-Y., "Implementation of an Inexact Approach to Solving Linear Semi-Infinite Programming Problems," *Journal of Computational and Applied Mathematics*, Vol. 61, 1995, pp. 87–103.

[29] Minoux, M., *Mathematical Programming*, Wiley-Interscience, New York, 1986.

[30] Murtagh, B.A., and Saunders, M.A., *MINOS 5.1 User's Guide*, Technical Report SOL 83-20R, Stanford University, Stanford, California, 1983.

[31] Peterson, E.L., "Geometric Programming," *SIAM Review*, Vol. 19, 1976, pp. 1–45.

[32] Rajasekera, J.R., and Fang, S.-C., "On the Convex Programming Approach to Linear Programming," *Operations Research Letters*, Vol. 10, 1991, pp. 309–312.

[33] Rajasekera, J. R., and Fang, S.-C., "Deriving an Unconstrained Convex Program for Linear Programming," *Journal of Optimization Theory and Applications*, Vol. 75, 1992, pp. 603–612.

[34] Renegar, J., "A Polynomial-time Algorithm Based on Newton's Method for Linear Programming," *Mathematical Programming*, Vol. 40, 1988, pp. 59–93.

[35] Rockafellar, R.T., *Convex Analysis,* Princeton University Press, Princeton, New Jersey, 1970.

[36] Saigal, R., *Linear Programming: A Modern Integrated Analysis*, Kluwer Academic Publishers, Norwell, Massachusetts, 1995.

[37] Shanno, D.F., and Bagchi, A., "Unified View of Interior Point Methods for Linear Programming," Rutcor Research Report No. 35-88, Rutgers University, New Brunswick, New Jersey, July 1988.

[38] Sheu, R.-L., and Fang, S.-C., "Insights into Interior-Point Methods," *Zeitschrift fur Operations Research*, Vol. 36, 1992, pp. 227–257.

[39] Vanderbei, R.J., *LOQO User's Manual*, Princeton University Press, Princeton, New Jersey, 1992.

[40] Wang, D., " Experiments on the Unconstrained Convex Programming Approach to Linear Programming," IE Report No. 90-12, North Carolina State University, Raleigh, North Carolina, 1990.

[41] Xu, Z.K., and Fang, S.-C., "Unconstrained Convex Programming Approach to Linear Programming," *Journal of Optimization Theory and Applications*, Vol. 86, 1993, pp. 745–752.

6

L_P-NORM PERTURBATION APPROACH : A GENERALIZATION OF ENTROPIC PERTURBATION

Solving a linear or nonlinear program by perturbing its primal objective function with a barrier or penalty function has attracted much attention recently in developing interior-point methods [3, 17, 4, 13, 18]. However, the idea of perturbing the feasible region has not been fully explored. This chapter focuses on this idea and discusses a particular approach involving the l_p-norm of a vector measure of constraint violation. Three topics will be discussed in this chapter:

(i) perturbing the dual feasible region of a standard-form linear program,

(ii) perturbing the primal feasible region of a linear program with inequality constraints, and

(iii) perturbing the dual feasible region of a convex quadratic program.

The related duality theory and convergence theorems will be developed for each of the three cases. We show that, at least in theory, the effect of perturbing the primal objective with an entropic function could be captured by a corresponding perturbation of the dual feasible region under the l_p-norm. In this sense, the entropic perturbation approach discussed in Chapter 5 corresponds to a special case of this l_p-norm perturbation scheme. An ϵ-optimality theory will also be developed accordingly.

6.1 PERTURBING THE DUAL FEASIBLE REGION OF STANDARD-FORM LINEAR PROGRAMS

Consider the following (primal) linear program in its standard form:

Program L:

$$\min \quad \mathbf{c}^T\mathbf{x} \tag{6.1}$$

$$\text{s.t.} \quad \mathbf{A}\mathbf{x} = \mathbf{b}, \tag{6.2}$$

$$\mathbf{x} \geq \mathbf{0}, \tag{6.3}$$

where $\mathbf{c}$ and $\mathbf{x}$ are n-dimensional column vectors, $\mathbf{A}$ is an $m \times n$ $(m \leq n)$ matrix, $\mathbf{b}$ is an m-dimensional column vector, and $\mathbf{0}$ is the n-dimensional zero vector. A linear dual of Program L is given as follows:

Program DL:

$$\max \quad \mathbf{b}^T\mathbf{w} \tag{6.4}$$

$$\text{s.t.} \quad \mathbf{A}^T\mathbf{w} \leq \mathbf{c}, \tag{6.5}$$

$$\mathbf{w} \in R^m. \tag{6.6}$$

By perturbing the primal objective function with an entropic barrier, Section 5.2 and previous work [2, 15, 4] showed that, for any $\mu > 0$,

Program L(μ):

$$\min \quad \mathbf{c}^T\mathbf{x} + \mu \sum_{j=1}^{n} x_j \ln x_j \tag{6.7}$$

$$\text{s.t.} \quad \mathbf{A}\mathbf{x} = \mathbf{b}, \tag{6.8}$$

$$\mathbf{x} \geq \mathbf{0}, \tag{6.9}$$

has an unconstrained geometric dual

Program DL(μ):

$$\max_{\mathbf{w} \in R^m} \left\{ \sum_{i=1}^{m} b_i w_i - \mu \sum_{j=1}^{n} \exp\{[(\sum_{i=1}^{m} a_{ij} w_i - c_j)/\mu] - 1\} \right\}. \tag{6.10}$$

Moreover, under some "standard" assumptions, Program L(μ) has a unique optimal solution $\mathbf{x}^*(\mu)$ which approaches an optimal solution $\mathbf{x}^*$ of Program L, as $\mu \to 0$. Similarly, Program DL(μ) has a unique optimal solution $\mathbf{w}^*(\mu)$ that tends to an optimal solution $\mathbf{w}^*$ of Program DL, as $\mu \to 0$.

In this section, we denote the feasible region of the dual program DL as

$$F \equiv \{\mathbf{w} | \mathbf{w} \in R^m, \mathbf{A}^T\mathbf{w} \leq \mathbf{c}\} \tag{6.11}$$

and investigate a new perturbation approach that perturbs this dual feasible region. Subsection 6.1.1 introduces the new approach. Subsection 6.1.2 shows how this approach generalizes the (primal) entropic perturbation approach. Concluding remarks are given in Subsection 6.1.3.

6.1.1 A Perturbed Dual Approach

Note that there are n constraints in Program DL. Given $\mathbf{w} \in R^m$, the jth constraint is satisfied, $j = 1, 2, \ldots, n$, if and only if

$$\exp\{\sum_{i=1}^{m} a_{ij}w_i - c_j\} \leq 1. \tag{6.12}$$

Therefore, we can define an n-dimensional positive "error vector"

$$\phi(\mathbf{w}) \equiv (\phi_1(\mathbf{w}), \ldots, \phi_n(\mathbf{w})), \tag{6.13}$$

for $\mathbf{w} \in R^m$, where

$$\phi_j(\mathbf{w}) \equiv \exp\{\sum_{i=1}^{m} a_{ij}w_i - c_j\} > 0, \; j = 1, 2, \ldots, n. \tag{6.14}$$

It is clearly seen that, for any given $\mathbf{w} \in R^m$, $\mathbf{w} \in F$ if and only if

$$||\phi(\mathbf{w})||_\infty \equiv \max_{j=1,\ldots,n} \{\phi_j(\mathbf{w})\} \leq 1. \tag{6.15}$$

Based on this fact, we would like to exploit the concept of perturbing the dual feasible domain under different norms. In particular, for $p \geq 1$, the l_p-norm of $\phi(\mathbf{w})$ is defined by

$$||\phi(\mathbf{w})||_p \equiv \left\{\sum_{j=1}^{n}(\phi_j(\mathbf{w}))^p\right\}^{1/p} = \left\{\sum_{j=1}^{n}\exp[p(\sum_{i=1}^{m}a_{ij}w_i - c_j)]\right\}^{1/p}. \quad (6.16)$$

Since the logarithmic function is strictly increasing, the l_p-norm is preserved by

$$\ln ||\phi(\mathbf{w})||_p = \frac{1}{p}\ln\left\{\sum_{j=1}^{n}\exp[p(\sum_{i=1}^{m}a_{ij}w_i - c_j)]\right\}. \quad (6.17)$$

A perturbed dual problem can be defined in the following form:

Program $\mathbf{DL}_p$:

$$\max \quad \mathbf{b}^T\mathbf{w} \quad (6.18)$$

$$\text{s.t.} \quad \frac{1}{p}\ln\left\{\sum_{j=1}^{n}\exp[p(\sum_{i=1}^{m}a_{ij}w_i - c_j)]\right\} \leq 0, \quad (6.19)$$

or, equivalently,

Program $\mathbf{DL}_p'$:

$$\max \quad \mathbf{b}^T\mathbf{w} \quad (6.20)$$

$$\text{s.t.} \quad ||\phi(\mathbf{w})||_p \leq 1. \quad (6.21)$$

We denote the feasible region of Program DL_p' by

$$F_p \equiv \{\mathbf{w} \mid \mathbf{w} \in R^m,\ ||\phi(\mathbf{w})||_p \leq 1\}. \quad (6.22)$$

Several properties of Program DL_p can be derived here.

Lemma 6.1. If $q \geq p \geq 1$, then the feasible domain F_p of Program DL_p is contained in the feasible domain F_q of Program DL_q, i.e., $F_p \subset F_q$.

Proof. Given $\mathbf{y} \equiv (y_1, \ldots, y_n)^T \geq \mathbf{0}$, we have $(\sum_{j=1}^n y_j)^{\sum_{j=1}^n y_j} \geq \prod_{j=1}^n y_j^{y_j}$. Therefore, $(\sum_{j=1}^n y_j)\ln(\sum_{j=1}^n y_j) \geq \sum_{j=1}^n y_j \ln y_j$. This implies that, for each

$\mathbf{w} \in R^m$ and $r \geq 1$, the partial derivative of the function $||\phi(\mathbf{w})||_r$ with respect to r is nonpositive. This in turn implies that $||\phi(\mathbf{w})||_p \geq ||\phi(\mathbf{w})||_q$, for $q \geq p \geq 1$. Hence,

$$\ln ||\phi(\mathbf{w})||_p \geq \ln ||\phi(\mathbf{w})||_q, \tag{6.23}$$

for $q \geq p \geq 1$, and Lemma 6.1 follows immediately.

□

Let us denote $\mathbf{w}_p^*$ as an optimal solution of Program DL$_p$, for $p \geq 1$, when it exists. As a direct consequence of Lemma 6.1, we have the following result:

Lemma 6.2. If Program DL$_1$ is feasible, then $\{\mathbf{b}^T \mathbf{w}_p^* \mid p \geq 1\}$ forms an increasing sequence in p.

Our next objective is to show that, in the sense of "liminf" ([1], page 71), F_p approaches the interior of F, denoted by int(F), as $p \rightarrow \infty$. First, for $\mathbf{w} \in R^m$, we define a function $h(\mathbf{w})$ by

$$h(\mathbf{w}) \equiv \max_{j=1,\ldots,n} \{\sum_{i=1}^{m} a_{ij} w_i - c_j\}. \tag{6.24}$$

We now prove the following results:

Lemma 6.3. For $p \geq 1$ and $\mathbf{w} \in R^m$, $\ln ||\phi(\mathbf{w})||_p$ decreases to $h(\mathbf{w})$ as p increases to infinity.

Proof. For any $\mathbf{w} \in R^m$, since

$$||\phi(\mathbf{w})||_\infty \equiv \lim_{p \rightarrow \infty} ||\phi(\mathbf{w})||_p = \max_{j=1,\ldots,n} \{\phi(\mathbf{w})\}, \tag{6.25}$$

we know that

$$\lim_{p \rightarrow \infty} \ln ||\phi(\mathbf{w})||_p = \ln \left(\max_{j=1,\ldots,n} \{\phi_j(\mathbf{w})\} \right) = \max_{j=1,\ldots,n} \{\ln \phi_j(\mathbf{w})\}. \tag{6.26}$$

Lemma 6.3 follows from Lemma 6.1.

□

Lemma 6.4. For $p \geq 1$ and $\mathbf{w} \in R^m$,

$$h(\mathbf{w}) \leq \ln ||\phi(\mathbf{w})||_p \leq h(\mathbf{w}) + (1/p) \ln n, \tag{6.27}$$

and hence $\ln ||\phi(\mathbf{w})||_p$ converges to $h(\mathbf{w})$ uniformly on $\mathbf{w} \in R^m$. Moreover, for any increasing sequence $\{p_i \mid p_i \geq 1, \ i = 1, 2, \ldots, \infty\}$, $\liminf F_{p_i} = \text{int}(F)$ as p_i approaches infinity.

Proof. For any $p \geq 1$ and $\mathbf{w} \in R^m$,

$$\begin{aligned} \ln ||\phi(\mathbf{w})||_p - h(\mathbf{w}) &= \frac{1}{p} \ln \left\{ \sum_{j=1}^{n} \exp \left[p(\sum_{i=1}^{m} a_{ij} w_i - c_j) \right] \right\} - \ln\{\exp \ h(\mathbf{w})\} \\ &= \frac{1}{p} \ln \left\{ \sum_{j=1}^{n} \exp \left\{ p \left[(\sum_{i=1}^{m} a_{ij} w_i - c_j) - h(\mathbf{w}) \right] \right\} \right\} \end{aligned} \tag{6.28}$$

Note that, for $j = 1, ..., n$,

$$\sum_{i=1}^{m} a_{ij} w_i - c_j \leq h(\mathbf{w}). \tag{6.29}$$

Hence, we have

$$\exp \left\{ p[(\sum_{i=1}^{m} a_{ij} w_i - c_j) - h(\mathbf{w})] \right\} \leq 1. \tag{6.30}$$

Consequently, for any $p \geq 1$ and $\mathbf{w} \in R^m$,

$$\ln ||\phi(\mathbf{w})||_p - h(\mathbf{w}) \leq (1/p) \ln n. \tag{6.31}$$

By using Lemma 6.3, we obtain Equation (6.27) and the rest follows.

□

Note that Lemma 6.4 suggests that Program DL_p becomes Program DL as p approaches infinity. Also note that when $\text{int}(F)$ is nonempty, it is easy to see that, for $p \geq 1$,

$$F_p \subset \text{int}(F). \tag{6.32}$$

All the lemmas derived earlier culminate in the following theorem:

Theorem 6.1. If Program L and Program DL_1 are feasible, then $\{\mathbf{b}^T\mathbf{w}_p^* \mid p \geq 1\}$ forms an increasing sequence which converges to the optimal value of Program DL. Moreover, any convergent subsequence of $\{\mathbf{w}_p^* \mid p \geq 1\}$ converges to an optimal solution of Program DL from the interior of F.

Note that the perturbed dual program DL_p is also equivalent to

Program DL_p'':

$$\max \quad \mathbf{b}^T\mathbf{w} \tag{6.33}$$

$$\text{s.t.} \quad \sum_{j=1}^{n} \exp[p(\sum_{i=1}^{m} a_{ij}w_i - c_j)] \leq 1. \tag{6.34}$$

It is clearly seen that Program DL_p'' is a convex program with a linear objective function and one convex C^∞ constraint. When p is large enough, the n constraints of Program DL are virtually aggregated into this single constraint.

To study its optimality conditions, let us define the following Lagrangian function for Program DL_p'':

$$L(\mathbf{w}, \lambda) \equiv \sum_{i=1}^{m} b_i w_i - \lambda \left\{ \sum_{j=1}^{n} \exp[p(\sum_{i=1}^{m} a_{ij}w_i - c_j)] - 1 \right\}, \tag{6.35}$$

where $\lambda \geq 0$ and $\mathbf{w} \in R^m$. From convex analysis, at a (finite) optimal solution $\mathbf{w}_p^*$ of Program DL_p'', Equation (6.34) turns out to be an equality constraint with a corresponding positive multiplier $\lambda^* > 0$. In this case, the following first-order optimality conditions must also be satisfied for optimality:

$$\frac{\partial L(\mathbf{w}, \lambda)}{\partial w_i} = b_i - \lambda^* p \left\{ \sum_{j=1}^{n} a_{ij} \exp[p(\sum_{l=1}^{m} a_{lj} w_{p,l}^* - c_j)] \right\} = 0. \tag{6.36}$$

Then, by defining

$$x_j(\lambda^*, \mathbf{w}_p^*) = \lambda^* p \exp[p(\sum_{i=1}^{m} a_{ij} w_{p,i}^* - c_j)], \quad j = 1, 2, ..., n, \tag{6.37}$$

we see that $x_j(\lambda^*, \mathbf{w}_p^*) \geq 0$, j=1,2,...,n, and

$$\sum_{j=1}^{n} a_{ij} x_j(\lambda^*, \mathbf{w}_p^*) = b_i, \; i = 1, \ldots, m. \tag{6.38}$$

This implies that $\mathbf{x}(\lambda^*, \mathbf{w}_p^*)$ is a feasible solution to Program L. Moreover, when p is sufficiently large, $\mathbf{w}_p^*$ and $\mathbf{x}(\lambda^*, \mathbf{w}_p^*)$ are "good" approximate optimal solutions to Program DL and Program L respectively.

6.1.2 Generalization of Entropic Perturbation

In this subsection, we make two observations. Both show that the proposed dual l_p-norm perturbation approach generalizes the primal entropic perturbation approach.

Observation 6.1:

Consider the dual problem of Karmarkar's standard-form linear program:

Program DKL:

$$\max \quad w_{m+1} \tag{6.39}$$

$$\text{s.t.} \quad \sum_{i=1}^{m} a_{ij} w_i + w_{m+1} \leq c_j, \quad j = 1, 2, ..., n. \tag{6.40}$$

For $p \geq 1$, the perturbed dual is defined as follows:

Program DKL$_p$:

$$\max \quad w_{m+1} \tag{6.41}$$

$$\text{s.t.} \quad \frac{1}{p} \ln \left\{ \sum_{j=1}^{n} \exp \left[p(\sum_{i=1}^{m} a_{ij} w_i + w_{m+1} - c_j) \right] \right\} \leq 0. \tag{6.42}$$

The constraint can be reduced to

$$w_{m+1} \leq -\frac{1}{p} \ln \left\{ \sum_{j=1}^{n} \exp \left[p(\sum_{i=1}^{m} a_{ij} w_i - c_j) \right] \right\}. \tag{6.43}$$

Consequently, Program DKL_p is equivalent to:

Program $\mathbf{DKL}_p^{'}$:

$$\max_{\mathbf{w} \in R^m} -\frac{1}{p} \ln \left\{ \sum_{j=1}^{n} \exp \left[p(\sum_{i=1}^{m} a_{ij} w_i - c_j) \right] \right\}. \tag{6.44}$$

By setting $\mu = 1/p$, we obtain the following program:

Program $\mathbf{DKL}_p^{''}$:

$$\min_{\mathbf{w} \in R^m} \mu \ln \left\{ \sum_{j=1}^{n} \exp \left[(\sum_{i=1}^{m} a_{ij} w_i - c_j)/\mu \right] \right\}. \tag{6.45}$$

This is exactly the unconstrained dual program as derived in Section 5.1.

Observation 6.2:

Consider the Lagrangian function defined by Equation (6.35). At the particular value of $\lambda = 1/pe$, where e = 2.71828... is the exponential constant, the corresponding Lagrangian dual maximization problem becomes:

$$\max_{\mathbf{w} \in R^m} \left\{ L(\mathbf{w}, \frac{1}{pe}) = \sum_{i=1}^{m} b_i w_i - \frac{1}{p} \left\{ \sum_{j=1}^{n} \exp[p(\sum_{i=1}^{m} a_{ij} w_i - c_j) - 1] - \frac{1}{e} \right\} \right\}.$$

Setting $\mu = 1/p$ and omitting the constant term, we obtain the following equivalent program:

$$\max_{\mathbf{w} \in R^m} \left\{ \sum_{i=1}^{m} b_i w_i - \mu \sum_{j=1}^{n} \exp \left\{ [(\sum_{i=1}^{m} a_{ij} w_i - c_j)/\mu] - 1 \right\} \right\}, \tag{6.46}$$

which is exactly the unconstrained dual program DL(μ) as derived in Section 5.2. Therefore, the perturbed dual DL_p'' can be viewed as a generalized version of the entropic perturbation approach.

6.1.3 Concluding Remarks

In this section, we have introduced a dual l_p-norm perturbation approach to solving standard-form linear programs, and established its relationship with the primal entropic perturbation approach discussed in Chapter 5. We have shown that this dual approach provides a more general setting DL_p'' than its counterpart DL(μ) in the framework of entropic perturbation of objective function.

6.2 SOLVING LINEAR PROGRAMS WITH INEQUALITY CONSTRAINTS VIA PERTURBATION OF FEASIBLE REGION

The previous section introduced a perturbation scheme for the dual feasible region of a standard-form linear program. In this section, we further investigate into solving linear programming problems with inequality constraints by directly perturbing its *primal* feasible region.

Consider the following linear program with inequality constraints:

Program LI:

$$\min \quad \mathbf{c}^T\mathbf{x} \tag{6.47}$$

$$\text{s.t.} \quad \mathbf{A}\mathbf{x} \geq \mathbf{b}, \tag{6.48}$$

$$\mathbf{x} \geq \mathbf{0}, \tag{6.49}$$

where $\mathbf{c}$ and $\mathbf{x}$ are n-dimensional column vectors, $\mathbf{A}$ is an $m \times n$ matrix, $\mathbf{b}$ is an m-dimensional column vector, and $\mathbf{0}$ is the n-dimensional zero vector. Without loss of generality, we may assume that $\mathbf{c} \neq \mathbf{0}$ in this section.

Its linear dual is

Program DLI:

$$\max \quad \mathbf{b}^T\mathbf{w} \tag{6.50}$$
$$\text{s.t.} \quad \mathbf{A}^T\mathbf{w} \leq \mathbf{c}\,, \tag{6.51}$$
$$\mathbf{w} \geq \mathbf{0}. \tag{6.52}$$

By perturbing the primal objective function with an entropic barrier function, it can be shown that, for any perturbation parameter $\mu > 0$, the following problem

Program LI (μ):

$$\min \quad \mathbf{c}^T\mathbf{x} + \mu\sum_{j=1}^{n} x_j \ln x_j \tag{6.53}$$
$$\text{s.t.} \quad \mathbf{A}\mathbf{x} \geq \mathbf{b}, \tag{6.54}$$
$$\mathbf{x} \geq \mathbf{0}, \tag{6.55}$$

has a geometric dual with only nonnegativity constraints:

Program DLI (μ):

$$\max_{\mathbf{w}\geq 0}\left\{\sum_{i=1}^{m} b_i w_i - \mu\sum_{j=1}^{n}\exp\{[(\sum_{i=1}^{m} a_{ij}w_i - c_j)/\mu] - 1\}\right\}. \tag{6.56}$$

Moreover, for any given tolerance level $\epsilon > 0$, a sufficiently small μ can be found so that $\mathbf{x}^*(\mu)$ is an ϵ-optimal solution to Program LI in the sense that $|\mathbf{c}^T\mathbf{x}^* - \mathbf{c}^T\mathbf{x}^*(\mu)| \leq \epsilon$. The proof is similar to that for Program L(μ), as discussed in detail in Section 5.2.

In this section, we define

$$F \equiv \{\mathbf{x} \mid \mathbf{x} \in R^n, \mathbf{A}\mathbf{x} \geq \mathbf{b}, \mathbf{x} \geq \mathbf{0}\}. \tag{6.57}$$

Then a perturbed feasible region will be defined by a smooth convex constraint which specifies a measure of constraint violation in terms of l_p-norms in Subsection 6.2.1. We show that the perturbed feasible region is contained in F

and expands to the interior of F as p approaches ∞. Subsection 6.2.2 studies the duality theory and provides a primal-to-dual conversion formula that transforms an optimal solution of the perturbed program into a feasible solution of Program DLI. It also establishes a convergence theory using the duality theory. Under appropriate assumptions, we show that the optimal solution of the perturbed problem approaches an optimal solution of Program LI, while the dual solution obtained through the conversion formula converges to an optimal solution of Program DLI. In Subsection 6.2.3, we introduce a "reference linear program" and use it to develop an ϵ-optimality theory for the proposed approach. Concluding remarks are given in Subsection 6.2.4.

6.2.1 Perturbing the Primal Feasible Region

We treat the regular constraints (6.48) and the nonnegativity constraints (6.49) separately. Note that the regular constraints, for $i = 1, ..., m$,

$$\sum_{j=1}^{n} a_{ij} x_j \geq b_i, \tag{6.58}$$

are equivalent to

$$\phi_{1i}(\mathbf{x}) \equiv \exp\{-\sum_{j=1}^{n} a_{ij} x_j + b_i\} \leq 1, \quad i = 1, \ldots, m. \tag{6.59}$$

Similarly, the nonnegative constraints are equivalent to

$$\phi_{2j}(\mathbf{x}) \equiv \exp\{-x_j\} \leq 1, \quad j = 1, \ldots, n. \tag{6.60}$$

Define a vector measure of constraint violation by

$$\phi^T(\mathbf{x}) \equiv (\phi_1^T(\mathbf{x}), \phi_2^T(\mathbf{x})), \tag{6.61}$$

where

$$\phi_1^T(\mathbf{x}) \equiv (\phi_{11}(\mathbf{x}), \ldots, \phi_{1m}(\mathbf{x})), \tag{6.62}$$

and

$$\phi_2^T(\mathbf{x}) \equiv (\phi_{21}(\mathbf{x}), \ldots, \phi_{2n}(\mathbf{x})). \tag{6.63}$$

Also define

$$h(\mathbf{x}) \equiv \max\{\max_{i=1,\ldots,m}\{-\sum_{j=1}^{n} a_{ij}x_j + b_i,\}, \max_{j=1,\ldots,n}\{-x_j\}\}, \tag{6.64}$$

$$||\phi(\mathbf{x})||_\infty \equiv \max\{\max_{i=1,\ldots,m}\{\phi_{1i}(\mathbf{x})\}, \max_{j=1,\ldots,n}\{\phi_{2j}(\mathbf{x})\}\} = \exp[h(\mathbf{x})]. \tag{6.65}$$

Note that $\mathbf{x} \in F$ if and only if

$$h(\mathbf{x}) \leq 0, \quad \text{or equivalently,} \quad \ln ||\phi(\mathbf{x})||_\infty \leq 0. \tag{6.66}$$

Therefore, Program LI is equivalent to

Program $\mathbf{LI}'$:

$$\min \quad \sum_{j=1}^{n} c_j x_j \tag{6.67}$$

$$\text{s.t.} \quad \ln ||\phi(\mathbf{x})||_\infty \leq 0. \tag{6.68}$$

Now, for $p \geq 1$, the l_p-norm of $\phi(\mathbf{x})$ is defined by

$$\begin{aligned} ||\phi(\mathbf{x})||_p &\equiv \left[\sum_{i=1}^{m}[\phi_{1i}(\mathbf{x})]^p + \sum_{j=1}^{n}[\phi_{2j}(\mathbf{x})]^p\right]^{1/p} \\ &= \left\{\sum_{i=1}^{m}\exp\left[p\left(-\sum_{j=1}^{n}a_{ij}x_j + b_i\right)\right] + \sum_{j=1}^{n}\exp\left[-px_j\right]\right\}^{1/p}. \end{aligned}$$

By using the l_p-norm to approximate l_∞-norm, we define the following perturbed program:

Program $\mathbf{LI}_p$:

$$\min \quad \sum_{j=1}^{n} c_j x_j \tag{6.69}$$

$$\text{s.t.} \quad \ln ||\phi(\mathbf{x})||_p \leq 0. \tag{6.70}$$

Denote the feasible region of Program LI_p by

$$F_p \equiv \{\mathbf{x} \mid \mathbf{x} \in R^n, \ \ln ||\phi(\mathbf{x})||_p \leq 0\}. \tag{6.71}$$

We now begin to establish some properties of Program LI_p.

Lemma 6.5. If $q \geq p \geq 1$, then the feasible domain F_p of Program LI_p is contained in the feasible domain F_q of Program LI_q, i.e., $F_p \subset F_q$.

The proof is similar to that of Lemma 6.1 and is omitted. We are now ready to prove that, in the sense of "liminf" as defined in [1], F_p approaches the interior of F, denoted by $int(F)$, as p tends to infinity.

Lemma 6.6. For $p \geq 1$ and $\mathbf{x} \in R^n$,

$$h(\mathbf{x}) \leq \ln ||\phi(\mathbf{x})||_p \leq h(\mathbf{x}) + (1/p)\ln(m+n). \tag{6.72}$$

Proof. For any $p \geq 1$ and $\mathbf{x} \in R^n$,

$$\begin{aligned} \ln ||\phi(\mathbf{x})||_p - h(\mathbf{x}) &= (1/p)\ln \left\{ \sum_{i=1}^{m} \exp\left\{ p[-\sum_{j=1}^{n} a_{ij}x_j + b_i - h(\mathbf{x})] \right\} \right. \\ &+ \left. \sum_{j=1}^{n} \exp\{p[-x_j - h(\mathbf{x})]\} \right\}. \end{aligned} \tag{6.73}$$

By the definition of $h(\mathbf{x})$, we have

$$\ln ||\phi(\mathbf{x})||_p - h(\mathbf{x}) \leq (1/p)\ln(m+n). \tag{6.74}$$

Since, for any $p > 1$ and $\mathbf{x} \in R^n$, $\ln ||\phi(\mathbf{x})||_p$ decreases to $h(\mathbf{x})$ as p increases to infinity, we obtain Inequality (6.72).

□

The following corollary is an immediate consequence of Lemma 6.6.

Corollary 6.1. $\ln ||\phi(\mathbf{x})||_p$ converges to $h(\mathbf{x})$ uniformly on $\mathbf{x} \in R^n$. Moreover, for any increasing sequence $\{p_i \geq 1 \mid i = 1, 2, \ldots, \infty\}$, $\liminf F_{p_i} = \text{int}(F)$ as p_i approaches infinity.

Note that Corollary 6.1 indicates that Program LI_p becomes Program LI as p approaches infinity. Also note that when $\text{int}(F)$ is nonempty, it is easy to see that, for $p \geq 1$,

$$F_p \subset \text{int}(F). \tag{6.75}$$

Let us now focus on the feasible region of Program LI_p, i.e., F_p, which is "strictly convex" in the sense that every boundary point of F_p is an extreme point. Here, a boundary point of F_p is defined to be a point $\mathbf{x}$ such that $\ln ||\phi(\mathbf{x})||_p = 0$ and an extreme point is defined to be one that cannot be expressed as a convex combination of two other points of F_p. The following lemma facilitates the proof of the strict convexity.

Lemma 6.7. Let $k > 1$, $\mathbf{y}$ and $\mathbf{z}$ be any k-dimensional column vectors such that $\mathbf{0} < \mathbf{y}$, $\mathbf{0} < \mathbf{z}$, $\mathbf{y} \neq \mathbf{z}$, $\sum_{i=1}^k y_i = 1$ and $\sum_{i=1}^k z_i = 1$. Then, for $0 < \lambda < 1$, $\sum_{i=1}^k y_i^\lambda z_i^{1-\lambda} < 1$.

Proof. Let us define $f(\lambda) \equiv \sum_{i=1}^k y_i^\lambda z_i^{1-\lambda}$. Since $\frac{d^2 f(\lambda)}{d\lambda^2} = \sum_{i=1}^k (\ln y_i - \ln z_i)^2 y_i^\lambda z_i^{1-\lambda} > 0$, $f(\lambda)$ is strictly convex. The lemma follows from the fact that $f(0) = f(1) = 1$.

□

Theorem 6.2. If F_p is nonempty, then every boundary point of F_p is an extreme point.

Proof. Note that when $m + n = 1$, i.e., $n = 1$ and $m = 0$, F_p has only one boundary point. This is a trivial case. Otherwise, let $\mathbf{x}^1$ and $\mathbf{x}^2$ denote any two distinct boundary points of F_p. Any point $\mathbf{x}^\lambda \equiv \lambda\mathbf{x}^1 + (1-\lambda)\mathbf{x}^2$, $0 < \lambda < 1$, cannot be a boundary point because, by Lemma 6.7,

$$\sum_{i=1}^m \exp[p(-\sum_{j=1}^n a_{ij} x_j^\lambda + b_i)] + \sum_{j=1}^n \exp[-p x_j^\lambda]$$

$$= \sum_{i=1}^{m}\{\exp[p(-\sum_{j=1}^{n} a_{ij}x_j^1 + b_i)]\}^{\lambda}\{\exp[p(-\sum_{j=1}^{n} a_{ij}x_j^2 + b_i)]\}^{1-\lambda}$$

$$+ \sum_{j=1}^{n}\{\exp(-px_j^1)\}^{\lambda}\{\exp(-px_j^2)\}^{1-\lambda} < 1. \tag{6.76}$$

□

By the "strict convexity" of F_p, we have the following immediate result regarding the existence and uniqueness of an optimal solution to Program $\mathbf{LI}_p$.

Corollary 6.2. For $p \geq 1$, if F_p is nonempty and bounded, then Program LI_p has a unique optimal solution.

Proof. The compactness of F_p implies the existence of an optimal solution. Since $\mathbf{c} \neq \mathbf{0}$ and F_p is a convex set, all optimal solutions lie on the boundary of F_p. Suppose that there exist two distinct optimal solutions. By the linearity of the objective function $\mathbf{c}^T\mathbf{x}$ of Program LI_p, any convex combination of the two distinct optimal solutions is also optimal. By Theorem 6.2, any such convex combination must not be a boundary point of F_p. But, this contradicts the fact that all optimal solutions lie on the boundary of F_p. Therefore, Program LI_p has a unique optimal solution.

□

We now illustrate the approach by considering the following linear program:

Program $\mathbf{LI^0}$:

$$\begin{aligned} \min \quad & -x_1 \\ \text{s.t.} \quad & -x_1 - x_2 \geq -5 \\ & -x_1 + x_2 \geq -3 \\ & x_1 \geq 0 \\ & x_2 \geq 0. \end{aligned}$$

The perturbed program is

Program LI_p^0:

$$\begin{array}{ll} \min & -x_1 \\ \text{s.t.} & e^{p(x_1+x_2-5)} + e^{p(x_1-x_2-3)} + e^{-px_1} + e^{-px_2} \le 1. \end{array}$$

The nested feasible regions of Programs LI_p^0, $p = 1, 2, 3$, and Program LI^0 are plotted in Figure 6.1.

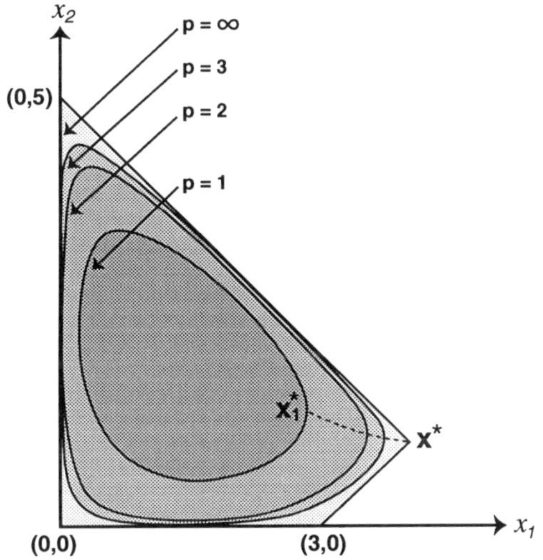

Figure 6.1 Perturbed feasible regions and convergence of the optimal solutions.

6.2.2 Duality and Convergence Theory

Although Program LI_p could be infeasible, unbounded, or bounded from below, to simplify our treatment of the duality and convergence theory, we assume throughout this subsection that F_1 is nonempty and F is bounded. The former ensures the existence of a feasible solution to Program LI_p for $p \ge 1$, while the latter ensures the existence of an optimal solution for Program LI_p as well as its uniqueness. Note that F_1 being nonempty implies that F has a nonempty interior. Let us denote $\mathbf{x}_p^*$ as the optimal solution of Program LI_p, for $p \geq 1$. As a consequence of Lemma 6.5, we have the following result:

Lemma 6.8. If F_1 is nonempty and F is bounded, then $\mathbf{c}^T\mathbf{x}_p^*$ is a decreasing function of p that converges to the optimal value of Program LI as $p \to \infty$.

Proof. Since F_1 is nonempty and F is bounded, $\mathbf{x}_p^*$ exists for $p \geq 1$, and is the unique optimal solution of Program LI_p. By Corollary 6.1, $\mathbf{c}^T\mathbf{x}_p^*$ is decreasing in p. Recall that F_1 being nonempty implies that F has a nonempty interior. This and Corollary 6.1 again imply that $\mathbf{c}^T\mathbf{x}_p^*$ converges to the optimal value of Program LI.

□

We will study the convergence of the optimal solution $\mathbf{x}_p^*$ later in conjunction with the duality theory. We now derive the duality theory for Program LI_p. In particular, we provide an effortless conversion formula transforming an optimal solution of Program LI_p to a feasible solution to Program DLI.

Define the following Lagrangian function for Program LI_p, with $\lambda \geq 0$:

$$L(\lambda, \mathbf{x}) = \sum_{j=1}^{n} c_j x_j + \lambda\{\sum_{i=1}^{m} \exp[p(-\sum_{j=1}^{n} a_{ij}x_j + b_i)] + \sum_{j=1}^{n} \exp[-px_j] - 1\}.$$

Given $\lambda \geq 0$, the minimum of the Lagrangian function is attained at a point $\mathbf{x}^*(\lambda)$ such that

$$c_j = \sum_{i=1}^{m} a_{ij} \left\{ \lambda p \,\exp\left[p(-\sum_{j=1}^{n} a_{ij}x_j^*(\lambda) + b_i) \right] \right\} + \lambda p \,\exp[-px_j^*(\lambda)]. \quad (6.77)$$

Denote the λ at which the minimum of the Lagrangian function is maximized by λ_p^*. In other words, λ_p^* denotes an optimal solution of the Lagrangian dual of Program LI_p. Therefore, $\mathbf{x}_p^* = \mathbf{x}^*(\lambda_p^*)$. Consider the following conversion formula:

$$w_{p,i} \;\equiv\; \lambda_p^* p \,\exp[p(-\sum_{j=1}^{n} a_{ij}x_{p,j}^* + b_i)], \quad i = 1, \ldots, m, \qquad (6.78)$$

$$s_{p,j} \equiv \lambda_p^* p \exp[-px_{p,j}^*], \quad j = 1, ..., n. \tag{6.79}$$

It will be shown in the following lemma that Equation (6.78) provides a feasible solution $\mathbf{w}_p$ to Program DLI. Also note that $\mathbf{s}_p$ defined in Equation (6.79) corresponds to a dual slack vector of Program DLI.

Lemma 6.9. If F_1 is nonempty and F is bounded, then the vector $\mathbf{w}_p$ defined by Equation (6.78) is feasible to Program DLI.

Proof. For $p \geq 1$, since $\lambda_p^* \geq 0$, $\mathbf{w}_p \geq \mathbf{0}$. Since $\lambda_p^* p \exp(-px_{p,j}^*) > 0$, $j = 1, 2, ..., n$, we have, by Equation (6.77), $\sum_{i=1}^m a_{ij} w_{p,i} \leq c_j$.

□

Theorem 6.3. Suppose that F_1 is nonempty, F is bounded with a nonempty interior, and Program DLI has a bounded feasible region. Then, the set $\{(\mathbf{x}_p^*, \mathbf{w}_p, \mathbf{s}_p) \mid p \geq 1\}$, where $\mathbf{x}_p^*$ is the optimal solution of Program LI_p and $\mathbf{w}_p$ and $\mathbf{s}_p$ are defined by Equations (6.78) and (6.79), respectively, has at least one cluster point. Let $\{p_i \geq 1 \mid i = 1, 2, \ldots, \infty\}$ be an increasing sequence of real numbers approaching infinity. If $\{\mathbf{x}_{p_i}^*, \mathbf{w}_{p_i}, \mathbf{s}_{p_i}\}$ is a convergent sequence, then $\mathbf{x}_{p_i}^*$ converges to an optimal solution of Program LI, and $\mathbf{w}_{p_i}$ converges to an optimal solution of Program DLI.

Proof. By the Bolzano-Weierstrass Theorem, the nonemptiness and boundedness assumptions ensure the existence of a cluster point. Now, consider any convergent sequence $\{\mathbf{x}_{p_i}^*, \mathbf{w}_{p_i}, \mathbf{s}_{p_i}\}$. Let $\mathbf{x}_\infty^*$ and $\mathbf{w}_\infty$ denote the limits as $p_i \to \infty$. We show that the Karush-Kuhn-Tucker (K.K.T.) conditions are met in the limit as $i \to \infty$ and $p_i \to \infty$. The feasibility of $\mathbf{x}_\infty^*$ and $\mathbf{w}_\infty$ is obvious. It suffices to show that the complementary conditions hold. The convergence of $\mathbf{w}_{p_i}$ and $\mathbf{s}_{p_i}$ implies the convergence of their sum $\mathbf{w}_{p_i} + \mathbf{s}_{p_i}$. Since the objective function is linear and nonzero, $\mathbf{x}_p^*$ always lies on the boundary of the feasible region F_p. Therefore, we have, for $p \geq 1$,

$$\sum_{i=1}^m \exp[p(-\sum_{j=1}^n a_{ij} x_{p,j}^* + b_i)] + \sum_{j=1}^n \exp[-px_{p,j}^*] = 1. \tag{6.80}$$

This, together with the convergence of $\mathbf{w}_{p_i} + \mathbf{s}_{p_i}$, imply that $\lambda_{p_i}^* p_i$ converges, as $i \to \infty$. Therefore, if $x_{\infty,j}^* > 0$, then, as $i \to \infty$,

$$s_{p_i,j} \equiv \lambda^*_{p_i} p_i \exp[-p_i x^*_{p_i,j}] \to 0, \tag{6.81}$$

for $j = 1, ..., n$. But Equation (6.77) further implies that, for $j = 1, ..., n$,

$$\sum_{i=1}^{m} a_{ij} w_{\infty,i} = c_j. \tag{6.82}$$

On the other hand, for $j = 1, ..., n$, if

$$\sum_{i=1}^{m} a_{ij} w_{\infty,i} < c_j, \tag{6.83}$$

then $x^*_{\infty,j}$ must be 0 because, otherwise, Inequality (6.83) would have been an equality.

□

6.2.3 ϵ-Optimality

In this subsection, we first define the following "reference linear program" R_p and then use it to establish an ϵ-optimality theory.

Program R_p:

$$\min \quad \sum_{j=1}^{n} c_j x_j \tag{6.84}$$

$$\text{s.t.} \quad \sum_{j=1}^{n} a_{ij} x_j \geq b_i + (1/p)\ln(m+n), \quad i = 1, ..., m, \tag{6.85}$$

$$x_j \geq (1/p)\ln(m+n), \quad j = 1, ..., n. \tag{6.86}$$

Let H_p denote the feasible region of Program R_p. We show that $H_p \subset F_p$.

Lemma 6.10. The feasible region H_p of Program R_p is a subset of the feasible region F_p of Program LI_p.

Proof. Let $\mathbf{x}$ be any feasible solution of Program R_p. We have

$$\max\{\max_{i=1,\ldots,m}\{-\sum_{j=1}^{n} a_{ij}x_j + b_i + (1/p)\ln(m+n)\},$$

$$\max_{j=1,\ldots,n}\{-x_j + (1/p)\ln(m+n)\}\} \leq 0. \tag{6.87}$$

Since the left-hand side is equivalent to $h(\mathbf{x}) + (1/p)\ln(m+n)$, we know

$$h(\mathbf{x}) \leq -(1/p)\ln(m+n). \tag{6.88}$$

This and Lemma 6.6 together imply that

$$\ln ||\phi(\mathbf{x})||_p \leq 0. \tag{6.89}$$

Hence, $\mathbf{x}$ is in F_p.

□

Although the exact shape of the feasible region of Program LI_p is not easy to visualize, it is clear that its boundary must be contained completely between the boundaries of Program LI and Program R_p. The quantity $(1/p)\ln(m+n)$ can be viewed as an index of maximum shrinkage. A smaller p results in a larger perturbation with a smoother constraint that defines the boundary of the corresponding perturbed program.

Unlike the technique of perturbing the dual feasible region as discussed in Section 6.1 and [6, 7], perturbing the primal feasible region, with the aid of the reference program R_p, has an intuitive and clear geometric interpretation. Also, to facilitate computation, one can rescale the variables as well as constraints so that the "shrinkage" index $(1/p)\ln(m+n)$ reflects a common acceptable imprecision or error on the right-hand side.

Now we focus on investigating the maximum possible difference between the optima of Program LI and Program LI_p as a function of p. By Lemma 6.10, this maximum possible difference is bounded above by the maximum difference between the optima of Program LI and Program R_p. Assume that Program LI has an interior-point solution and Program DLI is feasible. There must exist a $p \geq 1$ such that Program R_p has an interior-point solution. Define

$$p_0 \equiv \inf\{p \mid p \geq 1 \text{ and } R_p \text{ has an interior-point solution}\}. \tag{6.90}$$

Consider in the rest of this subsection those $p > p_0$. Since there exists no duality gap between Programs LI and DLI and between Programs R_p and its linear dual, denoted by DR_p, it suffices to investigate the maximum possible difference between the optima of Program DLI and Program DR_p in terms of p. Once this relationship is known, for any given $\epsilon > 0$, we can find an appropriate $p > 0$ such that $\mathbf{x}_p^*$ is an ϵ-optimal solution to Program LI.

Denoting $(1/p)\ln(m+n)$ by α, it can be shown that the dual of Program R_p is given by

Program DR_p:

$$\max \quad \sum_{i=1}^{m} w_i b_i + \alpha \left[\sum_{i=1}^{m} w_i + \sum_{j=1}^{n} \left(c_j - \sum_{i=1}^{m} a_{ij} w_i \right) \right] \tag{6.91}$$

$$\text{s.t.} \quad \sum_{i=1}^{m} a_{ij} w_i \leq c_j, \quad j = 1, \ldots, n, \tag{6.92}$$

$$w_i \geq 0, \quad i = 1, \ldots, m. \tag{6.93}$$

Note that Program DR_p is different from Program DLI only in the objective function, which has an additional term that is proportional to α. Since $\alpha > 0$, for any feasible solution of Program DR_p (and hence of Program DLI), the objective value of Program DR_p at $\mathbf{w}$ is greater than or equal to that of Program DLI. It is clearly seen that when p is sufficiently large, Program DR_p can be used to approximate Program DLI.

We now derive the relationship between a given tolerance level $\epsilon > 0$ and the perturbation parameter p. Let $\mathbf{w}^*$ denote an optimal solution to Program DLI and $\mathbf{u}^p$ an optimal solution to Program DR_p. Since they are optimal solutions, we have

$$\sum_{i=1}^{m} b_i w_i^* \geq \sum_{i=1}^{m} b_i u_i^p, \tag{6.94}$$

$$\sum_{i=1}^{m} u_i^p (b_i + \alpha - \alpha \sum_{j=1}^{n} a_{ij}) \geq \sum_{i=1}^{m} w_i^* (b_i + \alpha - \alpha \sum_{j=1}^{n} a_{ij}). \tag{6.95}$$

Consequently,

$$\alpha \left[\sum_{i=1}^{m} (1 - \sum_{j=1}^{n} a_{ij})(u_i^p - w_i^*) \right] \geq \sum_{i=1}^{m} b_i w_i^* - \sum_{i=1}^{m} b_i u_i^p \geq 0. \tag{6.96}$$

If Program DLI has a bounded feasible region, i.e., there exists a positive number M such that the dual feasible region is contained in the spheroid centered at the origin with a radius of M, then

$$2\alpha KM \geq \sum_{i=1}^{m} b_i w_i^* - \sum_{i=1}^{m} b_i u_i^p \geq 0, \tag{6.97}$$

where

$$K \equiv \sum_{i=1}^{m} |1 - \sum_{j=1}^{n} a_{ij}|. \tag{6.98}$$

Therefore, given an $\epsilon > 0$, if

$$p \geq 2 \ln(m+n) KM/\epsilon, \tag{6.99}$$

then $\mathbf{x}_p^*$ is an ϵ-optimal solution.

We summarize these results in the following theorem.

Theorem 6.4. Suppose that Program LI has an interior-point solution and Program DLI has a nonempty and bounded feasible region. If p is selected according to Equation (6.99) and $p > p_0$, then $\mathbf{x}_p^*$ is an ϵ-optimal solution to Program LI.

6.2.4 Concluding Remarks

In this section, we have developed an approach to solving linear programming problems with inequality constraints by perturbing its primal feasible region. Both primal convergence and dual convergence of the perturbed solutions have been proven. We have also derived a formula for choosing proper perturbation parameters to achieve ϵ-optimality. Numerical methods for solving Program LI_p should be investigated to gain computational experience.

6.3 PERTURBING DUAL FEASIBLE REGION OF CONVEX QUADRATIC PROGRAMS

Consider the following convex quadratic program:

Program GQ:

$$\min \quad (1/2)\mathbf{x}^T\mathbf{Q}\mathbf{x} + \mathbf{c}^T\mathbf{x} \tag{6.100}$$

$$\text{s.t.} \quad \mathbf{A}\mathbf{x} = \mathbf{b}, \tag{6.101}$$

$$\mathbf{x} \geq \mathbf{0}, \tag{6.102}$$

where $\mathbf{Q}$ is an $n \times n$ symmetric positive semi-definite matrix, $\mathbf{c}$ and $\mathbf{x}$ are n-dimensional column vectors, $\mathbf{A}$ is an $m \times n$ $(m \leq n)$ matrix, $\mathbf{b}$ is an m-dimensional column vector, and $\mathbf{0}$ is the n-dimensional zero vector. Note that Program GQ differs from Program Q, as defined in Section 5.3, in that the matrix $\mathbf{Q}$ of Program GQ is a general symmetric positive semi-definite matrix, while the matrix $\mathbf{D}$ of Program Q is a nonnegative diagonal matrix. In this section, we denote the feasible region of Program GQ by F.

The Lagrangian dual of Program GQ, which is also a convex quadratic program, is given by:

Program DGQ:

$$\max \quad -(1/2)\mathbf{v}^T\mathbf{Q}\mathbf{v} + \mathbf{b}^T\mathbf{w} \tag{6.103}$$

$$\text{s.t.} \quad -\mathbf{Q}\mathbf{v} + \mathbf{A}^T\mathbf{w} \leq \mathbf{c}, \tag{6.104}$$

$$\mathbf{v} \in R^n, \quad \mathbf{w} \in R^m. \tag{6.105}$$

We denote the feasible region of Program DGQ by G. The K.K.T. conditions for Program GQ are

$$\mathbf{Ax} = \mathbf{b}, \quad \mathbf{x} \geq \mathbf{0}, \tag{6.106}$$

$$-\mathbf{Qx} + \mathbf{A}^T\mathbf{w} + \mathbf{s} = \mathbf{c}, \quad \mathbf{s} \geq \mathbf{0}, \tag{6.107}$$

$$\mathbf{XSe} = \mathbf{0}, \tag{6.108}$$

where $\mathbf{X}$ and $\mathbf{S}$ are diagonal matrices formed by using the elements of $\mathbf{x}$ and $\mathbf{s}$ as diagonal elements, respectively, and $\mathbf{e}$ is an n-dimensional column vector of 1's.

When Program GQ has an interior feasible solution and attains its minimum, Program DGQ attains its maximum with no duality gap. If, in addition, matrix $\mathbf{Q}$ is positive-definite, then Program DGQ has a unique optimal solution $(\mathbf{v}^*, \mathbf{w}^*)$, and, more interestingly, $\mathbf{v}^*$ is also the optimal solution of Program GQ ([14], pages 288 - 291). Therefore, in this case, solving the dual program provides the unique solution to both the primal and dual programs.

If matrix $\mathbf{Q}$ is reduced to a diagonal matrix $\mathbf{D}$ with diagonal elements d_j's, as shown in Section 5.3, by perturbing the primal objective with an entropic function, for $\mu > 0$, the following problem

Program Q(μ):

$$\min \quad (1/2)\mathbf{x}^T\mathbf{Dx} + \mathbf{c}^T\mathbf{x} + \mu \sum_{j=1}^{n} x_j \ln x_j \tag{6.109}$$

$$\text{s.t.} \quad \mathbf{Ax} = \mathbf{b}, \tag{6.110}$$

$$\mathbf{x} \geq \mathbf{0}, \tag{6.111}$$

has an unconstrained dual problem

Program DQ(μ):

$$\max_{\mathbf{w} \in R^m} \left\{ -(1/2) \sum_{j=1}^{n} d_j g_j^2(\mathbf{w}) - \mu \sum_{j=1}^{n} g_j(\mathbf{w}) + \sum_{i=1}^{m} b_i w_i \right\}, \tag{6.112}$$

where $g_j(\mathbf{w}) > 0$, denoting a function defined on R^m, is the unique solution of the following equation

$$d_j g_j(\mathbf{w}) + \mu \ln g_j(\mathbf{w}) = \sum_{i=1}^{m} a_{ij} w_i - c_j - \mu. \tag{6.113}$$

Moreover, under some commonly used assumptions, Program Q(μ) has a unique optimal solution $\mathbf{x}^*(\mu)$ which approaches an optimal solution $\mathbf{x}^*$ of Program GQ, as $\mu \to 0$.

In this section, we investigate a dual perturbation approach which approximates the dual feasible region G by aggregating the n linear constraints (6.104) into a single l_p-norm constraint. The dual feasible region is perturbed by controlling the p value ($p \geq 1$). As $p \to \infty$, we show that the perturbed dual feasible region approaches G in the sense of "liminf" ([1], page 71) and hence the optimal solution to the perturbed dual program approaches an optimal solution of Program DGQ. Moreover, given the knowledge of an optimal solution to the perturbed dual program, we derive a closed-form dual-to-primal formula for obtaining a primal feasible solution, and show that such a primal solution converges to an optimal solution of Program GQ as $p \to \infty$. We also show that this approach generalizes the previous results obtained by the entropic perturbation approach.

This section is organized as follows. Subsection 6.3.1 presents the dual l_p-norm perturbation approach. Subsections 6.3.2 and 6.3.3 establish the dual and primal convergence results. Subsection 6.3.4 shows the relationship between the dual l_p-norm perturbation approach and the method of perturbing the primal objective with an entropic function. Concluding remarks are given in Subsection 6.3.5.

6.3.1 Perturbing the Dual Feasible Region

The feasible region G is defined by the following n constraints:

$$-\sum_{k=1}^{n} q_{jk}v_k + \sum_{i=1}^{m} a_{ij}w_i - c_j \leq 0, \ j = 1, \ldots, n. \tag{6.114}$$

Instead of dealing with this polyhedral set directly, we approximate it by a proper subset which has a smooth boundary defined by a single nonlinear differentiable function.

To do so, we measure the violation of the dual constraints (6.104) at $\mathbf{v} \in R^n$ and $\mathbf{w} \in R^m$ by the following n-dimensional "error vector"

$$\phi(\mathbf{v}, \mathbf{w}) \equiv (\phi_1(\mathbf{v}, \mathbf{w}), \ldots, \phi_n(\mathbf{v}, \mathbf{w})), \tag{6.115}$$

where

$$\phi_j(\mathbf{v}, \mathbf{w}) \equiv \exp\left[-\sum_{k=1}^{n} q_{jk}v_k + \sum_{i=1}^{m} a_{ij}w_i - c_j\right], \tag{6.116}$$

for $j = 1, ..., n$. It is clearly seen that, for any given $\mathbf{v} \in R^n$ and $\mathbf{w} \in R^m$, $\phi_j(\mathbf{v}, \mathbf{w}) > 0$ and $(\mathbf{v}, \mathbf{w}) \in G$ if and only if

$$||\phi(\mathbf{v}, \mathbf{w})||_\infty \equiv \lim_{p \to \infty} ||\phi(\mathbf{v}, \mathbf{w})||_p = \max_{j=1,\ldots,n} \{\phi_j(\mathbf{v}, \mathbf{w})\} \leq 1, \tag{6.117}$$

where

$$\begin{aligned} ||\phi(\mathbf{v}, \mathbf{w})||_p &\equiv \left[\sum_{j=1}^{n} (\phi_j(\mathbf{v}, \mathbf{w}))^p\right]^{1/p} \\ &= \left[\sum_{j=1}^{n} \exp[p(-\sum_{k=1}^{n} q_{jk}v_k + \sum_{i=1}^{m} a_{ij}w_i - c_j)]\right]^{1/p} \end{aligned} \tag{6.118}$$

is the l_p-norm of $\phi(\mathbf{v}, \mathbf{w})$, for $p \geq 1$. Consequently, for $p \geq 1$, we may define a perturbed dual program as follows:

Program $\mathbf{DGQ}_p$:

$$\max \quad -(1/2)\mathbf{v}^T\mathbf{Q}\mathbf{v} + \mathbf{b}^T\mathbf{w} \tag{6.119}$$

$$\text{s.t.} \quad ||\phi(\mathbf{v}, \mathbf{w})||_p \leq 1. \tag{6.120}$$

We denote its feasible region by G_p and an optimal solution by $(\mathbf{v}_p^*, \mathbf{w}_p^*)$, when it exists.

Since the logarithmic function is strictly increasing, the l_p-norm is preserved by

$$\ln ||\phi(\mathbf{v}, \mathbf{w})||_p = (1/p) \ln \left\{ \sum_{j=1}^{n} \exp[p(-\sum_{k=1}^{n} q_{jk} v_k + \sum_{i=1}^{m} a_{ij} w_i - c_j)] \right\}. \quad (6.121)$$

Therefore, Inequality (6.120) of Program DGQ_p can be replaced by

$$(1/p) \ln \left\{ \sum_{j=1}^{n} \exp[p(-\sum_{k=1}^{n} q_{jk} v_k + \sum_{i=1}^{m} a_{ij} w_i - c_j)] \right\} \leq 0. \quad (6.122)$$

Inequality (6.120) of Program DGQ_p can also be replaced by

$$\sum_{j=1}^{n} \exp[p(-\sum_{k=1}^{n} q_{jk} v_k + \sum_{i=1}^{m} a_{ij} w_i - c_j)] \leq 1. \quad (6.123)$$

It is clearly seen that Program DGQ_p is a convex program with a convex quadratic objective function and one convex C^∞ constraint. When p is sufficiently large, all of the n constraints of Program DGQ are virtually aggregated into this single constraint.

6.3.2 Dual Convergence

In this subsection we show that Program DGQ_p becomes Program DGQ as $p \to \infty$. The following lemma follows immediately from a well-known fact about the l_p-norm.

Lemma 6.11. If $q \geq p \geq 1$, then $\ln ||\phi(\mathbf{v}, \mathbf{w})||_p \geq \ln ||\phi(\mathbf{v}, \mathbf{w})||_q$ and $G_p \subset G_q$. Moreover, $\ln ||\phi(\mathbf{v}, \mathbf{w})||_p \to h(\mathbf{v}, \mathbf{w})$ as $p \to \infty$, where

$$h(\mathbf{v}, \mathbf{w}) \equiv \max_{j=1,\ldots,n} \{-\sum_{k=1}^{n} q_{jk} v_k + \sum_{i=1}^{m} a_{ij} w_i - c_j\}. \quad (6.124)$$

The following result is a direct consequence of Lemma 6.11.

Lemma 6.12. If Program DGQ_1 is feasible, then $-(1/2)\mathbf{v}_p^{*T}\mathbf{Q}\mathbf{v}_p^* + \mathbf{b}^T\mathbf{w}_p^*$, is nondecreasing in p, for $p \geq 1$.

We are now ready to show that, in the sense of "liminf", G_p approaches the interior of G, denoted by $\text{int}(G)$, as $p \to \infty$. Let E_n be any sequence of subsets of an arbitrary space Ω. Then, liminf $E_n \equiv \cup_{m=1}^{\infty} \cap_{n=m}^{\infty} E_n$ ([1], page 71).

Lemma 6.13. For $p \geq 1$, $\mathbf{v} \in R^n$, and $\mathbf{w} \in R^m$,

$$h(\mathbf{v}, \mathbf{w}) \leq \ln ||\phi(\mathbf{v}, \mathbf{w})||_p \leq h(\mathbf{v}, \mathbf{w}) + (1/p) \ln n, \tag{6.125}$$

and hence $\ln ||\phi(\mathbf{v}, \mathbf{w})||_p$ converges to $h(\mathbf{v}, \mathbf{w})$ uniformly, as $p \to \infty$. Moreover, if $\{p_i \mid p_i \geq 1, i = 1, 2, \ldots, \infty\}$ is an increasing sequence such that $p_i \to \infty$ as $i \to \infty$, then

$$\text{liminf } G_{p_i} = \text{int}(G), \text{ as } i \to \infty. \tag{6.126}$$

Proof. For any $p \geq 1$, $\mathbf{v} \in R^n$ and $\mathbf{w} \in R^m$,

$$\begin{aligned} &\ln ||\phi(\mathbf{v}, \mathbf{w})||_p - h(\mathbf{v}, \mathbf{w}) \\ &= (1/p) \ln \left\{ \sum_{j=1}^{n} \exp \left\{ p[(-\sum_{k=1}^{n} q_{jk} v_k + \sum_{i=1}^{m} a_{ij} w_i - c_j) - h(\mathbf{v}, \mathbf{w})] \right\} \right\}. \end{aligned}$$

Since, for $j = 1, ..., n$,

$$-\sum_{k=1}^{n} q_{jk} v_k + \sum_{i=1}^{m} a_{ij} w_i - c_j \leq h(\mathbf{v}, \mathbf{w}), \tag{6.127}$$

we have

$$\exp \left\{ p[(-\sum_{k=1}^{n} q_{jk} v_k + \sum_{i=1}^{m} a_{ij} w_i - c_j) - h(\mathbf{v}, \mathbf{w})] \right\} \leq 1. \tag{6.128}$$

Consequently, for $j = 1, ..., n$,

$$\ln ||\phi(\mathbf{v}, \mathbf{w})||_p - h(\mathbf{v}, \mathbf{w}) \leq (1/p) \ln n. \tag{6.129}$$

Combining Equation (6.129) with Lemma 6.11 gives Equation (6.125). The uniform convergence and the liminf convergence follow immediately.

□

Lemma 6.13 indicates that Program DGQ$_p$ becomes Program DGQ as p approaches infinity. All lemmas derived earlier culminate in the following lemma:

Lemma 6.14. If Program GQ and Program DGQ$_1$ are feasible, then, as $p \to \infty$, $-(1/2)\mathbf{v}_p^{*T}\mathbf{Q}\mathbf{v}_p^* + \mathbf{b}^T\mathbf{w}_p^*$ converges to the optimal value of Program DGQ, and any convergent sequence $\{(\mathbf{v}_{p_k}^*, \mathbf{w}_{p_k}^*) \mid 1 \leq p_k < p_{k+1} \to \infty,\ k = 1, 2, \ldots, \infty\}$ converges to an optimal solution of Program DGQ from the interior of G.

Since the l_p-norm (i.e., the left-hand side value of Inequality (6.123)) is determined by the slackness of the constraints of Program DGQ$_p$ (i.e., the left-hand side of Equation (6.114), some boundary points of G_p may not be extreme points. When $\mathbf{Q}$ is not positive-definite, this may lead to the existence of multiple optimal solutions for Program DGQ. Therefore, if $(\mathbf{v}_{p_k}^*, \mathbf{w}_{p_k}^*)$ is arbitrarily chosen among the multiple optimal solutions, the sequence cannot be guaranteed to be convergent. However, when $\mathbf{Q}$ is positive-definite, the next convergence result follows immediately:

Theorem 6.5. If Program GQ and Program DGQ$_1$ are feasible and $\mathbf{Q}$ is positive-definite, then Program DGQ$_p$ has a unique optimal solution $(\mathbf{v}_p^*, \mathbf{w}_p^*)$ which converges to the optimal solution of Program DGQ, as $p \to \infty$.

6.3.3 Primal Convergence

In this subsection, we use the optimality conditions of Program DGQ$_p$ to derive a dual-to-primal conversion formula which converts an optimal solution

$(\mathbf{v}_p^*, \mathbf{w}_p^*)$ of Program DGQ$_p$ into a feasible solution $\mathbf{x}_p$ of Program GQ. Then we show that such an $\mathbf{x}_p$ converges to an optimal solution of Program GQ, as $p \to \infty$.

To simplify our discussion, we make the following assumptions for the rest of this subsection:

(A1) $\mathbf{Q}$ is symmetric and positive semi-definite.

(A2) Program GQ has an interior feasible solution (relative to the nonnegativity constraints). Program DGQ also has an interior feasible solution.

(A3) All of the optimal solutions of Program DGQ lie on the boundary of G.

(A4) Program DGQ$_1$ is feasible.

(A1) is a basic assumption for convex quadratic programming. Assumption (A2) implies that Program GQ and Program DGQ attain a common finite optimum. Since Program DGQ is a convex quadratic program, its optimal solutions may fall within the interior of its feasible region, in which case the problem becomes trivial [3]. To avoid such triviality, we make the assumption (A3). Assumption (A4) is made for ease of discussion. Actually, the feasibility of Program DGQ$_p$ for any $p \geq 1$ suffices.

Consider the following Lagrangian function for Program DGQ$_p$ with the constraints expressed as in Inequality (6.123):

$$L(\lambda, \mathbf{v}, \mathbf{w}) \equiv -(1/2)\sum_{i=1}^{n}\sum_{j=1}^{n} v_i q_{ij} v_j + \sum_{i=1}^{m} b_i w_i$$
$$-\lambda\left\{\sum_{j=1}^{n} \exp[p(-\sum_{k=1}^{n} q_{jk} v_k + \sum_{i=1}^{m} a_{ij} w_i - c_j)] - 1\right\}, \qquad (6.130)$$

where $\lambda \geq 0$, $\mathbf{v} \in R^n$, and $\mathbf{w} \in R^m$.

Since all optimal solutions $(\mathbf{v}_p^*, \mathbf{w}_p^*)$ of Program DGQ$_p$ satisfy Inequality (6.123) and, by the assumption (A3), lie on the boundary of its feasible region, at these optimal solutions, Inequality (6.123) becomes an equality with a corresponding multiplier $\lambda_p^* > 0$. Moreover, the first-order optimality conditions must be satisfied. The partial derivative of L with respect to v_k, $k = 1, \ldots, n$, must vanish at $(\lambda_p^*, \mathbf{v}_p^*, \mathbf{w}_p^*)$. Therefore, we have

$$\sum_{j=1}^{n} q_{kj} v_{p,j}^* = \lambda_p^* p \sum_{j=1}^{n} q_{jk} \exp\left\{ p \left[-\sum_{i=1}^{n} q_{ji} v_{p,i}^* + \sum_{i=1}^{m} a_{ij} w_{p,i}^* - c_j \right] \right\}, \quad (6.131)$$

for $k = 1, \ldots, n$. Since the partial derivative of L with respect to w_i, $i = 1, \ldots, m$, must also vanish at $(\lambda_p^*, \mathbf{v}_p^*, \mathbf{w}_p^*)$, we obtain, for $i = 1, \ldots, m$,

$$b_i = \lambda_p^* p \left\{ \sum_{j=1}^{n} a_{ij} \exp\left[p(-\sum_{k=1}^{n} q_{jk} v_{p,k}^* + \sum_{k=1}^{m} a_{kj} w_{p,k}^* - c_j) \right] \right\}. \quad (6.132)$$

If we define, for $j = 1, ..., n$,

$$x_{p,j}(\lambda_p^*, \mathbf{v}_p^*, \mathbf{w}_p^*) = \lambda_p^* p \exp[p(-\sum_{k=1}^{n} q_{jk} v_{p,k}^* + \sum_{i=1}^{m} a_{ij} w_{p,i}^* - c_j)] \geq 0, \quad (6.133)$$

Equation (6.132) becomes

$$\sum_{j=1}^{n} a_{ij} x_{p,j}(\lambda_p^*, \mathbf{v}_p^*, \mathbf{w}_p^*) = b_i, \quad (6.134)$$

for $i = 1, ..., m$. This shows that $\mathbf{x}_p(\lambda^*, \mathbf{v}_p^*, \mathbf{w}_p^*)$ is a feasible solution to the primal program GQ. Therefore, solving Program DGQ$_p$ also results in a feasible solution $\mathbf{x}_p \equiv \mathbf{x}_p(\lambda_p^*, \mathbf{v}_p^*, \mathbf{w}_p^*)$ to Program GQ.

Theorem 6.6. Under the assumptions (A1) - (A4), if Program GQ has a bounded feasible region, then $\sum_{j=1}^{n} x_{p,j}(\lambda_p^*, \mathbf{v}_p^*, \mathbf{w}_p^*) = \lambda_p^* p$ and $\lambda_p^* \to 0$ as $p \to \infty$.

Proof. By Equation (6.133), we have

$$\sum_{j=1}^{n} x_{p,j}(\lambda_p^*, \mathbf{v}_p^*, \mathbf{w}_p^*) = \lambda_p^* p \sum_{j=1}^{n} \exp[p(-\sum_{k=1}^{n} q_{jk} v_{p,k}^* + \sum_{i=1}^{m} a_{ij} w_{p,i}^* - c_j)]. \quad (6.135)$$

Since the optimal solutions of Program DGQ lie on the boundary of its feasible region, we have

$$\sum_{j=1}^{n} \exp[p(-\sum_{k=1}^{n} q_{jk} v_{p,k}^* + \sum_{i=1}^{m} a_{ij} w_{p,i}^* - c_j)] = 1. \tag{6.136}$$

Consequently, Equation (6.135) reduces to

$$\sum_{j=1}^{n} x_{p,j}(\lambda_p^*, \mathbf{v}_p^*, \mathbf{w}_p^*) = \lambda_p^* p. \tag{6.137}$$

Since Program GQ has a bounded feasible region, the set $\{\lambda_p^* p \mid p \geq 1\}$ must be bounded. Therefore, $\lambda_p^* \to 0$ as $p \to \infty$.

□

Lemma 6.15. Under the assumptions (A1) - (A4), if $\{p_k \mid p_k \geq 1, k = 1, 2, \ldots, \infty\}$ is a strictly increasing sequence approaching infinity such that $(\mathbf{x}_{p_k}, \mathbf{w}_{p_k}^*)$ converges, then $\mathbf{x}_{p_k}$ converges to an optimal solution $\mathbf{x}^*$ of Program GQ.

Proof. Since $\mathbf{Q}$ is symmetric, by Equations (6.131) and (6.133), for $k = 1, 2, \ldots, n$, we have

$$\begin{aligned} \sum_{j=1}^{n} q_{kj} v_{p,j}^* &= \sum_{j=1}^{n} q_{kj} \lambda_p^* p \exp\left\{ p \left[-\sum_{i=1}^{n} q_{ji} v_{p,i}^* + \sum_{i=1}^{m} a_{ij} w_{p,i}^* - c_j \right] \right\} \\ &= \sum_{j=1}^{n} q_{kj} x_{p,j}(\lambda_p^*, \mathbf{v}_p^*, \mathbf{w}_p^*). \end{aligned} \tag{6.138}$$

Let us define, for $j = 1, \ldots, n$,

$$\alpha_{p,j} \equiv -\sum_{i=1}^{n} q_{ji} v_{p,i}^* + \sum_{i=1}^{m} a_{ij} w_{p,i}^* - c_j. \tag{6.139}$$

Recall that $\mathbf{x}_p \equiv \mathbf{x}_p(\lambda_p^*, \mathbf{v}_p^*, \mathbf{w}_p^*)$. By denoting the ith component of $\mathbf{x}_p$ by $x_{p,i}$, we obtain, in view of Equation (6.138),

$$\alpha_{p,j} = -\sum_{i=1}^{n} q_{ji} x_{p,i} + \sum_{i=1}^{m} a_{ij} w_{p,i}^* - c_j \leq 0, \tag{6.140}$$

for $j = 1, ..., n$. Since $(\mathbf{x}_{p_k}, \mathbf{w}_{p_k}^*)$ converges, $\alpha_{p_k,j}$ converges, as $k \to \infty$. Denote the limit of $\alpha_{p_k,j}$ by α_j as $k \to \infty$. By the convergence of $\mathbf{x}_{p_k}$ and Equation (6.137), $\lambda_{p_k} p_k$ converges. Therefore, from Equation (6.133), we know that if $\alpha_j \neq 0$, i.e., $\alpha_j < 0$, then $x_{p_k,j} \to 0$ as $k \to \infty$. On the other hand, if $x_{p_k,j} \to \beta_j > 0$ as $k \to \infty$, then $\alpha_j = 0$ as $k \to \infty$. (Otherwise, β_j would be 0). This shows the complementary slackness condition (6.108) at the limit. Together with Equation (6.133), (6.134), and dual feasibility, the K.K.T. conditions (6.106) - (6.108) for Programs GQ and DGQ are satisfied, as $k \to \infty$. This implies that $\mathbf{x}_{p_k}$ converges to an optimal solution of Program GQ.

□

Lemmas 6.14 and 6.15 basically say that, under suitable conditions, when p is sufficiently large, $(\mathbf{v}_p^*, \mathbf{w}_p^*)$ and $\mathbf{x}_p(\lambda_p^*, \mathbf{v}_p^*, \mathbf{w}_p^*)$ are "good" approximate solutions to Program DGQ and Program GQ, respectively. One additional result follows immediately.

Theorem 6.7. Under the assumptions (A1) - (A4), if $\mathbf{Q}$ is symmetric and positive-definite, then, for any $p \geq 1$, $\mathbf{x}_p = \mathbf{v}_p^*$, which converges to the unique optimal solution of Program GQ, as $p \to \infty$.

Proof. Equation (6.138) implies, for $k = 1, \ldots, n$,

$$\sum_{j=1}^{n} q_{kj} [v_{p,j}^* - x_{p,j}(\lambda_p^*, \mathbf{v}_p^*, \mathbf{w}_p^*)] = 0. \tag{6.141}$$

The positive-definiteness of $\mathbf{Q}$ implies its nonsingularity, which in turn implies $\mathbf{x}_p = \mathbf{v}_p^*$. The rest follows from Theorem 6.5 and from the fact that the primal optimal solution is embedded in the optimal dual solution, when $\mathbf{Q}$ is positive-definite [14].

□

Theorem 6.7 says that if $\mathbf{Q}$ is symmetric and positive-definite, then, under the stated assumptions, $\mathbf{v}_p^*$ is primal feasible. In other words, the conversion using Equation (6.133) to obtain a primal feasible solution is redundant. Note that $\mathbf{v}_p^*$ may not be primal feasible if $\mathbf{Q}$, instead, is only positive semi-definite, in which case Equation (6.133) provides an effortless means to obtain a primal feasible solution of Program GQ from an optimal solution $(\mathbf{v}_p^*, \mathbf{w}_p^*)$ of the perturbed dual Program DGQ_p.

6.3.4 Perturbing the Primal Objective

In this subsection, we show that the dual l_p-norm perturbation approach generalizes the primal entropic perturbation approach for convex quadratic programming discussed in Section 5.3 and [5].

Consider a special case of Program GQ where the matrix $\mathbf{Q}$ is diagonal with diagonal elements $q_{jj} \equiv d_j > 0$, for $j = 1, \ldots, n$. Given any $\lambda \geq 0$, the first-order optimality conditions include, in light of Equation (6.131),

$$-d_j v_j + \lambda p d_j \exp\{p[-d_j v_j + \sum_{i=1}^{m} a_{ij} w_i - c_j]\} = 0, \tag{6.142}$$

for $j = 1, ..., n$. Since $d_j > 0$, we have

$$d_j v_j + (1/p) \ln v_j = \sum_{i=1}^{m} a_{ij} w_i - c_j + \ln(\lambda p)/p. \tag{6.143}$$

Selecting a particular value $\lambda = 1/(pe)$ for the Lagrange multiplier, where e = 2.71828 is the exponential constant, we have

$$d_j v_j + (1/p) \ln v_j = \sum_{i=1}^{m} a_{ij} w_i - c_j - 1/p. \tag{6.144}$$

Denoting $\mu \equiv 1/p$ results in

$$d_j v_j + \mu \ln v_j = \sum_{i=1}^{m} a_{ij} w_i - c_j - \mu. \tag{6.145}$$

Note that, since v_j is a function of $\mathbf{w}$, Equation (6.145) is actually identical to Equation (6.113). By plugging Equation (6.144) into Equation (6.130), the corresponding Lagrangian maximization problem becomes

$$\max_{\mathbf{w}\in R^m} L(1/pe, \mathbf{v}, \mathbf{w}), \tag{6.146}$$

where

$$\begin{aligned}
&L(1/pe, \mathbf{v}, \mathbf{w}) \\
&= -(1/2)\sum_{j=1}^{n} d_j v_j^2 + \sum_{i=1}^{m} b_i w_i - \lambda \left\{ \sum_{j=1}^{n} \exp[p((1/p)\ln v_j + (1/p))] - 1 \right\} \\
&= -(1/2)\sum_{j=1}^{n} d_j v_j^2 - (1/p)\sum_{j=1}^{n} v_j + \sum_{i=1}^{m} b_i w_i + 1/(pe).
\end{aligned} \tag{6.147}$$

By the fact that $\mu \equiv 1/p$ and by treating $1/pe$ as a constant, it is clear that Equation (6.147) is equivalent to the objective function of Program DQ(μ) as defined by Equation (6.112). Therefore, the perturbed dual program DGQ_p can be viewed as a generalized version of Program DQ(μ), which was derived by perturbing the primal objective with an entropic function. When the matrix $\mathbf{Q} = \mathbf{0}$, the theory developed thus far specializes to its linear counterpart, as detailed in Section 6.1 and [6], which again can be viewed as a generalization of the entropic perturbation approach to linear programming.

6.3.5 Concluding Remarks

In this section, we have introduced a dual l_p-norm perturbation approach to solving convex quadratic programming problems. The feasible region of the Lagrangian dual program is approximated by a proper subset that is defined by a single smooth constraint involving the l_p-norm of a vector measure of constraint violation. The perturbed dual program becomes the dual program as $p \to \infty$ and, under some standard conditions, the optimal solution of the perturbed program converges to a dual optimal solution. A closed-form formula for converting an optimal solution of the perturbed dual program into a feasible solution of the primal convex quadratic program was also provided. Similar to its dual counterpart, such a primal feasible solution converges to an optimal primal solution.

We have also shown that the proposed dual approach provides a more general setting than the primal entropic perturbation approach. The theory specializes easily for linear programming. The potential computational advantages of the proposal approach is yet to be fully investigated. Various numerical methods, including the Augmented Lagrangian method [11, 17], could lead to some interesting computational results.

Since the feasible domain, optimal objective value, and optimal solution set of the perturbed program are all indexed by the perturbation parameter p, we can define such correspondences as point-to-set mappings. In this way, the dual convergence results of Lemma 6.13, Lemma 6.14, and Theorem 6.5 can actually be cast in the more general framework of "sensitivity and stability analysis" for mathematical programming via the concept of point-to-set mapping. Readers are referred to [12, 9, 10, 16].

REFERENCES

[1] Chung, K.L., *A Course in Probability Theory, 2nd Edition*, Academic Press, New York, 1974.

[2] Fang, S.-C., "A New Unconstrained Convex Programming Approach to Linear Programming," *Zeitschrift fur Operations Research*, Vol. 36, 1992, pp. 149–161.

[3] Fang, S.-C., and Puthenpura, S., *Linear Optimization and Extensions: Theory and Algorithms*, Prentice Hall, Englewood Cliffs, New Jersey, 1993.

[4] Fang, S.-C., and Tsao, H.-S.J., "Linear Programming with Entropic Perturbation," *Zeitschrift fur Operations Research*, Vol. 37, 1993, pp. 171–186.

[5] Fang, S.-C., and Tsao, H.-S.J., "An Unconstrained Convex Programming Approach to Solving Convex Quadratic Programming Problems," *Optimization*, Vol. 27, 1993, pp. 235–243.

[6] Fang, S.-C., and Tsao, H.-S.J., "A Dual Perturbation View of Linear Programming," *Mathematical Methods of Operations Research,* Vol. 44, 1996, pp. 1–9.

[7] Fang, S.-C., and Tsao, H.-S.J., "Perturbing Dual Feasible Region for Solving Convex Quadratic Programs," OR Technical Report No. 297, North Carolina State University, Raleigh, North Carolina, 1994, to appear in *Journal of Optimization Theory and Applications*, Vol. 94, 1997.

[8] Fang, S.-C., and Tsao, H.-S.J., "Solving Linear Programs with Inequality Constraints via Perturbation of Feasible Region," *Optimization*, Vol. 37, 1996, pp. 213–223.

[9] Fiacco, A.V., *Introduction to Sensitivity and Stability Analysis in Nonlinear Programming*, Academic Press, New York, 1983.

[10] Fiacco, A.V., and Liu, J., "On the Stability of General Convex Programs under Slater's Condition and Primal Solution Boundedness," *Optimization*, Vol. 32, 1995, pp. 291–300.

[11] Gill, P.E., Murray, T. A., and Wright, M.H., *Practical Optimization*, Academic Press, London, 1981.

[12] Hogan, W.W., "Point-to-set Maps in Mathematical Programming," *SIAM Review*, Vol. 15, 1973, pp. 591–603.

[13] Luenberger, D.G., *Linear and Nonlinear Programming, 2nd Edition*, Addison-Wesley, Reading, Massachusetts, 1984.

[14] Van De Panne, C., *Methods for Linear and Quadratic Programming*, North-Holland, Amsterdam, 1975.

[15] Rajasekera, J.R., and Fang, S.-C., "On the Convex Programming Approach to Linear Programming," *Operations Research Letters*, Vol. 10, 1991, pp. 309–312.

[16] Robinson, S.M., and Day, R., "A Sufficient Condition for Continuity of Optimal Sets in Mathematical Programming," *Journal of Mathematical Analysis and Applications*, Vol. 45, 1974, pp. 506–511.

[17] Saigal, R., *Linear Programming: A Modern Integrated Analysis*, Kluwer Academic Publishers, Norwell, Massachusetts, 1995.

[18] Zangwill, W.I., *Nonlinear Programming: A Unified Approach*, Prentice-Hall, Englewood Cliffs, New Jersey, 1969.

7

EXTENSIONS AND RELATED RESULTS

This chapter discusses extensions of the methods developed in previous chapters and discusses some closely related subjects, including

(i) Entropy optimization problems with a finite number of constraints but a countably infinite number of variables,

(ii) A relationship between entropy optimization and Bayesian statistical estimation,

(iii) Entropic regularization method for solving min-max problems, and

(iv) Entropic regularization method for solving semi-infinite min-max problems.

7.1 ENTROPY OPTIMIZATION WITH COUNTABLY MANY VARIABLES

Sections 3.2 and 5.2 developed theory and algorithms for solving linearly constrained entropy optimization problems based on a geometric dual approach. In this section, we extend the geometric dual approach for solving linearly constrained entropy optimization problems with countably many variables. We start with the Jensen's inequality [10] and show how a dual problem can be naturally derived. The geometric dual in this case turns out to be analogous to the Lagrangian dual obtained in [2] for continuous probability distributions.

It explains how the maximum likelihood principle becomes the dual of the minimum cross-entropy principle for the exponential family.

We will define our entropy optimization problem in Subsection 7.1.1, establish a dual approach based on the Jensen's inequality in Subsection 7.1.2, provide an example to illustrate the power of this duality theory in Subsection 7.1.3, and conclude our findings in Subsection 7.1.4.

7.1.1 Problem Statement

Let $\{x_j > 0 \mid j = 1, 2, \ldots\}$ be a probability distribution on a countably infinite sample space and $\{p_j > 0 \mid j = 1, 2, \ldots\}$ be a given prior distribution that estimates $\{x_j\}$. The "cross-entropy" (or the relative entropy) of $\{x_j\}$ with respect to $\{p_j\}$ is defined by $\sum_{j=1}^{\infty} x_j \ln(x_j/p_j)$ when the infinite sum exists. Note that the measure of cross-entropy is not symmetric with respect to the two probability distributions appearing in the definition. In this section, the cross-entropy is always defined with respect to a prior distribution.

When the values of certain linear combinations of the probabilities are given or known, i.e., $\sum_{j=1}^{\infty} a_{ij}x_j = b_i$ for $i = 1, \ldots, m$, we are interested in finding a probability distribution, among all feasible ones, attaining the minimum cross-entropy with respect to the given prior. Note that many useful statistical quantities can be expressed as such linear combinations, e.g., expected values of random variables defined on the sample space. Since the cross-entropy is defined as an infinite sum which does not always exist, we naturally confine our attention to those probability distributions $\mathbf{x} \equiv \{x_j > 0 \mid j = 1, 2, \ldots\}$ such that the infinite sum $\sum_{j=1}^{\infty} x_j \ln(x_j/p_j)$ exists. Denote the set of all such probability distributions by Γ. Then we solve the following entropy optimization problem:

Program EIV:

$$\inf \quad f(\mathbf{x}) = \sum_{j=1}^{\infty} x_j \ln(x_j/p_j) \tag{7.1}$$

$$\text{s.t.} \quad \sum_{j=1}^{\infty} a_{ij}x_j = b_i, \quad i = 1, 2, \ldots, m, \tag{7.2}$$

$$\sum_{j=1}^{\infty} x_j = 1, \tag{7.3}$$

$$x_j > 0, \quad j = 1, 2, ..., \infty, \tag{7.4}$$

$$\mathbf{x} \in \Gamma. \tag{7.5}$$

Note that Program EIV contains the explicitly the probability constraint. This program has many applications. A recent study formulated a GI/G/c queueing problem as a special case of Program EIV, where linear constraints were used to represent the average number of customers in the system and the conservation law of traffic [41]. Another study analyzed the performance of tandem queues with generalized blocking schemes after formulating the queueing problem as a special case of Program EIV [30]. For such special cases of Program EIV, researchers either relied on the use of Lagrange multipliers and certain asymptotic properties of the GI/G/c queues or used ad hoc inequalities to solve their problems. We will present a more rigorous and systematic dual approach to solving Program EIV.

7.1.2 Jensen's Inequality for Dual Approach

Based on the "geometric-arithmetic inequality," a dual approach for solving entropy optimization problems in a finite-dimensional space has been introduced in Chapters 3 and 4. Here we show that Jensen's inequality can be used to extend the results for solving such problems with countably many variables. Let us denote the set of natural numbers by N. We have the following lemma:

Lemma 7.1. If $\delta_j \geq 0$, $\omega_j > 0$, $\forall j \in N$, and $\sum_{j=1}^{\infty} \delta_j = 1$, then whenever $\sum_{j=1}^{\infty} \delta_j \omega_j < \infty$ and $\prod_{j=1}^{\infty} \omega_j^{\delta_j} < \infty$,

$$\sum_{j=1}^{\infty} \delta_j \omega_j \geq \prod_{j=1}^{\infty} \omega_j^{\delta_j}. \tag{7.6}$$

Moreover, the equality holds if and only if there exists a constant $C \in R$ such that

$$\omega_j = C, \ \forall \ j \text{ with } \delta_j > 0. \tag{7.7}$$

Proof. Define a random variable V on N with the probability distribution $Prob\{V = j\} = \delta_j$, and a function f on N with $f(j) = \omega_j$. Since the natural logarithm function is concave, the Jensen's inequality implies that

$$\ln E(f(V)) \geq E(\ln f(V)), \tag{7.8}$$

whenever $0 < E(f(V)) < \infty$ and $-\infty < E(\ln f(V)) < \infty$. Therefore,

$$\ln \sum_{j=1}^{\infty} \delta_j \omega_j \geq \sum_{j=1}^{\infty} \delta_j \ln \omega_j, \tag{7.9}$$

or, equivalently,

$$\sum_{j=1}^{\infty} \delta_j \omega_j \geq \prod_{j=1}^{\infty} \omega_j^{\delta_j}, \tag{7.10}$$

whenever $\sum_{j=1}^{\infty} \delta_j \omega_j < \infty$ and $\prod_{j=1}^{\infty} \omega_j^{\delta_j} < \infty$. Note that Jensen's inequality becomes an equality if and only if $f(V) = E(f(V))$ (with probability 1). Therefore, the equality in the lemma holds if and only if there exists a constant $C > 0$ such that $\omega_j = C$, $\forall j$ with $\delta_j > 0$.

□

In case $\delta_j > 0$, for $j \in N$, we can replace ω_j by ω_j/δ_j in Inequality (7.6) and obtain the next lemma.

Lemma 7.2. If $\delta_j > 0$, $\omega_j > 0$, $\forall\, j \in N$, $\sum_{j=1}^{\infty} \delta_j = 1$, $\sum_{j=1}^{\infty} \omega_j < \infty$, and $\prod_{j=1}^{\infty} (\omega_j/\delta_j)^{\delta_j} < \infty$, then

$$\sum_{j=1}^{\infty} \omega_j \geq \prod_{j=1}^{\infty} \left(\frac{\omega_j}{\delta_j}\right)^{\delta_j}. \tag{7.11}$$

The equality holds if and only if there exists a constant $C > 0$ such that $\omega_j/\delta_j = C$, $\forall\, j \in N$.

Now, let $\{p_j > 0 : j \in N\}$ and $\{x_j > 0 : i \in N\}$ be any two probability distributions as described in Program EIV. For any sequence of real numbers $\{y_j : j \in N\}$, since $p_j e^{y_j} > 0$, $\forall j$, we can apply Lemma 7.2 to obtain

$$\sum_{j=1}^{\infty} p_j e^{y_j} \geq \prod_{j=1}^{\infty} \left(\frac{p_j e^{y_j}}{x_j} \right)^{x_j}, \tag{7.12}$$

whenever both the infinite sum and the infinite product exist. Taking logarithm on both sides, we have

$$\begin{aligned} \ln \sum_{j=1}^{\infty} p_j e^{y_j} &\geq \sum_{j=1}^{\infty} x_j (\ln p_j e^{y_j} - \ln x_j) \\ &= \sum_{j=1}^{\infty} x_j (y_j + \ln p_j - \ln x_j) \\ &= -\sum_{j=1}^{\infty} x_j \ln \left(\frac{x_j}{p_j} \right) + \sum_{j=1}^{\infty} x_j y_j, \end{aligned} \tag{7.13}$$

whenever the infinite sums exist. Hence, we have the following result:

Corollary 7.1. Let $\{p_j > 0 : i \in N\}$ and $\{x_j > 0 : j \in N\}$ be two probability distributions. Then, for real numbers y_j, $j \in N$, the following inequality holds

$$\sum_{j=1}^{\infty} x_j \ln \left(\frac{x_j}{p_j} \right) \geq -\ln \left(\sum_{j=1}^{\infty} p_j e^{y_j} \right) + \sum_{j=1}^{\infty} x_j y_j, \tag{7.14}$$

whenever all three infinite sums exist.

Notice that since $x_j > 0$, $\forall\, j \in N$, by Lemma 7.2, we know that Inequality (7.12) becomes an equality if and only if

$$\frac{p_j e^{y_j}}{x_j} = C, \quad \text{for some } C > 0. \tag{7.15}$$

Since $\{x_j > 0 : j \in N\}$ is a probability distribution and $\sum_{j=1}^{\infty} p_j e^{y_j} < \infty$, we have

$$x_j = \frac{p_j e^{y_j}}{\sum_{k=1}^{\infty} p_k e^{y_k}}. \tag{7.16}$$

Therefore, when (7.16) holds, Inequality (7.14) becomes

$$-\sum_{j=1}^{\infty} x_j \ln\left(\frac{x_j}{p_j}\right) + \sum_{j=1}^{\infty} x_j y_j = \left\{ \ln \sum_{j=1}^{\infty} p_j e^{y_j} \right\}. \tag{7.17}$$

Summarizing the discussion, we have the next corollary.

Corollary 7.2. Under the conditions of Corollary 7.1, Inequality (7.14) of Corollary 7.1 becomes an equality if and only if, $\forall j$,

$$x_j = \frac{p_j e^{y_j}}{\sum_{k=1}^{\infty} p_k e^{y_k}}. \tag{7.18}$$

Moreover, in this case,

$$-\sum_{j=1}^{\infty} x_j \ln\left(\frac{x_j}{p_j}\right) + \sum_{j=1}^{\infty} x_j y_j = \left\{ \ln \sum_{j=1}^{\infty} p_j e^{y_j} \right\}. \tag{7.19}$$

We are now ready to derive a dual problem for Program EIV. Applying Corollary 7.1 and incorporating the first m constraints of Program EIV, we know that if $\mathbf{x}$ is primal feasible and the infinite sums exist, then

$$\sum_{j=1}^{\infty} x_j \ln\left(\frac{x_j}{p_j}\right) \geq -\ln \sum_{j=1}^{\infty} p_j e^{y_j} - \sum_{i=1}^{m} b_i w_i + \sum_{j=1}^{\infty} x_j y_j + \sum_{i=1}^{m} \sum_{j=1}^{\infty} a_{ij} x_j w_i, \tag{7.20}$$

where w_i, $i = 1, 2, ..., m$, are arbitrary real numbers. If w_i's satisfy the equations

$$y_j + \sum_{i=1}^{m} a_{ij} w_i = 0, \quad \forall\, j \in N, \tag{7.21}$$

then (7.20) becomes

$$\sum_{j=1}^{\infty} x_j \ln\left(\frac{x_j}{p_j}\right) \geq -\ln \sum_{j=1}^{\infty} p_j e^{y_j} - \sum_{i=1}^{m} b_i w_i. \tag{7.22}$$

Therefore, we can take the right-hand-side of Inequality (7.22) and combine it with Equation (7.21) to form a maximization problem as a dual problem for

Program EIV. Note that in Inequality (7.20) the existence of the cross-entropy is guaranteed by the restriction on the domain of infimum in Program EIV, while the existence of the last term on the right-hand-side is guaranteed by the feasibility of $\{x_j > 0 : j \in N\}$. Its existence and Equation (7.21) guarantee the existence of infinite sum $\sum_{j=1}^{\infty} x_j y_j$ in Inequality (7.20). The only remaining infinite sum is $\sum_{j=1}^{\infty} p_j e^{y_j}$. We let Ω be the set of all $\mathbf{y}$ such that this infinite sum exists, i.e.,

$$\Omega = \left\{ \mathbf{y} : \sum_{j=1}^{\infty} p_j e^{y_j} < \infty \right\}, \tag{7.23}$$

and denote $\{w_i : i = 1, 2, ..., m\}$ by $\mathbf{w}$. We are now ready to define a dual problem for Program EIV as follows:

Program DEIV:

$$\sup \quad d(\mathbf{y}, \mathbf{w}) \equiv - \ln \sum_{j=1}^{\infty} p_j e^{y_j} - \sum_{i=1}^{m} b_i w_i \tag{7.24}$$

$$\text{s.t.} \quad y_j + \sum_{i=1}^{m} a_{ij} w_i = 0, \quad j \in N, \tag{7.25}$$

$$\mathbf{y} \in \Omega. \tag{7.26}$$

By defining Ω' as the set of $\mathbf{w} \in R^m$ such that $\sum_{j=1}^{\infty} p_j \exp(-\sum_{i=1}^{m} a_{ij} w_i)$ exists, i.e.,

$$\Omega' = \left\{ \mathbf{w} : \sum_{j=1}^{\infty} p_j \exp(-\sum_{i=1}^{m} a_{ij} w_i) < \infty \right\}, \tag{7.27}$$

Program DEIV is equivalent to the following dual problem without explicit constraints:

Program DEIV$'$:

$$\sup_{\Omega'} \left\{ d'(\mathbf{w}) \equiv - \ln \sum_{j=1}^{\infty} p_j \exp\left\{ - \sum_{i=1}^{m} a_{ij} w_i \right\} - \sum_{i=1}^{m} b_i w_i \right\}. \tag{7.28}$$

The previous discussion results in the following "weak duality" theorem:

Theorem 7.1. If $\mathbf{x}$ is a feasible solution of Program EIV and $\mathbf{w} \in \Omega'$, then

$$f(\mathbf{x}) \geq d'(\mathbf{w}). \tag{7.29}$$

Note that the whole derivation procedure that we are proposing here is also applicable to the finite dimensional problem. In that case, the dual Program DEIV is exactly the geometric dual obtained before. Also notice that our dual program is analogous to the Lagrangian dual obtained in [2] for continuous probability distributions. Hence it explains how the maximum likelihood principle becomes the dual of the minimum cross-entropy principle for the exponential family in the countably infinite sample space.

In order to study the so-called "strong duality theorem," let us assume that Program DEIV$'$ attains its optimum value $d'(\mathbf{w}^*)$ at an optimal solution $\mathbf{w}^*$. For $j \in N$, according to Equation (7.25), we define

$$y_j{}^* = -\sum_{i=1}^{m} a_{ij} w_i^*. \tag{7.30}$$

Since $\sum_{j=1}^{\infty} p_j e^{y_j^*} < \infty$, we can define $\mathbf{x}^*$ as in Corollary 7.2 by

$$x_j{}^* = \frac{p_j e^{y_j^*}}{\sum_{k=1}^{\infty} p_k e^{y_k^*}} > 0. \tag{7.31}$$

Now, if $\mathbf{w}^*$ is an interior point of the set Ω', the first partial derivatives of $d'(\mathbf{w})$ must be zero at this point. In other words, $\forall\ i,\ i = 1, 2, ..., m$,

$$0 = \frac{-\frac{\partial}{\partial w_i} \sum_{j=1}^{\infty} p_j e^{-\sum_{k=1}^{m} a_{kj} w_k}}{\sum_{j=1}^{\infty} p_j e^{-\sum_{k=1}^{m} a_{kj} w_k}} - b_i \tag{7.32}$$

$$= \sum_{j=1}^{\infty} \left(\frac{p_j e^{-\sum_{k=1}^{m} a_{kj} w_k}}{\sum_{l=1}^{\infty} p_l e^{-\sum_{k=1}^{m} a_{kl} w_k}} \right) a_{ij} - b_i. \tag{7.33}$$

This shows that $\mathbf{x}^*$ is a feasible solution of Program EIV if the corresponding cross-entropy is well-defined. As a direct consequence of Corollary 7.2, we see that $\mathbf{x}^*$ is an optimal solution to Program EIV. Moreover, $f(\mathbf{x}^*) = d'(\mathbf{w}^*)$ in this case.

We now state the "strong duality theorem" as follows:

Theorem 7.2. If $\sup d'(\mathbf{w})$ is attained by $\mathbf{w}^*$, $\mathbf{w}^*$ is an interior point of Ω', and the cross-entropy associated with $\mathbf{x}^*$ defined by

$$x_j^* = \frac{p_j e^{y_j^*}}{\sum_{k=1}^{\infty} p_k e^{y_k^*}} \tag{7.34}$$

exists, then $\mathbf{x}^*$ solves Program EIV, and there is no duality gap between Programs EIV and DEIV$'$.

Note that Program EIV is a cross-entropy minimization problem, and we have developed related duality theory for it. By removing the prior distribution $\{p_j : i \in N\}$ from the definition of cross-entropy, the same theory applies to the entropy maximization problem.

7.1.3 Geometric Distribution

In this section, we use Theorem 7.2 to demonstrate that the commonly used geometric distribution is a "natural" solution of an maximal entropy model. Here we consider the following entropy optimization problem:

$$\inf \quad f(\mathbf{x}) = \sum_{j=1}^{\infty} x_j \ln x_j \tag{7.35}$$

$$\text{s.t.} \quad \sum_{j=1}^{\infty} j x_j = c, \tag{7.36}$$

$$\sum_{j=1}^{\infty} x_j = 1, x_j > 0, \quad j = 1, 2, \ldots, \infty, \tag{7.37}$$

$$\sum_{j=1}^{\infty} x_j \ln x_j < \infty. \tag{7.38}$$

The dual problem is

$$\sup_{\Omega'} \left\{ d'(w) \equiv -\ln \sum_{j=1}^{\infty} e^{-jw} - cw \right\}, \tag{7.39}$$

where $\Omega' = \{w : \sum_{j=1}^{\infty} exp(-jw) < \infty\}$. Let us take a closer look at the dual problem. If $w \leq 0$, the infinite sum does not exist and $d'(w)$ is not defined. When $w > 0$,

$$d'(w) = -\ln\left(\frac{e^{-w}}{1 - e^{-w}}\right) - cw. \tag{7.40}$$

If $c \leq 1$, the objective function is strictly increasing and $d'(w)$ tends to ∞. If $c > 1$, then the first-order optimality condition becomes $w^* = -\ln(1 - 1/c)$. By the concavity of the objective function, we know that w^* is an optimal solution. Now, since $w^* > 0$ is an interior point of Ω' and the entropy of $\mathbf{x}^*$ defined by

$$x_j^* \equiv \frac{e^{-jw^*}}{\sum_{k=1}^{\infty} e^{-kw^*}} = p(1-p)^{j-1}, \tag{7.41}$$

with $p \equiv 1/c$, exists, we can further conclude, by Theorem 7.2, that $\mathbf{x}^*$ is an optimal primal solution. Hence we obtain the geometric distribution with parameter $p = 1/c$ as a solution to the entropy maximization problem.

7.1.4 Concluding Remarks

We have extended the geometric dual approach to the entropy optimization problem with countably many variables through the use of Jensen's inequality. The dual program is analogous to the Lagrangian dual developed by Ben-Tal and Charnes [2] except that we deal with discrete probability distributions explicitly. The conversion mechanism from an infinite-dimensional optimization problem with explicit constraints to a finite-dimensional optimization problem without explicit constraints can be easily explained in this framework.

7.2 ENTROPY OPTIMIZATION AND BAYESIAN STATISTICAL ESTIMATION

Many mathematical similarities between the Bayesian approach and the maximum entropy methods have been observed by various researchers, e.g., Jaynes [22, 23, 24], Williams [40], van Campenhout & Cover [7], and Zellner

[44]. In particular, the connection between the linearly constrained minimum cross-entropy problem (LCMXE) and the Bayesian hypothesis testing has been established (see Kullback [26]). This section focuses on the relationship between the LCMXE method and the Bayesian estimation procedure in which the probability distribution associated with the statistical experiment belongs to a single-parameter exponential family [38]. In this section, both entropy and cross-entropy are defined for probability distributions.

In statistics, the following estimation problem often occurs. A statistical experiment (or random event) is of interest, where the outcome of that experiment is uncertain. Through statistical testing or some other inferences, the distribution of the random outcome may be judged to be a member of a particular family of probability distributions, e.g., the family of Normal distributions. However, it is uncertain which exact Normal distribution best describes the distribution of the random outcome. Members of a family of probability distributions are typically indexed by a parameter (or a vector of parameters), denoted by θ. Therefore, the uncertainty about the statistical experiment (or random event) is completely captured by the uncertainty about the corresponding parameter θ. In the framework of Bayesian statistical inference, uncertainty is expressed in terms of a probability distribution. For this reason, the parameter can be and is often thought of as a random variable.

In a Bayesian statistical estimation procedure, a "prior" distribution capturing the experimenter's knowledge about the parameter θ prior to the experiment is first obtained. Then, based on (i) this prior, (ii) the given family of probability distributions associated with the statistical experiment, and (iii) the actual observations made during the experiment, a "posterior" distribution is obtained and expressed as a function of the prior distribution and the observations via the Bayes formula.

Given a prior distribution, the LCMXE approach, on the other hand, produces the probability distribution that is the "closest" to the prior distribution while satisfying given properties of the random variable θ, which are expressed as the expected values of functions of the random variable. Therefore, the optimal solution of the LCMXE problem, when it exists, is a probability distribution, which is also a function of the prior distribution. Consequently, we may call the optimal solution of the LCMXE problem the "LCMXE posterior." Upon careful examination, the LCMXE posterior also has the form of the Bayes formula. This leads us to view the LCMXE method as a Bayesian estimation procedure. But the underlying statistical experiment performed between the prior and posterior is still missing. Once this *embedded* statistical experiment is discovered, an equivalence relationship, under fairly general conditions, between the LCMXE problem and the Bayesian estimation procedure should become apparent.

To study the embedded statistical experiment, a dual approach to understanding the LCMXE method is taken. The dual problem was studied by many researchers [2, 4, 12, 37, 10]. Basically, the dual was viewed as the maximum likelihood estimation for the exponential family [4]. In this section, a new meaning for the dual approach is provided. For the LCMXE problem, once the prior distribution and the functions of the random variable θ are determined, the expected values of these functions are going to determine the LCMXE posterior completely. As for the Bayesian estimation procedure, once the prior distribution and the statistical model is determined, the observations made in the embedded experiment provide the only input to it. It will be shown that the dual problem of an LCMXE turns out to be the process of "translating" the information input to the LCMXE problem, i.e., the expected values, into the equivalent information input to the Bayesian estimation procedure, i.e., the observations made in the embedded experiment.

In this section, for reasons of simplicity, the space of absolutely continuous distributions is chosen as the domain of the LCMXE problem. The prior distribution and all feasible solutions of the LCMXE, including the posterior, will possess a probability density function. However, the theory actually holds for discrete probability distributions too. As for the family of distributions associated with the statistical experiment, both absolutely continuous and discrete distributions are considered.

To set the stage for the Bayesian framework, denote the sample space for the unknown parameter θ by Θ. A sequence of statistical experiments X_j, j=1,2,...,n, is to be performed. X_j's are defined on a common sample space Ω and are assumed to be independent and identically distributed given the unknown parameter θ. In the Bayesian framework, θ and X_j's are considered interdependent random variables with an $(n+1)$-variate joint distribution. In Bayesian terminology, X_j's are conditionally independent given θ, and their common distribution is actually considered as the conditional distribution of X_j given θ. We will refer to this common distribution as the conditional distribution associated with the statistical experiments or simply conditional distribution in short. The case where the conditional distribution is either discrete or absolutely continuous is considered. The common probability density function (p.d.f.) or probability function (p.f.) of X_j's is denoted by $r(x|\theta)$. Assume that for all $\theta \in \Theta$, $r(x|\theta)$ belongs to an exponential family of distributions over Θ. In other words, $r(x|\theta) = I_\Omega(x)a(x)b(\theta)\exp(-\sum_{i=1}^m h_i(\theta)g_i(x))$, where (i) $I_\Omega(x)$ is the indicator function associated with the sample space Ω, (ii) $b(\theta)$ and $h_i(\theta)$'s are functions of θ, and (iii) $a(x)$ and $g_i(x)$'s are functions of x (so that $r(x|\theta)$ is indeed a p.d.f. or p.f. for all $\theta \in \Theta$). Notice that in order for $r(x|\theta)$ to be a probability (density) function for all $\theta \in \Theta$, $b(\theta)$ has to be strictly positive over its entire domain Θ. Therefore, $b(\theta)$ can be rewritten as $\exp(h_{m+1}(\theta)g_{m+1}(x))$ with $h_{m+1}(\theta) \equiv \ln(b(\theta))$ and $g_{m+1}(x) \equiv 1$. Consequently, without loss of generality, $r(x|\theta) = I_\Omega(x)a(x)\exp(-\sum_{i=1}^m h_i(\theta)g_i(x))$. This expression will be used throughout this section. Let $\mathbf{x}^n$ denote the set of n outcomes $(x_1, x_2, \ldots\ldots, x_n)$ of the experiments X_j, j=1,2,....,n, and define $p_0(\theta)$ as the prior distribution. Finally, define $p_1(\theta)$ as the posterior distribution obtained by updating $p_0(\theta)$ with the Bayes formula and $\mathbf{x}^n$.

In the definition of the exponential family, except for the indicator function, the variables x and θ play symmetric roles. In this section, the convention of using θ as the unknown parameter and x as the outcome of experiment is maintained. Many well-known families of distributions are exponential families of distributions, and any particular exponential family is characterized by specific $a(x)$, $h_i(\theta)$'s, and $g_i(x)$'s. The Bayesian estimation problem discussed in this section is to find the posterior distribution $p_1(\theta)$ by updating the prior $p_0(\theta)$ with the information contained in the experiment X_j, j=1,2,....,n, using the Bayes formula.

Given two probability density functions $q(\theta)$ and $p_0(\theta)$ over the same sample space Θ, the cross-entropy (or the relative entropy) of $q(\theta)$ with respect to $p_0(\theta)$ is defined to be $\int_\Theta q(\theta)\ln(q(\theta)/p_0(\theta))d\theta$, when the integral exists. Note again that cross-entropy is not symmetric with respect to the two probability distributions. In this section, cross-entropy will always be defined with respect to a prior distribution. Therefore, for convenience of discussion, the term "cross-entropy" will be used without explicit reference to the nonsymmetric roles of the two underlying distributions. Let Γ be the set of all probability density functions which have meaningful cross-entropy with respect to $p_0(\theta)$. The LCMXE problem *corresponding* to our Bayesian estimation problem is defined as follows:

Program ELC:

$$\inf \quad f(q(\theta)) = \int_{\Theta} q(\theta) \ln(q(\theta)/p_0(\theta)) d\theta \tag{7.42}$$

$$\text{s.t.} \quad \int_{\Theta} h_i(\theta)\, q(\theta) d\theta = b_i, \quad i = 1, 2, ..., m, \tag{7.43}$$

$$\int_{\Theta} q(\theta)\, d\theta \; = \; 1, \tag{7.44}$$

$$q(\theta) \; \in \Gamma. \tag{7.45}$$

One way to solve Program ELC is to solve its dual problem as proposed by many researchers including Ben-Tal et al. [2, 4] and Tsao et al. [37]. With the notation $\mathbf{w} = (w_1, w_2, ..., w_m)^T$ and $\Lambda = \{\mathbf{w} : \int_{\Theta} p_0(\theta) \exp(-\sum_{i=1}^{m} h_i(\theta) w_i) d\theta < \infty\}$, the dual problem of Program ELC can be stated as follows:

Program DELC:

$$\sup_{\mathbf{w} \in \Lambda} \left\{ d(\mathbf{w}) \; \equiv \; -\ln \int_{\Theta} p_0(\theta) \exp(-\sum_{i=1}^{m} h_i(\theta) w_i) d\theta - \sum_{i=1}^{m} b_i w_i \right\}. \tag{7.46}$$

Under fairly general conditions [2, 4], the extreme values can be attained and min(ELC) = max(DELC). In particular, should there be an almost everywhere (with respect to the Borel measure over the whole sample space Θ) strictly positive probability density function that satisfies all the constraints of Program ELC, the duality gap between Programs ELC and DELC vanishes. This condition is referred to as *strict feasibility*. Under this condition, the optimal solution to Program ELC is in the form of

$$q^*(\theta) = p_0(\theta)\exp(-\sum_{i=1}^{m} h_i(\theta)w_i^*)/\int_\Theta p_0(\phi)\exp(-\sum_{i=1}^{m} h_i(\phi)w_i^*)\,d\phi, \quad (7.47)$$

where $\mathbf{w}^*$ is the optimal solution to Program DELC. Note that if $h_i(\theta)$'s are fixed and b_i's allowed to vary, $\mathbf{w}^*$ becomes a function of b_i's.

Equation (7.47) can be viewed as an application of the Bayes formula if there exists a statistical experiment according to a conditional distribution $I_\Omega(x)a(x)\exp(-\sum_{i=1}^m h_i(\theta)g_i(x))$, and the experimental outcome $\mathbf{x}^n$ satisfies $\sum_{j=1}^n g_i(x_j) = w_i^*$. This statistical experiment will be referred to as an *embedded* experiment. It is clear that b_i's influence the optimal solution only through w_i^*'s and, with $h_i(\theta)$ being fixed, w_i^*'s depend exclusively on b_i's. Therefore, with predetermined $h_i(\theta)$, the mean values b_i's can be viewed as the information input to the process that leads to the LCMXE formulation, and w_i^* can be viewed as the information input (the sample statistics $\sum_{j=1}^n g_i(x_j)$) to the corresponding Bayesian estimation procedure. Furthermore, the dual program DELC defines the relationship between b_i's and w_i^*'s and actually "translates" the LCMXE information input of mean values into its equivalent Bayesian information input as sample statistics.

With the embedded experiment in mind, it can be seen that the process of reaching the posterior from the prior using the LCMXE approach is equivalent to the process of reaching the same posterior from the same prior using the corresponding Bayesian estimation procedure. Note that the exponential family associated with the experiment is not completely specified. Actually, the sample space Ω for the exponential family, the statistics $g_i(x)$'s associated with the exponential family, and the function $a(x)$ can all be arbitrary. This equivalence relation is formalized as the following theorem:

Theorem 7.3. If Program ELC is strictly feasible and the optimal solution of Program DELC is denoted by $\mathbf{w}^* = (w_1^*, w_2^*, \ldots\ldots, w_m^*)^T$, then any statistical experiment with (i) observations $\mathbf{x}^n$ according to a probability (density) function $I_\Omega(x)a(x)\exp(-\sum_{i=1}^m h_i(\theta)g_i(x))$ for some Ω and (ii) $g_i(x)$'s such that $\sum_{j=1}^n g_i(x_j) = w_i^*$ will produce, according to the Bayes formula, a posterior distribution that is identical to the optimal solution of Program ELC.

The converse relationship can be stated as follows:

Theorem 7.4. Given (i) a prior $p_0(\theta)$, (ii) a statistical experiment according to a single-parameter exponential family $I_\Omega(x)a(x)\exp(-\sum_{i=1}^m h_i(\theta)g_i(x))$, (iii) some observations $\mathbf{x}^n$, and (iv) the resulting Bayesian posterior distribution $p_1(\theta)$, if its corresponding Program ELC with $b_i \equiv \int_\Theta h_i(\theta)p_1(\theta)d\theta$ is strictly feasible, then $p_1(\theta)$ is optimal to this LCMXE problem.

As a direct consequence, we have the following result:

Corollary 7.3. If the prior $p_0(\theta)$ in Theorem 7.4 is almost everywhere positive over its entire domain Θ with respect to the Borel measure and $b_i \equiv \int_\Theta h_i(\theta)p_1(\theta)d\theta$ exists, then $p_1(\theta)$ is also optimal to the corresponding LCMXE problem.

Parallel to the fact that the maximum entropy principle can be used to help determine the prior of the Bayesian approach [23], this equivalence relationship can be used to help determine the proper exponential families as the family of conditional distributions for the Bayesian experiments. Note that in the context of LCMXE, a family of distributions is a conjugate family with respect to a LCMXE problem if, given any member of the family as the prior, the optimal solution is also a member in the same family. Therefore, Equation (7.47) can also be used to identify the exponential conjugate families associated with any LCMXE problem. Actually, for any given LCMXE conjugate family, there exist a collection of single-parameter exponential families of distributions that are conjugate, in the Bayesian sense, to this LCMXE conjugate family.

To illustrate this relationship, consider the following example:

Program ELC$_\gamma$:

$$\inf \quad f(q(\theta)) = \int_0^\infty q(\theta)\ln(q(\theta)/p_0(\theta))\, d\theta \tag{7.48}$$

$$\text{s.t.} \quad \int_0^\infty \theta\; q(\theta)d\theta = b_1, \tag{7.49}$$

$$\int_0^\infty \ln(\theta)\; q(\theta)d\theta = b_2, \tag{7.50}$$

$$\int_0^\infty q(\theta)\, d\theta \;=\; 1, \tag{7.51}$$

$$q(\theta) \;\in \Gamma. \tag{7.52}$$

The optimal solution, when it exists, is the following distribution:

$$q^*(\theta) = p_0(\theta)\exp(-\theta w_1^* - (\ln\theta)w_2^*)/\int_0^\infty p_0(\phi)\exp(-\phi w_1^* - (\ln\phi)w_2^*)d\phi,$$

where w_1^* and w_2^* are obtained by solving the following dual program:

$$\inf_{\mathbf{w}\in\Lambda}\left\{d(\mathbf{w}) \equiv -\ln\int_0^\infty p_0(\theta)\exp(-\theta w_1 - (\ln\theta)w_2)d\theta - b_1w_1 - b_2w_2\right\}.$$

For simplicity, we assume that only one observation is made. Many different well-known exponential families fit the requirements for equivalence to this LCMXE problem. For example, the Poisson distribution, Exponential distribution, and Normal distribution with known mean but unknown precision (reciprocal of variance) are all proportional to $\exp(-\theta w_1 - (\ln\theta)w_2)$ $= \theta^{-w_2}\exp(-w_1\theta)$. These three distributions become proper candidates when the sample spaces are specified as the nonnegative integers, the nonnegative real numbers, and the real numbers, respectively. Note that the Gamma distribution is a conjugate family with respect to LCMXE ELC_γ and it is also a conjugate family, in the Bayesian sense, to all three families mentioned above (see DeGroot [10]).

7.3 ENTROPIC REGULARIZATION FOR MIN-MAX PROBLEMS

A variety of scientific and engineering problems can be formulated as the following min-max problem [5, 8, 32]:

$$\min_{\mathbf{x}\in X}\max_{1\le i\le m}\{f_i(\mathbf{x})\}, \tag{7.53}$$

where X is a common domain of the component functions $f_i(\mathbf{x})$, $i = 1, ...m$, which are usually assumed to be twice differentiable.

One major difficulty encountered in developing solution methods is the nondifferentiability of the max function

$$F(\mathbf{x}) \equiv \max_{1 \leq i \leq m} \{f_i(\mathbf{x})\}, \ \textit{for } \mathbf{x} \in X. \tag{7.54}$$

A distinct feature of the recent development centers around the idea of developing "smooth algorithms" [11, 16, 32, 33, 39, 43]. Among them, a class called *regularization* methods has been developed based on approximating the max function $F(\mathbf{x})$ by certain smooth functions [5, 8, 16, 27, 33, 43].

Note that the min-max problem (7.53) has the following Lagrangian function:

$$L(\mathbf{x}, \boldsymbol{\lambda}) = \sum_{i=1}^{m} \lambda_i f_i(\mathbf{x}), \tag{7.55}$$

for each $\mathbf{x} \in X$, where the m-vector $\boldsymbol{\lambda}$ denotes the Lagrange multipliers which satisfy the nonnegativity and normality conditions; i.e., these multipliers are restricted to fall within the simplex

$$\Lambda \equiv \{\boldsymbol{\lambda} | \ \boldsymbol{\lambda} \geq \mathbf{0}; \ \ \sum_{i=1}^{m} \lambda_i = 1\}. \tag{7.56}$$

Under some regular assumptions, the max function $F(\mathbf{x})$ can be obtained as

$$F(\mathbf{x}) = \max_{\boldsymbol{\lambda} \in \Lambda} L(\mathbf{x}, \boldsymbol{\lambda}), \ \textit{for } \mathbf{x} \in X. \tag{7.57}$$

Unfortunately, the maximization of $\sum_{i=1}^{m} \lambda_i f_i(x)$ over $\boldsymbol{\lambda} \in \Lambda$ rarely has an explicit solution $\boldsymbol{\lambda}(\mathbf{x})$ for all $\mathbf{x} \in X$. Therefore, general regularization methods consider a substitute (of $L(\mathbf{x}, \boldsymbol{\lambda})$):

$$L_p(\mathbf{x}, \boldsymbol{\lambda}) = \sum_{i=1}^{m} \lambda_i f_i(x) + \frac{1}{p} R(\boldsymbol{\lambda}; \boldsymbol{\beta}), \tag{7.58}$$

where $p > 0$ is a control parameter, R is a regularization function, and $\boldsymbol{\beta}$ is an optional parameter vector of R. By carefully choosing the function R, maximizing the substitute (7.58) could result in a smooth approximation function

$$F_p(\mathbf{x}) \equiv \max_{\boldsymbol{\lambda} \in \Lambda} L_p(\mathbf{x}, \boldsymbol{\lambda}). \tag{7.59}$$

Some known regularization methods have used the following functions:

- Gigola and Gomez [16]: $R(\boldsymbol{\lambda};\boldsymbol{\beta}) = -\frac{1}{2}\sum_{i=1}^{m}(\lambda_i - \beta_i)^2$.
- Hiriart-Urruty and Lemarechal [19]: $R(\boldsymbol{\lambda};\boldsymbol{\beta}) = \sum_{i=1}^{m}\ln\lambda_i$.

However, no explicit expression of $F_p(\mathbf{x})$, for $\mathbf{x} \in X$, was given in the previous work.

In this section, a regularization method using entropy functions is introduced to derive a smooth function $F_p(\mathbf{x})$ that is defined explicitly and is a good approximation for the max function $F(\mathbf{x})$. It has a natural interpretation which fits the special structure of the min-max problem and applies to related problems. The rest of this section is arranged as follows. Subsection 7.3.1 describes the entropic regularization method and proves some desirable properties of $F_p(\mathbf{x})$. Subsection 7.3.2 discusses several related applications. Subsection 7.3.3 concludes the section by making some remarks.

7.3.1 Entropic Regularization Method

Given $\mathbf{x} \in X$, when the Shannon's entropy function [21] is chosen as the regularization function, i.e., $R(\boldsymbol{\lambda}) = -\sum_{i=1}^{m}\lambda_i\ln\lambda_i$, we define

$$L_p(\mathbf{x},\boldsymbol{\lambda}) \equiv \sum_{i=1}^{m}\lambda_i f_i(x) - \frac{1}{p}\sum_{i=1}^{m}\lambda_i\ln\lambda_i, \tag{7.60}$$

where $\boldsymbol{\lambda} \in \Lambda$. Since $L_p(\mathbf{x},\boldsymbol{\lambda})$ is strictly concave in $\boldsymbol{\lambda}$, maximizing it over Λ gives a unique optimal solution:

$$\lambda_i^*(\mathbf{x},p) = \frac{\exp[pf_i(\mathbf{x})]}{Z}, \quad 1 \le i \le m, \tag{7.61}$$

where

$$Z = \sum_{i=1}^{m}\exp[pf_i(\mathbf{x})]. \tag{7.62}$$

Substituting λ_i in (7.55) by $\lambda_i^*(\mathbf{x},p)$ for (7.59) results in an explicit expression

$$F_p(\mathbf{x}) = \frac{1}{p}\ \ln\{\sum_{i=1}^{m}\exp[pf_i(\mathbf{x})]\}, \tag{7.63}$$

which is a smooth function. It should be emphasized that the $F_p(\mathbf{x})$ function obtained here is indeed the maximum of $L_p(\mathbf{x},\boldsymbol{\lambda})$ over Λ, for each $\mathbf{x} \in X$.

The concept of using an entropic perturbation for smooth optimization can be traced back to Li's unpublished thesis [27]. A more general result can be found by using the concept of recession function [3]. Our motivation for defining $L_p(\mathbf{x}, \boldsymbol{\lambda})$ by Equation (7.55) is to find an optimal combination of $\lambda_i f_i(\mathbf{x})$ over Λ such that $F(\mathbf{x}) = \sum_{i=1}^m \lambda_i f_i(\mathbf{x})$. Since the simplex Λ can be viewed as a probability space, we may follow the maximum entropy principle to consider the following problem:

$$\max_{\boldsymbol{\lambda}\in\Lambda} \quad H(\boldsymbol{\lambda}) \equiv -k\sum_{i=1}^{m} \lambda_i \ln \lambda_i \tag{7.64}$$

$$\text{s.t.} \quad \sum_{i=1}^{m} \lambda_i f_i(\mathbf{x}) = F(\mathbf{x}), \tag{7.65}$$

$$\sum_{i=1}^{m} \lambda_i = 1, \tag{7.66}$$

$$\boldsymbol{\lambda} \geq \mathbf{0}, \tag{7.67}$$

where $H(\boldsymbol{\lambda})$ is the Shannon's entropy function [21], k is a positive constant, and $F(\mathbf{x})$ works as an expectation.

When the parameter p is sufficiently small, the entropy term in $L_p(\mathbf{x}, \boldsymbol{\lambda})$ dominates. In this case, all multipliers become almost the same for the component functions and thereby gather around the center of Λ. As p increases, the solution will gradually shift to the boundaries of Λ to get the best mixture for $f_i(\mathbf{x})$, $i = 1, ..., m$. Therefore, p can also be viewed as a homotopy parameter which bridges a maximum entropy problem with a known answer $\lambda_i = 1/m$, for $i = 1, ..., m$, to a solution of the problem $\max_{\boldsymbol{\lambda}\in\Lambda} L(\mathbf{x}, \boldsymbol{\lambda})$.

Lemma 7.3. $F_p(\mathbf{x})$ converges to $F(\mathbf{x})$ pointwisely on X, as p tends to infinity.

Proof. Given $\mathbf{x} \in X$, if $V(\mathbf{x})$ is defined to be a vector-valued function with components $v_i(\mathbf{x}) = \exp[f_i(\mathbf{x})], 1 \leq i \leq m$, then the l_p-norm of $V(\mathbf{x})$ becomes

$$||V(\mathbf{x})||_p \equiv \{\sum_{i=1}^{m} [v_i(\mathbf{x})]^p\}^{\frac{1}{p}} = \{\sum_{i=1}^{m} \exp[pf_i(\mathbf{x})]\}^{\frac{1}{p}}. \tag{7.68}$$

Hence,

$$F_p(\mathbf{x}) = \ln ||V(\mathbf{x})||_p. \tag{7.69}$$

Consequently,

$$\lim_{p \to \infty} \|V(\mathbf{x})\|_p = \max_{1 \le i \le m} v_i(\mathbf{x}) = \exp[F(\mathbf{x})], \tag{7.70}$$

and

$$\lim_{p \to \infty} F_p(\mathbf{x}) = F(\mathbf{x}). \tag{7.71}$$

□

Lemma 7.4. $F_p(\mathbf{x})$ converges to $F(\mathbf{x})$ uniformly on X, as p tends to infinity.

Proof. By the properties of l_p-norm, for $\mathbf{x} \in X$, $F_p(\mathbf{x})$ is a monotonically decreasing function in terms of p, i.e.,

$$F_s(\mathbf{x}) \le F_r(\mathbf{x}), \text{ for } r \le s. \tag{7.72}$$

Hence, from Lemma 7.3, for any $p > 0$, we have

$$\begin{aligned} 0 \le F_p(\mathbf{x}) - F(\mathbf{x}) &= \ln \|V(\mathbf{x})\|_p - \ln\{\exp[F(\mathbf{x})]\} && (7.73) \\ &= (1/p) \ln \sum_{i=1}^{m} \exp\{p[f_i(\mathbf{x}) - F(\mathbf{x})]\} && (7.74) \\ &\le (1/p) \ln m. && (7.75) \end{aligned}$$

In other words, for any $p > 0$,

$$F(\mathbf{x}) \le F_p(\mathbf{x}) \le F(\mathbf{x}) + \frac{\ln m}{p}, \tag{7.76}$$

Hence, $F_p(\mathbf{x})$ converges to $F(\mathbf{x})$ uniformly on X, as p tends to infinity.

□

Lemma 7.5. $F_p(\mathbf{x})$ is convex, if each $f_i(\mathbf{x})$, $i = 1, 2, ..., m$, is a convex function defined on a convex set X.

Proof. For any $\mathbf{x}, \mathbf{y} \in X$ and $\alpha \in (0, 1)$, we have

$$F_p(\alpha\mathbf{x} + (1-\alpha)\mathbf{y}) = \frac{1}{p} \ln \sum_{i=1}^{m} \exp[pf_i(\alpha\mathbf{x} + (1-\alpha)\mathbf{y})]$$

$$\leq \frac{1}{p}\ln\sum_{i=1}^{m}\exp[\alpha pf_i(\mathbf{x})+(1-\alpha)pf_i(\mathbf{y})]$$
$$= \frac{1}{p}\ln\sum_{i=1}^{m}\{\exp[pf_i(\mathbf{x})]\}^{\alpha}\{\exp[pf_i(\mathbf{y})]\}^{1-\alpha}.$$

The Holder inequality ([25], page 67) implies that

$$\sum_{i=1}^{m}\{\exp[pf_i(\mathbf{x})]\}^{\alpha}\{\exp[pf_i(\mathbf{y})]\}^{1-\alpha} \leq \{\sum_{i=1}^{m}\exp[pf_i(\mathbf{x})]\}^{\alpha}\{\sum_{i=1}^{m}\exp[pf_i(\mathbf{y})]\}^{1-\alpha}.$$

Consequently,

$$\begin{aligned} F_p(\alpha\mathbf{x}+(1-\alpha)\mathbf{y}) &\leq \frac{\alpha}{p}\ln\sum_{i=1}^{m}\exp[pf_i(\mathbf{x})]+\frac{1-\alpha}{p}\ln\sum_{i=1}^{m}\exp[pf_i(\mathbf{y})] \\ &= \alpha F_p(\mathbf{x})+(1-\alpha)F_p(\mathbf{y}). \end{aligned} \tag{7.77}$$

This shows that $F_p(\mathbf{x})$ is a convex function on X.

□

Summarizing all results derived thus far results in the following main theorem.

Theorem 7.5. Given that $X = R^n$ and $f_i(\mathbf{x})$ is convex, for $i = 1, 2, ..., m$, a near-optimal solution of the min-max problem (7.53) can be obtained by solving an unconstrained convex programming problem $\{\min_{\mathbf{x}} F_p(\mathbf{x})\}$ with a sufficiently large p.

Instead of using Jaynes' maximum entropy principle, the Kullback's minimum cross-entropy principle [26] leads to another entropic regularization:

$$F_p(\mathbf{x},\boldsymbol{\beta}) \equiv \max_{\boldsymbol{\lambda}\in\Lambda} L_p(\mathbf{x},\boldsymbol{\lambda},\boldsymbol{\beta}) = \sum_{i=1}^{m}\lambda_i f_i(x) - \frac{1}{p}\sum_{i=1}^{m}\lambda_i \ln\frac{\lambda_i}{\beta_i}, \tag{7.78}$$

where $\boldsymbol{\beta}$ is an additional control vector representing some known *a priori* information (distribution).

A simple calculation results in the following solution, in terms of $\mathbf{x}$ and $\boldsymbol{\beta}$:

$$\lambda_i^*(\mathbf{x},\boldsymbol{\beta}) = \frac{\beta_i\exp[pf_i(\mathbf{x})]}{Z_b}, \quad i = 1, 2, ..., m, \tag{7.79}$$

where

$$Z_b = \sum_{i=1}^{m} \beta_i \exp[pf_i(\mathbf{x})]. \tag{7.80}$$

Substituting (7.79) into (7.78), we have

$$F_p(\mathbf{x}, \boldsymbol{\beta}) = \frac{1}{p} \ln \sum_{i=1}^{m} \beta_i \exp[pf_i(\mathbf{x})]. \tag{7.81}$$

It is interesting to note that this approximation is exactly the one proposed by Bertsekas [5] by means of a tedious derivation using the multiplier methods with an exponential penalty function. In comparison with $F_p(\mathbf{x})$, $F_p(\mathbf{x}, \boldsymbol{\beta})$ has one more control vector but fewer definitive properties then those discussed earlier in this subsection.

7.3.2 Related Applications

A variety of numerical analysis and optimization problems can be converted to an equivalent min-max problem (7.53). Here we study systems of inequalities/equations, linear programs, and constrained min-max problems. Consider a system of inequalities in R^n specified by

$$f_i(\mathbf{x}) \leq 0, \quad i = 1, 2, ...m. \tag{7.82}$$

Defining

$$F(\mathbf{x}) \equiv \max_{1 \leq i \leq m} \{f_i(\mathbf{x})\}, \tag{7.83}$$

we know that system (7.82) has a solution if and only if $F(\mathbf{x})$ has a minimizer $\mathbf{x}^*$ with $F(\mathbf{x}^*) \leq 0$.

By using $F_p(\mathbf{x})$ in place of $F(\mathbf{x})$, the problem of solving a system of inequalities becomes an unconstrained minimization problem with a smooth objective function $F_p(\mathbf{x})$, whose gradient and Hessian matrix can be easily calculated according to the following formulae:

$$\nabla F_p(\mathbf{x}) = J(\mathbf{x})\boldsymbol{\lambda}, \tag{7.84}$$

and

$$\nabla^2 F_p(\mathbf{x}) = \sum_{i=1}^{m} \lambda_i \nabla^2 f_i(\mathbf{x}) + pJ(\mathbf{x})(\boldsymbol{\Gamma} - \boldsymbol{\lambda}\boldsymbol{\lambda}^T)J(\mathbf{x})^T, \tag{7.85}$$

where $J(\mathbf{x})$ is the Jacobian matrix $[\partial f_i/\partial x_j]$, $\boldsymbol{\lambda}$ is a vector with

$$\lambda_i = \exp[pf_i(\mathbf{x})]\{\sum_{i=1}^{m} \exp[pf_i(\mathbf{x})]\}^{-1}, \tag{7.86}$$

and $\boldsymbol{\Gamma}$ is a diagonal matrix formed by $\boldsymbol{\lambda}$, i.e., $\boldsymbol{\Gamma} = diag(\lambda_i)$.

For solving a system of linear inequalities:

$$f_i(\mathbf{x}) \equiv \mathbf{A}_i\mathbf{x} - b_i \leq 0, \quad i = 1, ..., m, \tag{7.87}$$

where $\mathbf{A}_i$ is the ith row vector of an $m \times n$ matrix $\mathbf{A}$, a corresponding approximation becomes

$$F_p(\mathbf{x}) = \frac{1}{p}\ln\{\sum_{i=1}^{m} \exp[p(\mathbf{A}_i\mathbf{x} - b_i)]\}. \tag{7.88}$$

Its gradient and Hessian are given by

$$\nabla F_p(\mathbf{x}) = \mathbf{A}^T\boldsymbol{\lambda}, \tag{7.89}$$

and

$$\nabla^2 F_p(\mathbf{x}) = p\mathbf{A}^T(\boldsymbol{\Gamma} - \boldsymbol{\lambda}\boldsymbol{\lambda}^T)\mathbf{A}, \tag{7.90}$$

where

$$\lambda_i = \frac{\exp[p(\mathbf{A}_i\mathbf{x} - b_i)]}{\sum_{k=1}^{m} \exp[p(\mathbf{A}_k\mathbf{x} - b_k)]}. \tag{7.91}$$

As compared with the previous work of [31, 42], this approach can easily incorporate the second-order information and adaptively adjust the (surrogate) multipliers $\boldsymbol{\lambda}$.

To solve a system of equations defined by

$$f_i(\mathbf{x}) = 0, \quad i = 1, 2, ...m, \tag{7.92}$$

we can either replace each equation by two inequalities $f_i(\mathbf{x}) \leq 0$ and $-f_i(\mathbf{x}) \leq 0$ or by one single inequality $f_i^2(\mathbf{x}) \leq 0$. The former increases the dimensionality of the problem, while the latter complicates the component functions.

We now turn our attention to the following linear programming problem:

$$\min_{\mathbf{x}} \quad f(\mathbf{x}) \;\; \equiv \;\; \mathbf{c}^T\mathbf{x} \tag{7.93}$$

$$\begin{aligned} s.t. \quad \mathbf{A}\mathbf{x} &= \mathbf{b}, && (7.94)\\ \mathbf{e}^T\mathbf{x} &= M, && (7.95)\\ \mathbf{x} &\geq \mathbf{0}, && (7.96) \end{aligned}$$

where $\mathbf{e}$ is a vector of all 1's and M is a constant. Its linear dual becomes

$$\min_{\mathbf{y}} \quad d(\mathbf{y}) \equiv -\mathbf{b}^T\mathbf{y} - My_{m+1} \tag{7.97}$$

$$s.t. \quad \sum_{i=1}^{m} a_{ij}y_i - c_j \leq -y_{m+1}, \quad j = 1, ..., n. \tag{7.98}$$

Apparently, this is equivalent to the following unconstrained minimization problem:

$$\min_{\mathbf{y}}\{F(\mathbf{y}) \equiv -\mathbf{b}^T\mathbf{y} + M \max_{1\leq j\leq n}\{\sum_{i=1}^{m} a_{ij}y_i - c_j\}\}. \tag{7.99}$$

Recall that the Karmarkar's standard-form linear program [13] assumes $\mathbf{b} = \mathbf{0}$ and $M = 1$. In this case, its dual becomes a min-max problem:

$$\min_{\mathbf{y}} \quad \{F(\mathbf{y}) \equiv \max_{1\leq j\leq n}\{\sum_{i=1}^{m} a_{ij}y_i - c_j\}\}. \tag{7.100}$$

The entropic regularization method leads to the following problem:

$$\min_{\mathbf{y}} \quad \{F_p(\mathbf{y}) \equiv \frac{1}{p}\ln\sum_{j=1}^{n}\exp[p(\sum_{i=1}^{m} a_{ij}y_i - c_j)]\}, \tag{7.101}$$

which, with $p \equiv 1/\mu$, is exactly the Program DKL(μ) defined in Section 5.1, resulting from the (primal) entropic perturbation of a linear program in Karmarkar's standard form.

We now study the following constrained min-max problem:

$$\min_{\mathbf{x}} \quad \{f(\mathbf{x}) \equiv \max_{1\leq i\leq m}\{f_i(\mathbf{x})\}\} \tag{7.102}$$

$$\text{s.t.} \quad g_j(\mathbf{x}) \leq 0, \quad j = 1, ..., l. \tag{7.103}$$

Note that when $m = 1$, it is an ordinary nonlinear programming problem with l constraints.

The problem above is equivalent to the following:

$$\min_{\mathbf{x},\alpha} \quad \alpha \tag{7.104}$$

$$\text{s.t.} \quad f_i(\mathbf{x}) - \alpha \leq 0, \quad 1 \leq i \leq m, \tag{7.105}$$

$$g_j(\mathbf{x}) \leq 0, \quad 1 \leq j \leq l. \tag{7.106}$$

When the "method of centres" [18] is applied, the major task at each iteration is to find the center of a level set

$$G_\alpha(\mathbf{x}) \equiv \{\mathbf{x} \mid f_i(\mathbf{x}) - \alpha \leq 0,\ 1 \leq i \leq m;\ g_j(\mathbf{x}) \leq 0,\ 1 \leq j \leq l\}. \tag{7.107}$$

By defining a distance function

$$d(\mathbf{x}) \equiv \min\{-f_i(\mathbf{x}) + \alpha,\ 1 \leq i \leq m;\ -g_j(\mathbf{x}),\ 1 \leq j \leq l\}, \tag{7.108}$$

the center of $G_\alpha(\mathbf{x})$ can be found by maximizing $d(\mathbf{x})$, or equivalently, by minimizing $-d(\mathbf{x})$ via its smooth approximation

$$d_p(\mathbf{x}) = \frac{1}{p} \ln\{\sum_{i=1}^{m} \exp[p(f_i(\mathbf{x}) - \alpha)] + \sum_{j=1}^{l} \exp[pg_j(\mathbf{x})]\}. \tag{7.109}$$

One advantage of using a max-type distance function, instead of the traditional logarithmic barrier function, lies in the fact that its domain is not restricted by nonnegativity.

7.3.3 Concluding Remarks

1. The existence of a smooth and uniform approximation of the max function not only simplifies the min-max problem (7.53) to an unconstrained optimization problem but also sets up a framework for us to study related problems.

2. The regularization by means of entropic functions provides a linkage to previously known results.

3. The entropic regularization method allows us to use any unconstrained nonlinear minimizer to solve a min-max related problem. The required software development effort is minimal.

7.4 SEMI-INFINITE MIN-MAX PROBLEMS

This section extends the entropic regularization method developed in the previous section to solve the following min-max problem:

$$\min_{x\in X} \max_{t\in T} \{f_t(x)\}, \tag{7.110}$$

where X is a compact convex subset of R^n, T is a compact and subset of R^m, and $f_t(x)$ is assumed to be twice differentiable with respect to x on X for each $t \in T$. Note that if T is finite with cardinality m, then the min-max problem defined above reduces to the problem discussed in Section 7.3. Note that both x and t are vectors, but, for notational simplicity, they are not expressed in boldface as they should be. Also $f_t(x)$ in general is a function of both x and t, which could be expressed as $f(x,t)$. But we denote t in the subscript form. In this way, we stay close to the derivations of the previous section.

As pointed out in Section 7.3, one major difficulty encountered in developing solution methods is the nondifferentiability of the max function

$$F(x) \equiv \max_{t\in T} \{f_t(x)\} \tag{7.111}$$

over X. To overcome this problem, we show how the "entropic regularization" approach can be extended.

This section is organized as follows. The entropic regularization method will be introduced in Subsection 7.4.1. The major properties of the smooth function that approximates the max function $F(x)$ will be derived in Section 7.4.2. The min-max theory will be applied to solving linear semi-infinite programming problems in Subsection 7.4.3. Some concluding remarks are given in Subsection 7.4.4.

7.4.1 Entropic Regularization Method

To develop a theory that generalizes the findings in the previous section, we start with a simple framework by defining the following function:

$$L(x, \lambda) = \int_T f_t(x)\lambda(t)dt, \tag{7.112}$$

for each $x \in X$ and

$$\lambda \in \Lambda \equiv \{\lambda(t) \mid \lambda(t) \in L^1(T), \lambda(t) \geq 0, \ a.e., \ \text{and} \int_T \lambda(t)dt = 1\}.$$

Under some regular assumptions, the max function $F(x)$ of Equation (7.111) can be obtained as

$$F(x) = \max_{\lambda \in \Lambda} L(x, \lambda), \quad \forall x \in X. \tag{7.113}$$

Unfortunately, the maximization of $\int_T f_t(x)\lambda(t)dt$ over $\lambda \in \Lambda$ rarely has an explicit solution $\lambda^*(t)$ for $x \in X$. Therefore, general regularization methods consider a substitute (of $L(x, \lambda)$):

$$L_p(x, \lambda) = \int_T f_t(x)\lambda(t)dt + \frac{1}{p}R(\lambda), \tag{7.114}$$

where $p > 0$ is a control parameter and R is a regularization function.

Given $x \in X$, here we choose an entropic regularization function $R(\lambda) = -\int_T \lambda(t) \ln \lambda(t)dt$. Then,

$$L_p(x, \lambda) = \int_T f_t(x)\lambda(t)dt - \frac{1}{p}\int_T \lambda(t) \ln \lambda(t)dt, \tag{7.115}$$

where $\lambda \in \Lambda$. Similar to [20], for each $t \in T$, we let

$$\lambda^*(t) = \exp[pf_t(x)] / \int_T \exp[pf_t(x)]dt. \tag{7.116}$$

Substituting λ in Equation (7.115) by λ^*, we have

$$L_p(x, \lambda^*) = \frac{1}{p} \ln(\int_T \exp[pf_t(x)]dt), \tag{7.117}$$

which is a smooth function. Now we define that, for each $x \in X$,

$$F_p(x) \equiv L_p(x, \lambda^*). \tag{7.118}$$

If we can show that $F_p(x)$ uniformly converges to $F(x)$ on X, as p approaches infinity, then we can find an approximate solution of the min-max problem (7.110) by solving a smooth minimization problem $min_{x \in X} F_p(x)$ with p being sufficiently large.

7.4.2 Properties of $F_p(x)$

Lemma 7.6. $F_p(x)$ is convex, if $f_t(x)$ is a convex function defined on a convex set X for each $t \in T$.

Proof. For any $x, y \in X$ and $\delta \in (0,1)$, we have

$$\begin{aligned} F_p(\delta x + (1-\delta)y) &= \frac{1}{p}\ln(\int_T \exp[pf_t(\delta x + (1-\delta)y)]dt) \\ &\leq \frac{1}{p}\ln(\int_T \exp[\delta p f_t(x) + (1-\delta)pf_t(y)]dt) \\ &= \frac{1}{p}\ln(\int_T \{\exp[pf_t(x)]\}^{\delta}\{\exp[pf_t(y)]\}^{1-\delta}dt). \end{aligned}$$

The Holder inequality ([36]) implies that

$$\begin{aligned} &\int_T \{\exp[pf_t(x)]\}^{\delta}\{\exp[pf_t(y)]\}^{1-\delta}dt \\ \leq\ & \{\int_T ((\exp[pf_t(x)])^{\delta})^{1/\delta}dt\}^{\delta}\{\int_T ((\exp[pf_t(y)])^{1-\delta})^{1/(1-\delta)}dt\}^{1-\delta} \\ =\ & \{\int_T \exp[pf_t(x)]dt\}^{\delta}\{\int_T \exp[pf_t(y)]dt\}^{1-\delta}. \end{aligned} \tag{7.119}$$

Consequently,

$$\begin{aligned} F_p(\delta x + (1-\delta)y) &\leq \frac{\delta}{p}\ln(\int_T \exp[pf_t(x)]dt) + \frac{1-\delta}{p}\ln(\int_T \exp[pf_t(y)]dt) \\ &= \delta F_p(x) + (1-\delta)F_p(y). \end{aligned} \tag{7.120}$$

This shows that $F_p(x)$ is a convex function on X.

□

In order to prove that $F_p(x)$ converges to $F(x)$ uniformly on X, $f_t(x)$ will be assumed to satisfy the following conditions in the rest of this section [1]:

(Condition 7.1): $f_t(x)$ is uniformly bounded, i.e., there exists an $M > 0$, such that $|f_t(x)| \leq M$, $\forall x \in X$ and $\forall t \in T$. (Condition 7.2): $f_t(x)$ is "super-uniformly continuous," i.e., $\forall \epsilon > 0$, there exists $\delta > 0$, such that if for any $x, y \in X$ and $|x - y| < \delta$, then $|f_t(x) - f_t(y)| < \epsilon$, $\forall t \in T$.

The following lemma will lead to our main result.

Lemma 7.7. Under Conditions 7.1 and 7.2, for the compact convex subset $X \subset R^n$ and any $0 < l < 1$, there exists a sufficiently large $P > 0$ such that

$$\frac{\max_{j=1,\dots,p} \min_{t \in T_j} exp[f_t(x)]}{l \cdot \max_{t \in T} exp[f_t(x)]} > 1, \quad \forall x \in X,$$

holds for any integer $p > P$ and any partition $\{T_1, \dots, T_p\}$ of T such that $\max_{j=1,\dots,p} d(T_j) \to 0$, as $p \to \infty$, where $d(T_j)$ represents the largest diameter of T_j.

Proof. If our claim is false, then, for the compact convex subset X, there exists a constant $0 < l < 1$, a convergent sequence $\{x^i\} \to x^*$, and for each i, there corresponds a p_i with $p_i \to \infty$ as $i \to \infty$, such that

$$\frac{\max_{j=1,\dots,p_i} \min_{t \in T_j} exp[f_t(x^i)]}{l \cdot \max_{t \in T} exp[f_t(x^i)]} \leq 1, \tag{7.121}$$

for some partition $\{T_1, T_2, \dots, T_{p_i}\}$ of T with $\max_{j=1,\dots,p_i} d(T_j) \to 0$, as $i \to \infty$.

For each $t \in T$, by Taylor's expansion, we have

$$exp[f_t(x^i)] = exp[f_t(x^*)] + \nabla exp[f_t(\hat{x}^i)] \cdot (x^i - x^*)$$

where $\hat{x}^i$ lies between x^i and x^*.

Since $x^i \to x^*$ as $i \to \infty$, there exists a closed ball $B(x^*, r)$ such that $x^i, \hat{x}^i \in B(x^*, r)$, when i is sufficiently large. Since X is convex and Condition 7.1 holds, there exists $M^* > 0$ such that, for each T_j, $j = 1, \dots, p_i$,

$$\begin{aligned} \min_{t \in T_j} exp[f_t(x^*)] - M^*\|x^i - x^*\| &\leq \min_{t \in T_j} exp[f_t(x^i)] \\ &\leq \min_{t \in T_j} exp[f_t(x^*)] + M^*\|x^i - x^*\|. \end{aligned}$$

Hence

$$\begin{aligned} & \lim_{i \to \infty} \big(\max_{j=1,\dots,p_i} \min_{t \in T_j} exp[f_t(x^i)] \big) \\ \leq\ & \lim_{i \to \infty} \big(\max_{j=1,\dots,p_i} \min_{t \in T_j} exp[f_t(x^*)] + M^*\|x^i - x^*\| \big) \end{aligned}$$

and

$$\lim_{i\to\infty}(\max_{j=1,\ldots,p_i}\min_{t\in T_j} exp[f_t(x^i)]) \tag{7.122}$$
$$\geq \lim_{i\to\infty}(\max_{j=1,\ldots,p_i}\min_{t\in T_j} exp[f_t(x^*)] - M^*\|x^i - x^*\|).$$

For the partition $\{T_1, \ldots, T_{p_i}\}$ of T with $\max_{j=1,\ldots,p_i} d(T_j) \to 0$, as $p_i \to \infty$, we have

$$\lim_{i\to\infty}(\max_{j=1,\ldots,p_i}\min_{t\in T_j} exp[f_t(x^i)]) = \max_{t\in T} exp[f_t(x^*)]. \tag{7.123}$$

Due to Condition 7.2, we also have

$$\lim_{i\to\infty}(l \cdot \max_{t\in T} exp[f_t(x^i)]) = l \cdot \max_{t\in T} exp[f_t(x^*)]. \tag{7.124}$$

Equations (7.123) and (7.124) result in

$$\lim_{i\to\infty} \frac{\max_{j=1,\ldots,p_i}\min_{t\in T_j} exp[f_t(x^i)]}{l \cdot \max_{t\in T} exp[f_t(x^i)]} = \frac{1}{l} > 1.$$

This contradicts Inequality (7.121) and completes the proof.

□

Theorem 7.6. If Conditions 7.1 and 7.2 hold, then $F_p(x)$ converges to $F(x)$ uniformly on X, as $p \to \infty$.

Proof. By basic integration theory, if T is divided into p equal sections, then, $\forall x \in X$,

$$\begin{aligned}
&(\int_T exp[pf_t(x)]dt)^{\frac{1}{p}} \\
\leq\ & (\frac{\text{Vol}(T)}{p}(\max_{t\in T_1} exp[pf_t(x)] + \cdots + \max_{t\in T_p} exp[pf_t(x)]))^{\frac{1}{p}} \\
=\ & (\frac{\text{Vol}(T)}{p}(\frac{\max_{t\in T_1} exp[pf_t(x)]}{\max_{t\in T} exp[pf_t(x)]} + \cdots + \frac{\max_{t\in T_p} exp[pf_t(x)]}{\max_{t\in T} exp[pf_t(x)]}))^{\frac{1}{p}} \cdot \max_{t\in T} exp[f_t(x)] \\
\leq\ & (\frac{\text{Vol}(T)}{p} \cdot p)^{\frac{1}{p}} \max_{t\in T} exp[f_t(x)] \\
=\ & (\text{Vol}(T))^{\frac{1}{p}} \max_{t\in T} exp[f_t(x)].
\end{aligned}$$

By Lemma 7.7, for $0 < l < 1$, there exists $P > 0$, such that for $p > P$ and $x \in X$,

$$\begin{aligned}
&(\int_T exp[pf_t(x)]dt)^{\frac{1}{p}} \\
\geq\ & (\frac{\mathrm{Vol}(T)}{p}(\min_{t\in T_1} exp[pf_t(x)] + \cdots + \min_{t\in T_p} exp[pf_t(x)]))^{\frac{1}{p}} \\
=\ & (\frac{\mathrm{Vol}(T)}{p}(\frac{\min_{t\in T_1} exp[pf_t(x)]}{\max_{t\in T} exp[pf_t(x)]} + \cdots + \frac{\min_{t\in T_p} exp[pf_t(x)]}{\max_{t\in T} exp[pf_t(x)]}))^{\frac{1}{p}} \cdot \max_{t\in T} exp[f_t(x)] \\
\geq\ & (\frac{\mathrm{Vol}(T)}{p}(\frac{\max_{j=1,\ldots,p} \min_{t\in T_j} exp[pf_t(x)]}{l \cdot \max_{t\in T} exp[f_t(x)]})^p)^{\frac{1}{p}}(l \cdot \max_{t\in T} exp[f_t(x)]) \\
\geq\ & (\frac{\mathrm{Vol}(T)}{p})^{\frac{1}{p}}(l \cdot \max_{t\in T} exp[f_t(x)]).
\end{aligned}$$

This means that, for $0 < l < 1$, there exists a sufficiently large P such that for $p > P$ and $x \in X$,

$$\begin{aligned}
&[(\frac{\mathrm{Vol}(T)}{p})^{\frac{1}{p}} l - 1] \cdot \max_{t\in T} exp[f_t(x)] \\
\leq\ & (\int_T exp[pf_t(x)]dt)^{\frac{1}{p}} - \max_{t\in T} exp[f_t(x)] \\
\leq\ & [(\mathrm{Vol}(T))^{\frac{1}{p}} - 1] \cdot \max_{t\in T} exp[f_t(x)].
\end{aligned}$$

Since $\max_{t\in T} exp[f_t(x)]$ is bounded on X, for any $\epsilon > 0$, there exists $0 < \bar{l} < 1$ such that there exists a sufficiently large P and for $p > P$,

$$\begin{aligned}
-\epsilon\ \leq\ & [(\frac{\mathrm{Vol}(T)}{p})^{\frac{1}{p}} \bar{l} - 1] \cdot \max_{t\in T} exp[f_t(x)] \\
\leq\ & (\int_T exp[pf_t(x)]dt)^{\frac{1}{p}} - \max_{t\in T} exp[f_t(x)] \\
\leq\ & [(\mathrm{Vol}(T))^{\frac{1}{p}} - 1] \cdot \max_{t\in T} exp[f_t(x)] \\
\leq\ & \epsilon
\end{aligned}$$

Hence, for any $\epsilon > 0$, there exists P, such that for $p > P$ and any $x \in X$, we have

$$|(\int_T exp[pf_t(x)]dt)^{\frac{1}{p}} - \max_{t\in T} exp[f_t(x)]| \leq \epsilon.$$

This means that $(\int_T exp[pf_t(x)]dt)^{\frac{1}{p}}$ converges to $\max_{t \in T} exp[f_t(x)]$ uniformly on X.

Since $(\int_T exp[pf_t(x)]dt)^{\frac{1}{p}} \geq (\frac{\text{Vol}(T)}{p})^{\frac{1}{p}}(l \cdot \max_{t \in T} exp[f_t(x)])$, for $p > P$ and $f_t(x)$ satisfies Condition 7.1, there exist $K_1 > 0$ and P^* such that

$$(\int_T exp[pf_t(x)]dt)^{\frac{1}{p}} \geq K_1, \qquad \forall p > P^* \quad \text{and } x \in X. \tag{7.125}$$

Similarly, for any $i = 1, ..., P^*$ and $x \in X$, there exists $K_2 > 0$, such that

$$(\int_T exp[if_t(x)]dt)^{\frac{1}{i}} \geq K_2. \tag{7.126}$$

Let $K \equiv min(K_1, K_2)$. Since $(\int_T exp[pf_t(x)]dt)^{\frac{1}{p}} \leq (\frac{\text{Vol}(T)}{p})^{\frac{1}{p}} \cdot \max_{t \in T} exp[f_t(x)]$ and $f_t(x)$ satisfies Condition 7.1, there exist $K_1' > 0$ and P' such that

$$(\int_T exp[pf_t(x)]dt)^{\frac{1}{p}} \leq K_1', \qquad \forall p > P', \quad x \in X. \tag{7.127}$$

Similarly, for any $i = 1, ..., P'$ and $x \in X$, there exists $K_2' > 0$, such that

$$(\int_T exp[if_t(x)]dt)^{\frac{1}{i}} \leq K_2'. \tag{7.128}$$

Let $K' \equiv max(K_1', K_2')$, then

$$K \leq (\int_T exp[pf_t(x)]dt)^{\frac{1}{p}} \leq K', \tag{7.129}$$

for any $p > 0$ and $x \in X$. Again, due to Condition 7.1, there exist $M > 0$ and $M' > 0$, such that

$$M \leq \max_{t \in T} exp[pf_t(x)] \leq M', \quad \forall x \in X. \tag{7.130}$$

Now let $L \equiv min(K, M)$ and $L' \equiv max(K', M')$. Since the logarithmic function is uniformly continuous on $[L, L']$, we know that $\ln(\int_T exp[pf_t(x)]dt)^{\frac{1}{p}}$ converges to $\max_{t \in T} f_t(x)$ uniformly on X. Hence $F_p(x)$ converges to $F(x)$ uniformly on X.

□

7.4.3 Solving Linear Semi-Infinite Programs

Consider the following linear semi-infinite programming problem [17] :

Program LSIP:

$$\begin{array}{lllr}
\min & \int_T c(t)x(t)dt & & (7.131)\\
\text{s.t.} & \int_T f_i(t)x(t)dt & = b_i, \quad 1 \le i \le m, & (7.132)\\
& \int_T x(t)dt & = 1, & (7.133)\\
& x(t) & \ge 0, \quad a.e. \text{ and } x(t) \in L^1[T], & (7.134)
\end{array}$$

where T is a compact set in R^m, $f_i(t)$, $i = 1, ..., m$, and $c(t)$, with respect to the Lebesgue measure, are continuous function on T.

Its linear dual problem becomes the following problem:

Program DLSIP:

$$\begin{array}{llllr}
\max & \sum_{1\le i\le m} b_i u_i & + & u_{m+1} & (7.135)\\
\text{s.t.} & \sum_{1\le i\le m} f_i(t)u_i & + & u_{m+1} \le c(t), \quad \forall t \in T, & (7.136)\\
& u_i \in R, & & i = 1, ..., m+1. & (7.137)
\end{array}$$

Or, equivalently, we have

$$\begin{array}{llllr}
\min & -\sum_{1\le i\le m} b_i u_i & -u_{m+1} & & (7.138)\\
\text{s.t.} & \sum_{1\le i\le m} f_i(t)u_i & -c(t) \le -u_{m+1}, & \forall t \in T, & (7.139)\\
& u_i \in R, & i = 1, ..., m+1. & & (7.140)
\end{array}$$

This can be reformulated as the following unconstrained minimization problem:

$$\min_{u\in R^m} [-\sum_{1\le i\le m} b_i u_i + \max_{t\in T}(\sum_{1\le i\le m} f_i(t)u_i - c(t))]. \qquad (7.141)$$

When $b_i = 0$, for each i, is assumed, the dual problem defined in Equation (7.141) becomes a min-max problem:

$$\min_{u\in R^m} \max_{t\in T}(\sum_{1\le i\le m} f_i(t)u_i - c(t)). \qquad (7.142)$$

Writing u as x, if we further assume that the optimal solution of Program DLSIP lies in a compact convex set X, then we only have to deal with the following problem:

$$\min_{x \in X} \max_{t \in T} \left(\sum_{1 \leq i \leq m} f_i(t) x_i - c(t) \right). \tag{7.143}$$

Since the objective function of Equation (7.143) satisfies the conditions of Theorem 7.6, the entropic regularization method leads to the following problem:

$$\min_{x \in X} \{ F_p(x) \equiv \frac{1}{p} \ln \int_T exp[p(\sum_{1 \leq i \leq m} f_i(t) x_i - c(t))] dt \}, \tag{7.144}$$

which is exactly the same as the result obtained through an alternative approach (unconstrained convex programming) to linear semi-infinite programming proposed in the literature [28, 29].

7.4.4 Concluding Remarks

The derivation of Subsection 7.4.1 can be generalized to the following framework of probability measure and general integration [35, 6]. As before, suppose that, for every $x \in X$, $f_t(x)$ is Lebesque integrable over T. Rather than focusing on only those cases where the probability density functions $\lambda(t)$ can be defined over T, one can derive similar but more general results using probability measures p_j over T, instead of $\lambda(t)$. In this general setting, Equation (7.112) can be defined, through the general integration with respect to a probability measure, as

$$L(x, p_j) = \int_T f_t(x) dp_j, \tag{7.145}$$

for every $x \in X$ and every probability measure p_j over T. Results derived in this framework can be readily specialized to those obtained in this section as well as in the previous section.

REFERENCES

[1] Apostol, A.M., *Mathematical Analysis,* Addison-Wesley Publishing, Redwood City, CA, 1977.

[2] Ben-Tal, A., and Charnes, A., "A Dual Optimization Framework for Some Problems of Information Theory and Statistics," *Problems of Control and Information Theory*, Vol. 8, 1979, pp. 387–401.

[3] Ben-Tal, A., and Teboulle, M., "A Smooth Technique for Nondifferentiable Optimization Problems," *Optimization – Fifth French-German Conference*, Castel Novel, 1988, *Lecture Notes in Mathematics 1405*, Springer Verlag, 1989, pp. 1–11.

[4] Ben-Tal, A., Teboulle, M., and Charnes, A., "The Role of Duality in Optimization Problems involving Entropy Functionals with Applications to Information Theory," *Journal of Optimization Theory and Applications*, Vol. 58, 1988, pp. 209–223.

[5] Bertsekas, D., *Constrained Optimization and Lagrange Multiplier Methods*, Academic Press, New York, 1989.

[6] Billingsley, P., *Probability and Measure*, John Wiley, New York, 1979.

[7] van Campenhout, J. and Cover, T.M., "Maximum Entropy and Conditional Probability," *IEEE Transactions on Information Theory*, Vol. 27, 1981, pp. 483–489.

[8] Charalambous, C., and Conn, A.R., "An Efficient Method to Solve the Minimax Problem Directly," *SIAM Journal on Numerical Analysis*, Vol. 15, 1978, pp. 162–187.

[9] Charnes, A., Cooper, W.W., and Seiford, L., "Extremal Principles and Optimization Qualities for Khinchin-Kullback-Leibler Estimation," *Mathematische Operationsforschung und Statistik, Series Optimization*, Vol. 9, 1978, pp. 21–29.

[10] DeGroot, M.H., *Optimal Statistical Decisions*, McGraw-Hill, New York, 1970.

[11] Di Pillo, G., Grippo, L., and Lucidi, S., "A Smooth Method for the Finite Minimax Problem," *Mathematical Programming*, Vol. 60, 1993, pp. 187–214.

[12] Fang, S.-C., and Rajasekera, J.R., "Quadratically Constrained Minimum Cross-Entropy Analysis," *Mathematical Programming*, Vol. 44, 1989, pp. 85–96.

[13] Fang, S.-C., and Puthenpura, S., *Linear Optimization and Extensions: Theory and Algorithms*, Prentice Hall, Englewood Cliffs, New Jersey, 1993.

[14] Fang, S.-C., and Wu, S.-Y., "Solving Min-Max Problems and Linear Semi-Infinite Programs," *Computers and Mathematics with Applications*, Vol. 32, 1996, pp. 87–93.

[15] Fletcher, R., *Practical Methods of Optimization*, Vol. 2, John Wiley, New York, 1981.

[16] Gigola, C., and Gomez, S., "A Regularization Method for Solving the Finite Convex Min-Max Problem," *SIAM Journal on Numerical Analysis*, Vol. 27, 1990, pp. 1621–1634.

[17] Hettich, R., and Kortanek, K.O., "Semi-Infinite Programming : Theory, Method and Applications," *SIAM Review*, Vol. 35, 1993, pp. 380–429.

[18] Huard, P., "Resolution of Mathematical Programming with Nonlinear Constraints by the Method of Centers," in *Nonlinear Programming*, edited by J. Abadie, North-Holland, Amsterdam, 1967, pp. 207–219.

[19] Hiriart-Urruty, J.-B., and Lemarechal, C., *Convex Analysis and Minimization Algorithm*, Springer-Verlag, Berlin, 1993.

[20] Li, X.S., and Fang, S.-C., "On the Entropic Regularization Method for Solving Min-Max Problems with Applications," to appear in *Mathematical Methods of Operations Research*, Vol. 46, 1997.

[21] Jaynes, E.T., "Information Theory and Statistical Mechanics," *Physics Review*, Vol. 106, 1957, pp. 620–630.

[22] Jaynes, E.T., "Information Theory and Statistical Mechanics II," *Physics Review*, Vol. 108, 1957, pp. 171–190.

[23] Jaynes, E.T., "Prior Probabilities," *IEEE Transactions. Systems Science and Cybernetics*, Vol. 4, 1968, pp. 227–241.

[24] Jaynes, E.T., "The Relation of Bayesian and Maximum Entropy Methods," *Maximum-Entropy and Bayesian Methods in Science and Engineering, Volume 1: Foundations*, edited by G. J. Erickson and C. R. Smith, Kluwer Academic Publishers, Dordrecht, The Netherlands, 1988, pp. 25–29.

[25] Kazarinoff, N.D., *Analytic Inequalities*, Holt, Rinehart and Winston, New York, 1961.

[26] Kullback, S., *Information and Statistics*, John Wiley, New York, 1959.

[27] Li, X.-S., *Entropy and Optimization*, Ph.D. Thesis, University of Liverpool, United Kingdom, 1987.

[28] Lin, C.J., Fang, S.-C., and Wu, S.Y., *Parametric Linear Semi-Infinite Programming, Applied Mathematics Letters*, Vol. 9, 1996, pp. 89–96.

[29] Lin, C.J., Fang, S.-C., and Wu, S.Y., *An Unconstrained Convex Programming Approach to Linear Semi-Infinite Programming*, OR Technical Report No. 296, North Carolina State University, Raleigh, North Carolina, 1994, submitted to *SIAM Journal on Optimization.*

[30] Mishra, S., and Fang, S.-C., "A Maximum Entropy Optimization Approach to Tandem Queues with Generalized Blocking," *Performance Evaluation*, Vol. 702, 1997, pp. 1–25.

[31] Oko, S.O., "Surrogate Methods for Linear Inequalities," *Journal of Optimization Theory and Applications*, Vol. 72, 1992, pp. 247–268.

[32] Polak, E., Mayne, D.Q., and Higgins, J.E., "Superlinearly Convergent Algorithm for Min-Max Problems," *Journal of Optimization Theory and Applications*, Vol. 69, 1991, pp. 407–439.

[33] Polyak, R.A., "Smooth Optimization Methods for Minimax Problems," *SIAM Journal of Control and Optimization*, Vol. 26, 1988, pp. 1274–1286.

[34] Rajasekera, J.R., and Fang, S.-C., "Deriving an Unconstrained Convex Program for Linear Programming," *Journal of Optimization Theory and Applications*, Vol. 75, 1992, pp. 603–612.

[35] Royden, H.L., *Real Analysis, 2nd Edition*, The Macmillan Company, New York, 1972.

[36] Rudin, W., *Principles of Mathematical Analysis*, McGraw Hill, New York, 1976.

[37] Tsao, H.-S.J., Fang, S.-C., and Lee, D.N., "On the Optimal Entropy Analysis," *European Journal of Operational Research*, Vol. 59, 1992, pp. 324–329.

[38] Tsao, H.-S.J., Fang, S.-C., and Lee, D.N., "A Bayesian Interpretation of the Linearly-Constrained Cross-Entropy Minimization Problem," *Engineering Optimization*, Vol. 22, 1993, pp. 65–75.

[39] Vardi, A., "New Minmax Algorithm," *Journal of Optimization Theory and Applications*, Vol. 75, 1992, pp. 613–633.

[40] Williams, P.M., "Bayesian Conditionalisation and the Principle of Minimum Information," *The British Journal for the Philosophy of Science*, Vol. 31, 1980, pp. 131–144.

[41] Wu, J.-S., and Chan, W.C., "Maximum Entropy Analysis of Multiple-Server Queueing Systems," *Journal of Operational Research Society*, Vol. 40, 1989, pp. 815–826.

[42] Yang, K., and Murty, K.G., "New Iterative Methods for Linear Inequalities," *Journal of Optimization Theory and Applications*, Vol. 72, 1992, pp. 163–185.

[43] Zang, I., "A Smoothing Out Technique for Min-Max Optimization," *Mathematical Programming*, Vol. 19, 1980, pp. 61–77.

[44] Zellner, A., "Optimal Information Processing and Bayes' Theorem," *The American Statistician*, Vol. 42, 1988, pp. 278–284.

BIBLIOGRAPHY

Agmon, N., Alhassid, Y., and Levine, R.D., "An Algorithm for Determining the Lagrange Parameters in the Maximal Entropy formalism," *The Maximum Entropy Formalism*, edited by R. D. Levine and M. Tribus, MIT Press, Cambridge, Massachusetts, 1979, pp. 207–209.

Alonso, W., "A Theory of the Urban Land Market," *Papers of the Regional Science Association*, Vol. 6, 1960, pp. 146–158.

Alonso, W., *Location and Land Use*, Harvard University Press, Cambridge, Massachusetts, 1964.

Anas, A., "Discrete Choice Theory, Information Theory and the Multinomial Logit and Gravity Models," *Transportation Research Part B*, Vol. 17B, 1983, pp. 13–23.

Anderson, E. J., and Nash, P., *Linear Programming in Infinite-Dimensional Spaces*, John Wiley, Chichester, New York, 1987.

Apostol, A.M., *Mathematical Analysis,* Addison-Wesley Publishing, Redwood City, CA, 1977.

Baierlein, R., "How Entropy Got Its Name," *American Journal of Physics*, Vol. 60, 1992, p. 1151.

Bazaraa, M. S., and Shetty, C. M., *Nonlinear Programming: Theory and Algorithms*, John Wiley, New York, 1979.

Bellman, R.E., *Mathematical Methods in Medicine*, World Scientific, Singapore, 1983.

Ben-Tal, A., and Charnes, A., "A Dual Optimization Framework for Some Problems of Information Theory and Statistics," *Problems of Control and Information Theory*, Vol. 8, 1979, pp. 387–401.

Ben-Tal, A., Melman, A., and Zowe, J., "Curved Search Methods for Unconstrained Optimization," *Optimization*, Vol. 21, 1990, pp. 669–695.

Ben-Tal, A., and Teboulle, M., "A Smooth Technique for Nondifferentiable Optimization Problems," *Optimization – Fifth French-German Conference*,

Castel Novel, 1988, *Lecture Notes in Mathematics 1405*, Springer Verlag, 1989, pp. 1–11.

Ben-Tal, A., Teboulle, M., and Charnes, A., "The Role of Duality in Optimization Problems involving Entropy Functionals with Applications to Information Theory," *Journal of Optimization Theory and Applications*, Vol. 58, 1988, pp. 209–223.

Bertsekas, D., *Constrained Optimization and Lagrange Multiplier Methods*, Academic Press, New York, 1989.

Billingsley, P., *Probability and Measure*, John Wiley, New York, 1979.

Bishop, Y.M.M., "Full Contingency Tables, Logits, and Split Contingency Tables," *Biometrika*, Vol. 25, 1969, pp. 339–383.

Boltzmann, L. *Vorlesungen uber Gastheorie*, J. A. Barth, Leipzig, 1896.

Borwein, J.M., and Lewis, A.S., "Duality Relationships for Entropy-Like Minimization Problems," *SIAM Journal of Control and Optimization*, Vol. 29, 1991, pp. 325–338.

Borwein, J.M., and Lewis, A.S., "Partially-Finite Programming in L_1 and the Existence of Maximum Entropy Estimates," *SIAM Journal of Optimization*, Vol. 3, 1993, pp. 248–267.

Bregman, L.M., "The Relaxation Method of Finding the Common Point of Convex Sets and Its Application to the Solution of Problems in Convex Programming," *U.S.S.R. Computational Mathematics and Mathematical Physics*, Vol. 7, 1967, pp. 200–217.

Brice, S., "Derivation of Nested Transport Models within a Mathematical Programming Framework," *Transportation Research Part B*, Vol. 23B, 1989, pp. 19–28.

Burch, S.F., Gull, S.F., and Skilling, J.K., "Image Restoration by a Powerful Maximum Entropy Method," *Computer Vision, Graphics, and Image Processing*, Vol. 23, 1983, pp. 113–128.

van Campenhout, J., and Cover, T.M., "Maximum Entropy and Conditional Probability," *IEEE Transactions on Information Theory*, Vol. 27, 1981, pp. 483–489.

Censor, Y., "Row-Action Methods for Huge and Sparse Systems and Their Applications," *SIAM Review*, Vol. 23, 1981, pp. 444–466.

Censor, Y., "On Linearly Constrained Entropy Maximization," *Linear Algebra and Its Applications*, Vol. 80, 1986, pp. 191–195.

Censor, Y., Elfving, T., and Herman, G.T., "Special Purpose Algorithms for Linearly Constrained Optimization," *Maximum-Entropy and Bayesian Spectral Analysis and Estimation Problems*, edited by C.R. Smith and G.J. Erickson, D. Reidel Publishing Company, 1987, pp. 241–254.

Censor, Y., and Herman, G.T., "On Some Optimization Techniques in Image Reconstruction," *Applied Numerical Mathematics*, Vol. 3, 1987, pp. 365–391.

Censor, Y., and Lent, A., "An Iterative Row-Action Method for Interval Convex Programming," *Journal of Optimization Theory and Applications*, Vol. 34, 1981, pp. 321–353.

Censor, Y., De Pierro, A.R., Elfving, T., Herman, G.T., and Iusem, A.N., "On Iterative Methods for Linearly Constrained Entropy Maximization," *Banach Center Publications*, Vol. 24, 1990, pp. 145–163.

Censor, Y., De Pierro, A.R., Elfving, T., Herman, G.T., and Iusem, A.N., "On Maximization of Entropies and Generalization of Bregman's Method of Convex Programming," Technical Report MIPG 113, Medical Image Processing Group, Department of Radiology, Hospital of the University of Pennsylvania, Philadelphia, PA., 1986.

Charalambous, C., "Nonlinear Least pth Optimization and Nonlinear Programming," *Mathematical Programming*, Vol. 12, 1977, pp. 195–225.

Charalambous, C., "Acceleration of the Least pth Algorithm for Minimax Optimization with Engineering Applications," *Mathematical Programming*, Vol. 17, 1979, pp. 270–297.

Charalambous, C., and Conn, A.R., "An Efficient Method to Solve the Minimax Problem Directly," *SIAM Journal on Numerical Analysis*, Vol. 15, 1978, pp. 162–187.

Charnes, A., Cooper, W.W., and Seiford, L., "Extremal Principles and Optimization Qualities for Khinchin-Kullback-Leibler Estimation," *Mathematische Operationsforschung und Statistik, Series Optimization*, Vol. 9, 1978, pp. 21–29.

Chung, K.L., *A Course in Probability Theory*, 2nd Edition, Academic Press, New York, 1974.

Clausius, R., "Ueber Verschiedene fur die Anwendung Bequeme Formen der Hauptgleichungen der Mechanischen Warmetheorie," *Annalen der Physik und Chemie*, Vol. 125, 1865, pp. 353–400.

Cottle, R.W., "Application of a Block Successive Overrelaxation Method to a Class of Constrained Matrix Problems," *Proceedings of the International Congress on Mathematical Programming*, edited by R.W. Cottle, M.L. Kelmanson, and B. Korte, Rio de Janeiro, Brazil, April, 1981, pp. 89–103.

Cozzolino, J.M., and Zahner, M.J., "The Maximum Entropy Distribution of the Future Market Price of a Stock," *Operations Research*, Vol. 21, 1973, pp. 1200–1211.

Darroch, J.N., and Ratcliff, D., "Generalized Iterative Scaling for Log-linear Models," *Annals of Mathematical Statistics*, Vol. 43, 1972, pp. 1470–1480.

Dennis, J.E., and Schnabel, R.B., *Numerical Methods for Unconstrained Optimization and Nonlinear Equations*, Prentice Hall, Englewood Cliffs, New Jersey, 1983.

DeGroot, M.H., *Optimal Statistical Decisions*, McGraw-Hill, New York, 1970.

Dinkel, J.J., and Kochenberger, G.A., "Constrained Entropy Models: Solvability and Sensitivity," *Management Science*, Vol. 25, 1979, pp. 555–564.

Dinkel, J.J., Kochenberger, G.A., and Wong, S.N., "Entropy Optimization and Geometric Programming," *Environment and Planning*, Vol. 9, 1977, pp. 419–427.

Dinkel, J.J., and Wong, D.S., "External Zones in Trip Distribution Models: Characterization and Solvability," *Transportation Science*, Vol. 18, 1984, pp. 253–266.

Di Pillo, G., Grippo, L., and Lucidi, S., "A Smooth Method for the Finite Minimax Problem," *Mathematical Programming*, Vol. 60, 1993, pp. 187–214.

Duffin, R.J., Peterson, E.L., and Zener, C., *Geometric Programming : Theory And Applications*, John Wiley, New York, 1967.

El-Affendi, M.A., and Kouvatsos, D.D., "A Maximum Entropy Analysis of the M/G/1 and G/M/1 Queueing System at Equilibrium," *Acta Informatica*, Vol.19, 1983, pp. 339–355.

Elfving, T., "On Some Methods for Entropy Maximization and Matrix Scaling," *Linear Algebra and its Applications*, Vol. 34, 1980, pp. 321–339.

Elton, E.D., and Gruber, M.J., *Modern Portfolio Theory and Investment Analysis, 4th Edition*, John Wiley, New York, 1991.

Eriksson, J.R., "A Note on Solution of Large Sparse Maximum Entropy Problems with Linear Equality Constraints," *Mathematical Programming*, Vol. 18, 1980, pp. 146–154.

Erickson, J.R., "Algorithms for Entropy and Mathematical Programming," Ph.D. Thesis, Department of Mathematics, Linkoping University, Sweden, 1981.

Erickson, J.R., "An Iterative Primal-Dual Algorithm for Linear Programming," Report LitH-MAT-R-1985-10, Department of Mathematics, Linkoping University, Linkoping, Sweden, 1985.

Erlander, S., "Accessibility, Entropy and the Distribution and Assignment of Traffic," *Transportation Research*, Vol. 11, 1977, pp. 149–153.

Erlander, S., "Entropy in Linear Programming," *Mathematical Programming*, Vol. 21, 1981, pp. 137–151.

Erlander, S., "Efficient Population Behavior and the Simultaneous Choices of Origins, Destinations and Routes," *Transportation Research Part B*, Vol. 24B, 1990, pp. 363–373.

Erlander, S., and Stewart, N.F., *The Gravity Model in Transportation Analysis*, VSP, The Netherlands, 1990.

Erlang, A.K., "Theory of Probabilities and Telephone Conversations," *Nyt Tidsskrift Matematik*, Vol. B-20, 1909, pp. 33–39.

Fang, S.-C., "A New Unconstrained Convex Programming Approach to Linear Programming," *Zeitschrift fur Operations Research*, Vol. 36, 1992, pp. 149–161.

Fang, S.-C., Lin, C.-J., and Wu, S.-Y., "On Solving Convex Quadratic Semi-Infinite Programming Problems," *Optimization*, Vol. 31, 1994, pp. 107–125.

Fang, S.-C., Peterson, E.L., and Rajasekera, J.R., "Controlled Dual Perturbations for Posynomial Programs," *European Journal of Operational Research*, Vol. 35, 1988, pp. 111–117.

Fang, S.-C., Peterson, E.L., and Rajasekera, J.R., "Minimum Cross-Entropy Analysis with Entropy-Type Constraints," *Journal of Computational and Applied Mathematics*, Vol. 39, 1992, pp. 165–178.

Fang, S.-C., and Puthenpura, S., *Linear Optimization and Extensions: Theory and Algorithms*, Prentice Hall, Englewood Cliffs, New Jersey, 1993.

Fang, S.-C., and Rajasekera, J.R., "Controlled Perturbations for Quadratically Constrained Quadratic Programs," *Mathematical Programming*, Vol. 36, 1986, pp. 276–289.

Fang, S.-C., and Rajasekera, J.R., "Controlled Dual Perturbations for l_p-Programming," *Zeitschrift fur Operations Research*, Vol. 30 1986, pp. A29–A42.

Fang, S.-C., and Rajasekera, J.R., "Quadratically Constrained Minimum Cross-Entropy Analysis," *Mathematical Programming*, Vol. 44, 1989, pp. 85–96.

Fang, S.-C., and Rajasekera, J.R., "Minimum Cross-Entropy Analysis with Convex Constraints," *Information and Computation*, Vol. 116, 1995, pp. 304–311.

Fang, S.-C., and Tsao, H.-S.J., "Linear Programming with Entropic Perturbation," *Zeitschrift fur Operations Research*, Vol. 37, 1993, pp. 171–186.

Fang, S.-C., and Tsao, H.-S.J., "An Unconstrained Convex Programming Approach to Solving Convex Quadratic Programming Problems," *Optimization*, Vol. 27, 1993, pp. 235–243.

Fang, S.-C., and Tsao, H.-S.J., "A Quadratically Convergent Global Algorithm for the Linearly-Constrained Minimum Cross-Entropy Problem," *European Journal of Operational Research*, Vol. 79, 1994, pp. 369–378.

Fang, S.-C., and Tsao, H.-S.J., "Perturbing Dual Feasible Region for Solving Convex Quadratic Programs," OR Technical Report No. 297, North Carolina State University, Raleigh, North Carolina, 1994, to appear in *Journal of Optimization Theory and Applications*, Vol. 94, 1997.

Fang, S.-C., and Tsao, H.-S.J., "Linear-Constrained Entropy Maximization Problem with Quadratic Cost and Its Applications to Transportation Planning Problems," *Transportation Science*, Vol. 29, 1995, pp. 353–365.

Fang, S.-C., and Tsao, H.-S.J., "On the Entropic Perturbation and Exponential Penalty Methods for Linear Programming," *Journal of Optimization Theory and Applications*, Vol. 89, 1996, pp. 461–466.

Fang, S.-C., and Tsao, H.-S.J., "A Dual Perturbation View of Linear Programming," *Mathematical Methods of Operations Research,* Vol. 44, 1996, pp. 1–9.

Fang, S.-C., and Wu, S.-Y., "An Inexact Approach to Solving Linear Semi-Infinite Programming Problems," *Optimization*, Vol. 28, 1994, pp. 291–299.

Ferdinand, A.E., "A Statistical Mechanics Approach to System Analysis," *IBM Journal of Research and Development*, Vol. 14, 1970, pp. 539–547.

Ferris, M.C., and Philpott, A.B., "An Interior Point Algorithm for Semi-Infinite Linear Programming," *Mathematical Programming*, Vol. 43, 1989, pp. 257–276.

Ferris, M.C., and Philpott, A.B., "On the Affine Scalingand Semi-Infinite Programming," *Mathematical Programming*, Vol. 56, 1992, pp. 361–364.

Fiacco, A.V., *Introduction to Sensitivity and Stability Analysis in Nonlinear Programming*, Academic Press, New York, 1983.

Fiacco, A.V., and Kortanek, K.O., *Semi-Infinite Programming and Applications*, Lecture Notes in Economics and Mathematical Systems No. 215, Springer-Verlag, New York, 1983.

Fiacco, A.V., and Liu, J., "On the Stability of General Convex Programs under Slater's Condition and Primal Solution Boundedness," *Optimization*, Vol. 32, 1995, pp. 291–300.

Fiacco, A.V., and McCormick, G.P., *Nonlinear Programming: Sequential Unconstrained Minimization Techniques*, John Wiley, New York, 1968.

Fletcher, R., *Practical Methods of Optimization*, Vol. 2, John Wiley, New York, 1981.

Gigola, C., and Gomez, S., "A Regularization Method for Solving the Finite Convex Min-Max Problem," *SIAM Journal on Numerical Analysis*, Vol. 27, 1990, pp. 1621–1634.

Gill, P.E., Murray, W., Saunders, M.A., Tomlin, J.A., and Wright, M.H., "On Projected Barrier Methods for Linear Programming and an Equivalence to Karmarkar's Projective Method," *Mathematical Programming*, Vol. 36, 1986, pp. 183–209.

Gill, P.E., Murray, T. A., and Wright, M.H., *Practical Optimization*, Academic Press, London, 1981.

Glashoff, K., and Gustafson, S.A., *Linear Optimization and Approximation*, Springer-Verlag, New York, 1982.

Golan, A., Judge, G., and Robinson, S., "Recovering Information from Incomplete or Partial Multi-sectoral Economic Data," *Review of Economics and Statistics*, Vol. LXXVI, 1994, pp. 541–549.

Goldfarb, D., and Todd, M.J., "Linear Programming," Cornell University, School of OR and IE, Tech. Report No. 777, 1988.

Gordon, R., Bender, R., and Herman, G. T., "Algebraic Reconstruction Techniques (ART) for Three Dimensional Electron Microscopy and X-ray Photography," *Journal of Theoretical Biology*, Vol. 29, 1970, pp. 471–481.

Grandy, W.T. Jr., and Schick, L.H., editors, *Proceedings of the 10th International Workshop on Maximum Entropy and Bayesian Methods*, Kluwer Academic Publishers, Dordrecht, The Netherlands, 1991.

Gross, D., and Harris, C.M., *Fundamentals of Queueing Theory*, 2nd Edition, John Wiley, New York, 1985.

Guiasu, S., *Information Theory with Applications*, McGraw-Hill, New York, 1977.

Guiasu, S., "Maximum Entropy Condition in Queueing Theory," *Journal of Operational Research Society*, Vol. 37, 1986, pp. 293–301.

Gustafson, S.A., and Kortanek, K.O., "Numerical Treatment of A Class of Semi-Infinite Programming Problems," *Naval Research Logistics Quarterly*, Vol. 20, 1973, pp. 473–504.

Herbert, D.J., and Stevens, B.H., "A Model for the Distribution of Residential Activity in Urban Areas," *Journal of Regional Science*, Vol. 2, 1960, pp. 21–36.

Herman, G.T., "A Relaxation Method for Reconstructing Objects from Noisy X-Rays," *Mathematical Programming*, Vol. 8, 1975, pp. 1–19.

Herman, G.T., editor, *Image Reconstruction from Projections: Implementation and Applications*, Springer-Verlag, Berlin, 1979.

Den Hertog, D., Roos, C., and Terlaky, T., "Inverse Barrier Methods for Linear Programming," Report of the Faculty of Technical Mathematics and Informatics, No. 91-27, Delft University of Technology, The Netherlands, 1991.

Hettich, R., and Kortanek, K., "Semi-Infinite Programming: Theory, Method and Applications," *SIAM Review*, Vol. 35, 1993, pp. 380–429.

Hiriart-Urruty, J.-B., and Lemarechal, C., *Convex Analysis and Minimization Algorithm*, Springer-Verlag, Berlin, 1993.

Hogan, W.W., "Point-to-set Maps in Mathematical Programming," *SIAM Review*, Vol. 15, 1973, pp. 591–603.

Hoshino, K., Kumar, V., Kumar, U., "On Optimizing of Generalized Iterative Scaling Method," Working Paper No. 89-22, School of Business, Carleton University, Ontario, Canada, 1989.

Huard, P., "Resolution of Mathematical Programming with Nonlinear Constraints by the Method of Centers," in *Nonlinear Programming*, edited by J. Abadie, North-Holland, Amsterdam, 1967, pp. 207–219.

Jan, G.W., and Fang, S.-C., "A New Variant of the Primal Affine Scaling Algorithm for Linear Programs," *Optimization*, Vol. 22, 1991, pp. 681–715.

Jaynes, E.T., "Information Theory and Statistical Mechanics," *Physics Review*, Vol. 106, 1957, pp. 620–630.

Jaynes, E.T., "Information Theory and Statistical Mechanics II," *Physics Review*, Vol. 108, 1957, pp. 171–190.

Jaynes, E.T., "Prior Probabilities," *IEEE Transactions. Systems Science and Cybernetics*, Vol. 4, 1968, pp. 227–241.

Jaynes, E.T., "The Relation of Bayesian and Maximum Entropy Methods," *Maximum-Entropy and Bayesian Methods in Science and Engineering, Volume 1: Foundations*, edited by G. J. Erickson and C. R. Smith, Kluwer Academic Publishers, Dordrecht, The Netherlands, 1988, pp. 25–29.

Jefferson, T.R., and Scott, C.H., "The Analysis of Flexible Entropy Models of Spatial Interaction Using Geometric Programming," *Proceedings of the OR Society of New Zealand*, Vol. 1, 1978, pp. 39–47.

Jefferson, T.R., and Scott, C.H., "Entropy and Geometric Programming in Transportation Planning," *Opsearch*, Vol. 15, 1978, pp. 22–34.

Jefferson, T.R., and Scott, C.H., "The Analysis of Entropy Models with Equality and Inequality Constraints," *Transportation Research*, Vol. 138, 1979, pp. 123–132.

Ji, J., and Potra, F.A., "Interior-Point Algorithm for Quadratically Constrained Entropy Minimization Problems," *Journal of Optimization Theory and Applications*, Vol. 77, 1993, pp. 79–95.

Johnson, R.W., and Shore, J.E., "Minimum Cross-entropy Spectral Analysis of Multiple Signals," *IEEE Transactions on Acoustics, Speech, and Signal Processing*, Vol. ASSP-31, 1983, pp. 574–582.

Johnston, R.J., and Pattie, C.J., "Using Entropy-Maximizing Procedure to Estimate Territorial Social Indicators: An Introduction and Illustration," *Social Indicators Research*, Vol. 27, 1992, pp. 235–256.

Jornsten, K.O., "An Algorithm for the Combined Distribution and Assignment Problem," *Transportation Research Part B*, Vol. 15B, 1981, pp. 21–33.

Jornsten, K.O., and Lundgren, J.T., "An Entropy-Based Model Split Model," *Transportation Research Part B*, Vol. 23B, 1989, pp. 345–359.

Kapur, J.N., *Maximum-Entropy Models in Science and Engineering*, Wiley Eastern Ltd., New Delhi, 1989.

Kapur, J.N., and Kesavan, H.K., *Entropy Optimization Principles with Applications*, Academic Press, Boston, 1992.

Karmarkar, N., "A New Polynomial Time Algorithm for Linear Programming," *Combinatorica*, Vol. 4, 1984, pp. 373–395.

Kazarinoff, N.D., *Analytic Inequalities*, Holt, Rinehart and Winston, New York, 1961.

Kortanek, K.O., Potra, F., and Ye, Y., "On Some Efficient Interior Point Methods for Nonlinear Convex Programming," *Linear Algebra and Its Applications*, Vol. 152, 1991, pp. 169–189.

Kumar, V., Hoshino, K., and Kumar, U., "An Application of the Entropy Maximization Approach in Shopping Area Planning," *International Journal of General Systems*, Vol. 16, 1989, pp. 25–42.

Kullback, S., *Information and Statistics*, John Wiley, New York, 1959.

Kullback, S., and Libler, R.A., "On Information and Sufficiency," *Annals of Mathematical Statistics*, Vol. 22, 1951, pp. 79–86.

Lagarias, J.C., and Bayer, D.A., "Karmarkar's Linear Programming Method and Newton's Method," *Bell Laboratories Technical Report*, 11218-870810-22TM, AT&T Bell Laboratories, Murray Hill, New Jersey, Aug 10, 1987.

Lai, H.C., and Wu, S.Y., "On Linear Semi-Infinite Programming Problems, an Algorithm," *Numerical Functional Analysis and Optimization*, Vol. 13, 1992, pp. 287–304.

Lamond, B., and Stewart, N.F., "Bregman's Balancing Method," *Transportation Research Part B*, Vol. 15B, 1981, pp. 239–248.

Lent, A., "Maximum Entropy and MART," in *Image Analysis and Evaluation*, SPSE Conference Proceedings, edited by R. Shaw, Toronto, Canada, 1976, pp. 249–257.

Lent, A., "A Convergent Algorithm for Maximum Entropy Image Restoration with a Medical X-Ray Application," in *Image Analysis and Evaluation*, edited by R. Shaw, Society of Photographic Scientists and Engineers (SPSE), Washington, D.C., 1977, pp. 249–257.

Leontief, W., "Quantitative Input-Output Relations in the Economic System of the United States," *Review of Economic and Statistics*, Vol. 18, 1936, pp. 105–125.

Leontief, W., *Input-Output Economics*, 2nd Edition, Oxford University Press, New York, 1986.

Li, X.-S., *Entropy and Optimization*, Ph.D. Thesis, University of Liverpool, United Kingdom, 1987.

Li, X.S., and Fang, S.-C., "On the Entropic Regularization Method for Solving Min-Max Problems with Applications," to appear in *Mathematical Methods of Operations Research*, Vol. 46, 1997.

Lin, C.J., Fang, S.-C., and Wu, S.Y., "An Unconstrained Convex Programming Approach to Linear Semi-Infinite Programming," OR Technical Report No. 296, North Carolina State University, Raleigh, North Carolina, 1994, submitted to *SIAM Journal on Optimization.*

Lin, C.J., Fang, S.-C., and Wu, S.Y., "Parametric Linear Semi-Infinite Programming," *Applied Mathematics Letters*, Vol. 9, 1996, pp. 89–96.

Luenberger, D.G., *Linear and Nonlinear Programming*, 2nd Edition, Addison-Wesley, Reading, Massachusetts, 1984.

Macgill, S.M., "Theoretical Properties of Bi-proportional Matrix Adjustments," *Environment and Planning A*, Vol. 9, 1977, pp. 687–701.

Markowitz, H.M., *Portfolio Selection: Efficient Diversification of Investments*, John Wiley, New York, 1959.

Mehrotra, S., and Sun, J., "An Algorithm for Convex Quadratic Programming That Requires $O(n^3.5L)$ Arithmetic Operations," *Mathematics of Operations Research*, Vol. 15, 1990, pp. 342–363.

Miller, R.E., and Blair, P.D., *Input-Output Analysis: Foundations and Extensions*, Prentice-Hall, Englewood Cliffs, New Jersey, 1985.

Minoux, M., *Mathematical Programming*, Wiley-Interscience, New York, 1986.

Mishra, S., and Fang, S.-C., "A Maximum Entropy Optimization Approach to Tandem Queues with Generalized Blocking," *Performance Evaluation*, Vol. 702, 1997, pp. 1–25.

Murtagh, B.A., and Saunders, M.A., *MINOS 5.1 User's Guide*, Technical Report SOL 83-20R, Stanford University, Stanford, California, 1983.

Oko, S.O., "Surrogate Methods for Linear Inequalities," *Journal of Optimization Theory and Applications*, Vol. 72, 1992, pp. 247–268.

Van De Panne, C., *Methods for Linear and Quadratic Programming*, North-Holland, Amsterdam, 1975.

Peterson, E.L., "Geometric Programming," *SIAM Review*, Vol. 19, 1976, pp. 1–45.

De Pierro, A.R., and Iusem, A.N., "A Relaxed version of Bregman's Method for Convex Programming," *Journal of Optimization Theory and Applications*, Vol. 51, 1986, pp. 421–440.

Polak, E., Mayne, D.Q., and Higgins, J.E., "Superlinearly Convergent Algorithm for Min-Max Problems," *Journal of Optimization Theory and Applications*, Vol. 69, 1991, pp. 407–439.

Polyak, R.A., "On the Best Convex Chebyshev Approximation," *Soviet Mathematics Doklady*, Vol. 12, 1971, pp. 1441–1444.

Polyak, R.A., "Smooth Optimization Methods for Minimax Problems," *SIAM Journal of Control and Optimization*, Vol. 26, 1988, pp. 1274–1286.

Potra, F., and Ye, Y., "A Quadratically Convergent Polynomial Algorithm for Solving Entropy Optimization Problems," *SIAM Journal on Optimization*, Vol. 3, 1993, pp. 843–860.

Rajasekera, J.R., and Fang, S.-C., "On the Convex Programming Approach to Linear Programming," *Operations Research Letters*, Vol. 10, 1991, pp. 309–312.

Rajasekera, J.R., and Fang, S.-C., "Deriving an Unconstarined Convex Program for Linear Programming," *Journal of Optimization Theory and Applications*, Vol. 75, 1992, pp. 603–612.

Renegar, J., "A Polynomial-time Algorithm Based on Newton's Method for Linear Programming," *Mathematical Programming*, Vol. 40, 1988, pp. 59–93.

Robinson, S.M., and Day, R., "A Sufficient Condition for Continuity of Optimal Sets in Mathematical Programming," *Journal of Mathematical Analysis and Applications*, Vol. 45, 1974, pp. 506–511.

Rockafellar, R.T., *Convex Analysis,* Princeton University Press, Princeton, New Jersey, 1970.

Royden, H.L., *Real Analysis*, 2nd Edition, The Macmillan Company, New York, 1972.

Rudin, W., *Principles of Mathematical Analysis*, McGraw Hill, New York, 1976.

Safwat, K.N.A., and Magnanti, T.L., "A Combined Trip Generation, Trip Distribution, Modal Split, and Trip Assignment Model," *Transportation Science*, Vol. 22, 1988, pp. 14–30.

Saigal, R., *Linear Programming: A Modern Integrated Analysis*, Kluwer Academic Publishers, Norwell, Massachusetts, 1995.

Schneider, M.H., and Zenios, S.A., "Comparative Study of Algorithms for Matrix Balancing," *Operations Research*, Vol. 38, 1990, pp. 439–455.

Scott, C.H., and Jefferson, T.R., "Entropy Maximizing Models of Residential Location via Geometric Programming," *Geographical Analysis*, Vol. 9, 1977, pp. 181–187.

Sen, A., and Smith, T.E., *Gravity Models of Spatial Interaction Behavior*, Springer-Verlag, Berlin, 1995.

Sengupta, J.K., "Mixed Strategy and Information Theory in Optimal Portfolio Choice," *International Journal of Systems Science*, Vol. 20, 1989, pp. 215–227.

Senior, M.L., and Wilson, A.G., "Exploration and Synthesis of Linear Programming and Spatial Interaction Models of Residential Location," *Geographical Analysis*, Vol. 6, 1974, pp. 209–238.

Shanno, D.F., and Bagchi, A., "Unified View of Interior Point Methods for Linear Programming," Rutcor Research Report No. 35-88, Rutgers University, New Brunswick, New Jersey, July 1988.

Shannon, C.E., "A Mathematical Theory of Communication," *Bell Systems Technical Journal*, Vol. 27, 1948, pp. 379–423.

Shannon, C.E., and Weaver, W., *The Mathematical Theory of Communication*, University of Illinois Press, Urbana, Illinois, 1962.

Sheu, R.L., and Fang, S.-C., "On the Generalized Path-Following Methods for Linear Programming," *Optimization*, Vol. 30, 1994, pp. 235–249.

Sheu, R.-L., and Fang, S.-C., "Insights into Interior-Point Methods," *Zeitschrift fur Operations Research*, Vol. 36, 1992, pp. 227–257.

Shore, J.E., "Minimum Cross-entropy Spectral Analysis," *IEEE Transactions on Acoustics, Speech, and Signal Processing*, Vol. ASSP-29, 1981, pp. 230–237.

Shore, J.E., "Information Theoretic Approximations for M/G/1 and G/G/1 Queueing Systems," *Acta Informatica*, Vol. 17, 1982, pp. 47–61.

Shore, J.E., and Johnson, R.W., "Axiomatic Derivation of Principle of Maximum Entropy and Minimum Cross-entropy," *IEEE Transactions on Information Theory*, Vol. IT-26, 1980, pp. 26–37.

Smith, C.R., and Grandy, W.T., editors, *Maximum-Entropy and Bayesian Methods in Inverse Problems*, D. Reidel Publishing Company, Dordrecht, Holland, 1985.

Stark, H., editor, *Image Recovery: Theory and Application*, Academic Press, Orlando, Florida, 1987.

Takahashi, Y., and Takami, Y., "A Numerical Method for the Steady-State Probabilities of a G1/G/C Queue in a General Class," *Journal of Operations Research Society of Japan*, Vol. 19, 1976, pp. 147–157.

Teboulle, M., "A Simple Duality Proof for Quadratically Constrained Entropy Functionals and Extension to Convex Constraints," *SIAM Journal on Applied Mathematics*, Vol. 49, 1989, pp. 1845–1850.

Todd, M.J., and Ye, Y., "A Centered Projective Algorithm for Linear Programming," *Mathematics of Operations Research*, Vol. 15, 1990, pp. 508–529.

Tomlin, J.A., "A Mathematical Programming Model for the Combined Distribution-Assignment of Traffic," *Transportation Science*, Vol. 5, 1971, pp. 122–140.

Tomlin, J.A., and Tomlin, S.G., "Traffic Distribution and Entropy," *Nature*, Vol. 220, 1968, pp. 974–976.

Tribus, M., *Rational Descriptions, Decisions, and Designs*, Pergamon Press, New York, 1969.

Tribus, M., "An Engineer Looks at Bayes," in *Maximum-Entropy and Bayesian Methods in Science and Engineering, Vol. 1: Foundations,* edited by Gary J. Erickson and C. Ray Smith, Kluwer Academic Publishers, Dordrecht, The Netherlands, 1988, pp. 31–52.

Tsao, H.-S.J., and Fang, S.-C., "Linear Programming with Inequality Constraints via Entropic Perturbation," *International Journal of Mathematics and Mathematical Sciences*, Vol. 19, 1996, pp. 177–184.

Tsao, H.-S.J., Fang, S.-C., and Lee, D.N., "On the Optimal Entropy Analysis," *European Journal of Operational Research*, Vol. 59, 1992, pp. 324–329.

Tsao, H.-S.J., Fang, S.-C., and Lee, D.N., "A Bayesian Interpretation of the Linearly-Constrained Cross-Entropy Minimization Problem," *Engineering Optimization*, Vol. 22, 1993, pp. 65–75.

Vanderbei, R.J., *LOQO User's Manual*, Princeton University Press, Princeton, New Jersey, 1992.

Vardi, A., "New Minmax Algorithm," *Journal of Optimization Theory and Applications*, Vol. 75, 1992, pp. 613–633.

Viel, V.Q., "Practices in Input-Output Table Compilation," *Regional Science and Urban Economics*, Vol. 24, 1994, pp. 27–54.

Wang, D., "Experiments on the Unconstrained Convex Programming Approach to Linear Programming," IE Report No. 90-12, North Carolina State University, Raleigh, North Carolina, 1990.

Wang, Y., and Lu, W., "Multicriterion Maximum Entropy Image Reconstruction from Projections," *IEEE Transactions on Medical Imaging*, Vol. 11, 1992, pp. 70–75.

Williams, P.M., "Bayesian Conditionalisation and the Principle of Minimum Information," *The British Journal for the Philosophy of Science*, Vol. 31, 1980, pp. 131–144.

Wilson, A.G., "A Statistical Theory of Spatial Distribution Models," *Transportation Research*, Vol. 1, 1967, pp. 253–269.

Wilson, A.G., *Entropy in Urban and Regional Planning*, Pion, London, 1970.

Wilson, A.G., *Urban and Regional Models in Geography and Planning*, John Wiley, London, 1974.

Wu, J-S., "Maximum Entropy Analysis of Open Queueing Networks with Group Arrivals," *Journal of Operational Research Society*, Vol. 43, 1992, pp. 1063–1078.

Wu, J.-S., and Chan, W.C., "Maximum Entropy Analysis of Multiple-Server Queueing Systems," *Journal of Operational Research Society*, Vol. 40, 1989, pp. 815–826.

Xu, Z.K., and Fang, S.-C., "Unconstrained Convex Programming Approach to Linear Programming," *Journal of Optimization Theory and Applications*, Vol. 86, 1993, pp. 745–752.

Yamada, M., and Rajasekera, J.R., "Portfolio Re-balancing with the Entropy Criteria," Report No. 310, QUICK Research Institute Corp., Tokyo, 1993.

Yang, K., and Murty, K.G., "New Iterative Methods for Linear Inequalities," *Journal of Optimization Theory and Applications*, Vol. 72, 1992, pp. 163–185.

Ye, Y., "Interior Algorithms for Linear, Quadratic, and Linearly Constrained Programming," Ph.D. Thesis, Department of Engineering-Economic Systems, Stanford University, Stanford, California, 1987.

Zang, I., "A Smoothing Out Technique for Min-Max Optimization," *Mathematical Programming*, Vol. 19, 1980, pp. 61–77.

Zangwill, W.I., *Nonlinear Programming: A Unified Approach*, Prentice-Hall, Englewood Cliffs, New Jersey, 1969.

Zellner, A., "Optimal Information Processing and Bayes' Theorem," *The American Statistician*, Vol. 42, 1988, pp. 278–284.

Zhang, J., and Brockett, P.L., "Quadratically Constrained Information Theoretic Analysis," *SIAM Journal on Applied Mathematics*, Vol. 47, 1987, pp. 871–885.

Zhu, J., and Ye, Y., "A Path-Following Algorithm for a Class of Convex Programming Problems," Working Paper No. 90-14, College of Business Administration, The University of Iowa, Iowa City, Iowa, 1990.

Zhuang, X., Haralick, R.M., and Zhao, Y., "Maximum Entropy Image Reconstruction," *IEEE Transactions on Signal Processing*, Vol. 39, 1991, pp. 1478–1480.

Zhuang, X., Ostevold, E., and Haralick, R.M., "The Principle of Maximum Entropy in Image Recovery," in *Image Recovery: Theory and Application*, edited by Henry Stark, Academic Press, Orlando, Florida, 1987, pp. 157–193.

INDEX